# Experimental Approaches in Biochemistry and Molecular Biology

# Experimental Approaches in Biochemistry and Molecular Biology

Henry M. Zeidan
Life College


William V. Dashek
Forest Products Laboratory

**WCB** **Wm. C. Brown Publishers**

Dubuque, IA  Bogota  Boston  Buenos Aires  Caracas  Chicago  Guilford, CT
London  Madrid  Mexico City  Seoul  Singapore  Sydney  Taipei  Tokyo  Toronto

 **Wm. C. Brown Publishers**

President and Chief Executive Officer   *Beverly Kolz*
Vice President, Publisher   *Kevin Kane*
Vice President, Director of Sales and Marketing   *Virginia S. Moffat*
Vice President, Director of Production   *Colleen A. Yonda*
National Sales Manager   *Douglas J. DiNardo*
Advertising Manager   *Janelle Keeffer*
Production Editorial Manager   *Renée Menne*
Publishing Services Manager   *Karen J. Slaght*
Royalty/Permissions Manager   *Connie Allendorf*

**A Times Mirror Company**

Copyediting, design, and production by Shepherd, Inc.



Library of Congress Catalog Card Number:   95-83500

ISBN 0-697-16735-6



Certain laboratory exercises in this manual cannot be taught by an inexperienced instructor. To perform all the exercises, the instructor must be fully trained and familiar with NIH guidelines for recombinant DNA, as well as with the use and disposal of hazardous compounds and radionuclides. Most academic institutions possess an environmental health and safety office as well as a radiation safety office. These individuals are familiar with OSHA and NIH as well as Nuclear Regulatory Agency Guidelines for disposal of hazardous and radioactive wastes, respectively. In addition, the above institutions have guidelines for the use of electron microscope and animal care facilities. The latter involve an animal care committee. The institution where the exercises are performed must be equipped with chemical hoods that have been certified to pull at 100 FPM. In addition, the cloning protocols will require hepafiltered hoods, such as those manufactured by NuAire Flow.

Some of the laboratory experiments included in this text may be hazardous if materials are handled improperly or if procedures are conducted incorrectly. Safety precautions are necessary when you are working with chemicals, glass test tubes, hot water baths, sharp instruments, and the like, or for any procedures that generally require caution. Your school may have set regulations regarding safety procedures that your instructor will explain to you. Should you have any problems with materials or procedures, please ask your instructor for help.

Printed in the United States of America by Times Mirror Higher Education Group, Inc.
2460 Kerper Boulevard, Dubuque, IA 52001

10   9   8   7   6   4   3   2   1

# Dedication

Dr. Zeidan dedicates this manual to his wife, Nabila, and his children, Iman, 16, Angie 14, Ronnie 8, Amanda 7, and Nora 5, who patiently dealt with his need for years of scholarship. He renews his sincere thanks to the following for his intellectual development: Dr. Kerry T. Yasunobu, emeritus Professor of Biochemistry and Chair, University of Hawaii, School of Medicine, Department of Biochem and Biophysics; Dr. Lawrence H. Piette, Dean of the Graduate School at Utah State University; Dr. Lawerence J. Berliner and his colleagues; and in particular Dr. Simmon Kwok, Director of Reproductive Biology at the Medical Center of Albert Einstein, Philadelphia, Pennsylvania. Dr. Zeidan also thanks Dr. K. Watanabe and H. Ishizkai.

Dr. Zeidan is indebted to the National Institutes of Health and National Science Foundation for financial support enabling certain of the monoamine and plasma amine oxidase experiments described herein. Dr. Zeidan is grateful to Dr. Craig Hill of Emory University, Dr. Keith Patel of University of Texas and Smith Kline of French Research Laboratories for facilities. Dr. Zeidan wishes to acknowledge the assistance of all graduate and professional students who contributed to his biomedical research activities. Dr. Zeiden also acknowledges Mr. K. Abbound for his outstanding graphic work.

Dr. Dashek dedicates this manual to his children, Kristin Ann Simpson and Karin Ann Waddil, who patiently dealt with his need for years of scholarship. He thanks the following for his intellectual development: Dr. Wm. F. Millington, Dr. Walter G. Rosen, Dr. D. T. A. Lamport, Dr. J. E. Varner, Dr. Lynn Margulis, and Dr. M. Schein. Dr. Dashek is indebted to the United States Department of Energy for financial support enabling certain of the polyphenol oxidase experiments described herein. Dr. Dashek extends his gratitude to Ms. Doris Donnell of the Morehouse School of Medicine for her patience and thoughtful clerical assistance. This manual would never have seen "light of day" without her dedication. He also thanks Ms. Kim Chancey of the University of Georgia for additional assistance.

Finally, Dr. Dashek is grateful to Dr. Terry L. Highley, Project Director, Biodeterioration of Wood, Forest Products Laboratory, USDA-Forest Service, for facilities and encouragement. He also expresses his appreciation to Dr. D. E. McMillin, Associate Professor of Biology at Clark Atlanta University in Georgia, for the preparation of Module 8.

We also thank Rick Hecker of Shepherd, Inc., and Mr. D. Bruflodt and Ms. Robin Steffek of Wm. C. Brown Co. for their thoughtful assistance. We are especially grateful to Ms. Steffek for her many creative suggestions and attention to detail.

# Table of Contents

# 6 ANTIBODY PRODUCTION

# 10  THE USE OF RADIOLABELED COMPOUNDS IN BIOCHEMISTRY: *IN VITRO* TRANSLATION    165

# 11 Bioenergetics: Transformation of Oxidation-Reduction Energy into Bioluminescence    187



# 12 TLC and GLC of Monosaccharides from Membrane-Bound Glycoproteins    193



# Appendixes    205



# Index    217

# Preface

This laboratory manual is intended for undergraduate and beginning graduate students who are currently enrolled but have not taken an introductory course in biochemistry/biophysics. In addition, the manual is designed around modules that could be utilized as separates by students enrolled in traditional biochemistry/biophysics courses or alternatively those in basic health-oriented biochemistry courses. The latter is evident, as basic biomedical information is included in certain modules, thereby rendering this manual different from standard available biochemistry laboratory manuals. Furthermore, this manual presents biochemical techniques in application to "cutting-edge" research problems in biotechnology, life sciences, and biomedical sciences. In this connection, molecular biology methodologies are included, and thus this manual bridges the gap between traditional biochemistry/biophysics on the one hand and molecular biology on the other. This lab manual will be well-suited for biochemistry lab courses that are oriented toward biotechnology, since the use of antibodies, recombinant DNA, and batch culture techniques is presented here; the quantitative aspects of these disciplines have not been ignored.

The underlying theme centers about the purification, physiochemical properties, and overproduction of enzymes, certain of which are relevant to biomedical research. This manual will be particularly relevant to schools possessing preprofessional undergraduate and/or graduate professional curricula. With the proper selection of topics, depending upon the equipment available, this manual may also be used in two-year colleges or those four-year colleges lacking a research thrust. However, it should be emphasized that this manual is "geared" toward problem-solving oriented programs. An alternative outline that could be performed in a one-semester course that emphasizes spectrophotometry, electrophoresis, enzyme kinetics, and recombinant DNA techniques with less concern for antibodies and radioisotopes, and the more sophisticated biophysical approaches could involve selected modules as highlighted in the Table of Contents.

To test the student's comprehension and retention of the contents of the modules in relation to the theme, review/self-study questions are included. The manual is keyed to the up-to-date versions of certain traditional biochemistry textbooks such as Lehninger, Stryer, Zubay, and others. Relevant references are included. Certain two-year schools will not be able to perform the physiochemical experiments; however, the majority of the schools could perform the biochemistry/ molecular biology experiments.

Finally, the strengths of this manual are 1) an exposure and appreciation for the problem-solving, research-oriented approach; 2) the integration of traditional biochemistry/molecular biology and biophysics with basic medical approach, as well as an appreciation for some relevant environmental problems; 3) an introduction to classical and highly contemporary biochemical/biophysical techniques, as well as "state-of-the art" molecular biology methodologies centering about recombinant DNA techniques; and finally 4) an appreciation for the quantitative aspects of biochemistry. Relevant recent references, as well as enduring references from the research literature, are included.

In summary, students will find this manual a unique resource book throughout their careers, and instructors can avail themselves to a manual that fills the gap between traditional biochemistry/biophysics and biotechnology.

This edition was conceived while H. Zeidan and W. V. Dashek were Associate Professors of Biochemistry and Biology, respectively, in the Biomedical Sciences Program at Clark Atlanta University. The edition, which was written while H. Zeidan was on academic leave at the University of Texas, and W. Dashek was a visiting research botanist at the University of Georgia, will be followed by subsequent editions incorporating recent advances in the topics covered in the laboratory modules and other biomedical research topics.

***Henry M. Zeidan Ph.D., FACB***
*Associate Professor of Biochemistry*
*Life College*
*Marietta, GA*

***William V. Dashek, Ph.D.***
*Visiting Scientist*
*Biodeterioration and Its Control*
*Forest Products Laboratory*
*Madison, WI*

# Experimental Approaches in Biochemistry and Molecular Biology

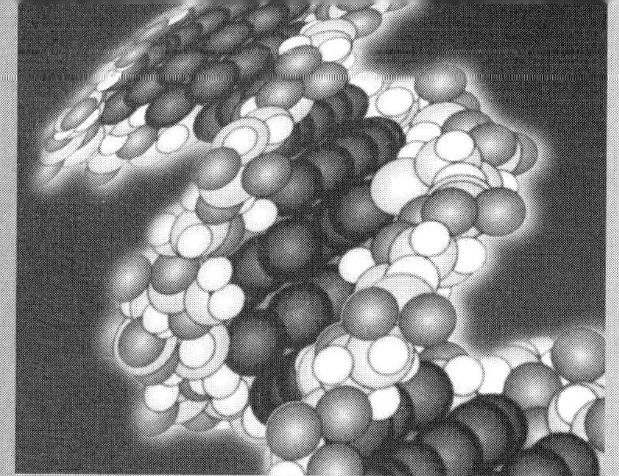

# Biochemical Calculations and Solution Preparation

## Outline of Module

Exponents
Experimental measurements
Biostatistical approaches
Solution preparation
    Concentrations based on volume
    Concentrations based on weight
    Acids and bases
    Ionization of weak acids
Laboratory buffers
    Buffer capacity

# Introduction

The objectives of this module are twofold. The first is to introduce the student to the mathematics and statistics used in biochemistry. A clear understanding of arithmetic and simple algebra is an essential component of biochemistry and molecular biology.[1-4] Therefore, sections I, II, and III discuss exponents, logarithms, experimental measurements, algebraic equations, and biostatistical approaches. Since biochemists are interested in studying reactions in solution, a survey of the various ways for expressing and interconverting concentrations of solutions forms the second objective. Section IV is devoted to these topics.

## I. Exponents (Scientific Notation)

An exponent is a number that indicates how many times the base appears as a factor. For example, $8 \times 8$ can be indicated by $8^2$; the exponent 2 appears as a superscript on the base 8. The base 8 is raised to the second power.

In scientific research, exponents are very useful, and the representation of either very large or very small numbers can be easily written. Any number can be represented in an exponential form. The ten (using the decimal system) is normally the base to use. The number 10,000 can be written as $1 \times 10^4$, 600,000 as $6 \times 10^5$, 0.00008 as $8 \times 10^{-5}$, and so on. Numbers that are integral (whole) as well as nonintegral powers of 10 can be represented. Generally, the exponential form is chosen so that it is the product of a number between 1 and 10 to some power. This exponential form is called scientific notation. The student can use the following two steps for writing a number in scientific notation.

1. Shift the decimal point so that the resulting number is between 1 and 10.
2. Determine the power of ten. To do this, count the number of positions the decimal point needs to be shifted to restore it to the original position. If this shift is to the right, the exponent will be positive, and if it is to the left, it will be negative.

***Example 1:***  Convert 0.0000983 to scientific notation.

***Solution:***
1. Write as a number between 1 and 10:

$$9.83$$

2. The decimal point can be restored by shifting 5 places to the left. The exponent on 10 is –5, and thus the number in scientific notation is

$$9.83 \times 10^{-5}$$

Scientific notation is very useful in the mathematical operations of addition, subtraction, multiplication, and division provided that the student follows the following rules.

### Addition and Subtraction

All the terms must have the same exponent. It is most common to convert all terms to the most positive exponent that appears. This gives the sum in scientific notation directly.

***Example 2:***  Perform the following addition:

$$8.0 \times 10^{-4} + 6.5 \times 10^{-2} + 9.4 \times 10^{-6}$$

***Solution:***

$$0.08 \times 10^{-2} + 6.5 \times 10^{-2} + 0.00094 \times 10^{-2}$$

Remove the common factor, $10^{-2}$, to yield

$$(0.08 + 6.5 + 0.00094) \times 10^{-2} = 6.580 \times 10^{-2}$$

***Example 3:***  Perform the following subtraction:

$$5.94 \times 10^8 - 4.2 \times 10^6$$

***Solution:***  Convert to the largest positive exponent, which is 8:

$$5.94 \times 10^8 - 0.042 \times 10^8$$

Remove the common factor, $10^8$, to give

$$(5.94 - 0.042) \times 10^8 = 5.898 \times 10^8$$
$$= 5.90 \times 10^8$$

### Multiplication

For multiplication of exponential numbers, the numerical portion is treated as usual, and the exponential part uses the law for multiplication of exponents. For the base 10 (decimal), it becomes

$$10^m \times 10^n = 10^{m+n}$$

The exponents are added algebraically to obtain the product in exponential form.

***Example 4:***  Multiple $900 \times 4000$.

***Solution:***  Convert to exponential numbers:

$$(9.0 \times 10^2) \times (4.0 \times 10^3)$$

Regroup the product:

$$9.0 \times 4.0 \times 10^2 \times 10^3$$

Multiply the parts:

$$36.0 \times 10^5$$

And convert to scientific notation:

$$3.6 \times 10^6$$

***Example 5:***   Multiply $0.0085 \times 3500$.

***Solution:***   Convert to scientific notation:

$$8.5 \times 10^{-3} \times 3.5 \times 10^{3}$$

Regroup the product:

$$8.5 \times 3.5 \times 10^{-3} \times 10^{3}$$

$$29.75 \times 10^{0} = 29.75$$

### Division

For division of exponential numbers, the quotient for the numerical part is obtained by the rule for division of exponents:

$$\frac{10^{m}}{10^{n}} = 10^{m-n}$$

***Example 6:***   Divide 6440 by 0.0025.

***Solution:***   Convert all factors to exponential numbers:

$$\frac{6.44 \times 10^{3}}{2.5 \times 10^{-3}} = 2.576 \times 10^{0} = 2.58$$

## II. Experimental Measurements

**Significant Figures.** The measurement of a physical quantity, no matter how precise, is still unreliable in a mathematical sense. A certain amount of error is associated with it. If, for example, the weight of a sample of a certain chemical is found to be 0.864 g and 0.868 g in two successive measurements, the answer is unreliable in the third decimal place. The measurement is said to contain three significant figures; that is, three of the digits are obtained with some reliability. The average weight obtained for the sample is 0.866 g, so the answer rounded off to the proper number of significant digits is 0.870 g. This problem illustrates the principle that the results of a calculation cannot be more reliable than the least reliable number used. Thus, we must drop digits that are not reliable. This process, which is called "rounding off," is governed by three rules.

1. If the digit to be dropped is less than 5, leave the last significant digit unchanged.
2. If it is more than 5, increase the last significant digit by one.
3. If the digit to be dropped is 5 followed by zero, the last remaining significant digit is left even.

***Example 1:***   Indicate the number of significant digits in the following numbers:

$$1560.0, \ 16.8, \ 2.520, \ 0.0268, \ 0.02460$$

***Solution:***

| Number | Significant Digits |
|---|---|
| 1560.0 | 5 |
| 16.8 | 3 |
| 2.520 | 4 |
| 0.0268 | 3 |
| 0.02460 | 4 |

***Example 2:***   Multiply $4.7 \times 400$. Give the answer to the proper number of significant digits.

***Solution:***

$$4.7 \times 400 = 1880$$

The minimum number of significant digits is 2, so that the product can be expressed to only 2 significant digits; thus,

$$4.7 \times 400 = 1900$$

The student may question how a number such as 1900 has only two significant figures. The answer to this question can easily be explained when the student writes this number in scientific notation, i.e., $19 \times 10^{2}$. This illustrates that only 1 and 9 are significant, and thus clarifies this point. In general, however, we assume the final zeros before the decimal point to be significant unless otherwise indicated. Final zeros after the decimal point are significant digits and should not be carelessly omitted. Thus, a reading of 10.0 g indicates that the weight is known to tenths of grams. Zeros appearing ahead of a number are not significant even though they may be to the right of the decimal point.

## III. Biostatistical Approaches

**Standard Deviation.** Standard deviation, or root-mean-square error, is preferred for some types of measurements. The standard deviation, $\sigma$, is defined as:

$$\sigma = \pm\sqrt{\frac{\left(x_i - \text{a.m.}\right)^{2}}{n-1}}$$

To understand the significance of standard deviation we need to examine the nature of the distribution function that determines the curve in Figure 1.1. Its

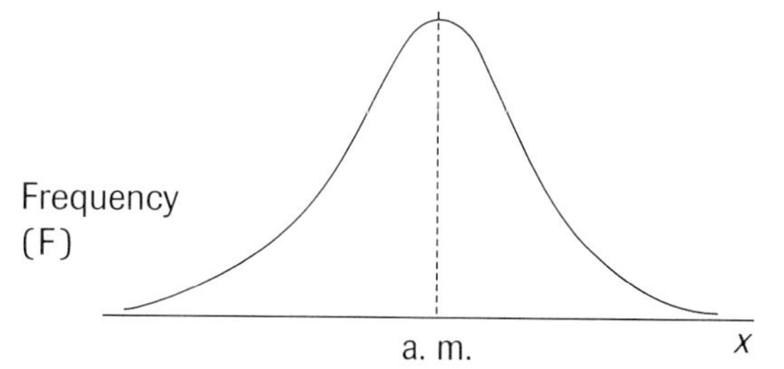


**FIGURE 1.1**   Gaussian/curve distribution.

mathematical representation for a large number of measurements is generally called the Gaussian distribution function,

$$F = \frac{1}{\sigma(2\pi)^{1/2}} e^{-\frac{(x_1 - \text{a.m.})^2}{2\sigma^2}}$$

where $x_1$, a.m. arithmetic mean and $\sigma$ is as defined earlier, and $F$ is the frequency of occurrence of a particular observation. Every experimental measurement that is made lies somewhere on this curve. The most probable value is the true value.

For a given range of $X_1$ about the true value, the probability that a measured value will fall in this range is proportional to the area under the curve. Since the standard deviation is a range of values spanning the true value, it is thus related to the probabilities for the Gaussian function, using a large number of measurements, as follows:

$$\text{a.m.} \pm Q \quad 68.3\%$$

$$\text{a.m.} \pm 2Q \quad 95.4\%$$

$$\text{a.m.} \pm 3Q \quad 99.7\%$$

Thus, 95.4% of the measurements should be within two standard deviations, and 99.7% of the measurements should be within three standard deviations.

***Example 1:***  The following data were gathered for the pH of a given buffer solution on six successive readings: 6.71, 6.75, 6.74, 6.77, 6.73, 6.74.

***Solution:***  Evaluate the arithmetic mean and the sum of the square of the deviations from the arithmetic mean and the sum of the square of the deviations from the arithmetic mean.

| **Observations** | **$[X_1 - \text{a.m.}]$** | **$[X - \text{a.m.}]^2$** |
|---|---|---|
| 6.71 | 0.03 | 0.0009 |
| 6.75 | 0.01 | 0.0001 |
| 6.74 | 0.00 | 0.0000 |
| 6.77 | 0.03 | 0.0009 |
| 6.73 | 0.01 | 0.0001 |
| 6.74 | 0.00 | 0.0000 |
| $\Sigma X_i = 40.44$ | $[x_i - \text{a.m.}] = 0.08$ | $\Sigma[x_i\ \text{a.m.}]^2 = 0.0020$ |

The arithmetic mean is

$$\text{a.m.} = \frac{x_i}{n}$$

Substitute

$$\text{a.m.} = \frac{40.44}{6} = 6.74$$

The standard deviation is calculated by:

$$\sigma = \pm\sqrt{\frac{(x_i - \text{a.m.})^2}{n-1}}$$

Substitute

$$\sigma = \pm\frac{0.002}{5}$$

$$\sigma = \pm 0.0004$$

$$\sigma = \pm 2.0 \times 10^{-2}$$

The pH of the solution can then be expressed as

$$6.74 + 0.020$$

This means that if more measurements were made, then 68.3% would fall between a pH of 6.72 and 6.76.

# Review Questions

1. Express the results of the following operations to the proper number of significant figures.
   a) $\dfrac{3.1 \times 3.57 \times 62.5}{14.6}$
   b) $0.0071 + 0.00364 + 0.00118$
   c) $3120.1 + 610.3 + 56$
2. Six separate determinations of the concentration of a solution of HCl gave these results: 0.1021, 0.1017, 0.1013, 0.1020, and 0.1016. Calculate the average deviation and standard deviation.
3. Calculate the average deviation and standard deviation for a radioactive sample that when counted gave the following data for 10 five-minute counts: 22,700, 21,650, 22,200, 23,100, 23,900, 22,000, 21,400, 22,300, 21,700, and 23,050.
4. The weight in grams of a sample of hexokinase on six successive weighings showed the following variations: 1.3146, 1.3131, 1.3137, 1.3135, 1.3141, and 1.3138. Express the average reading to the proper number of significant figures.
5. Five determinations of the molecular weight of the enzyme ribonuclease by ultracentrifugation gave the following results: 13,100, 13,640, 13,400, 13,250, and 13,790. Calculate the average deviation in molecular weight to the appropriate number of significant figures.

## IV. Solution Preparation

### *Concentration Based on Volume*
Concentrations based on the amount of dissolved solute per unit volume are the most widely used in biochemistry laboratories. The most common conventions are defined below.

Molarity (M) = the number of moles of solute per liter of solution.

To calculate M, we need to know the weight of dissolved solute and its molecular weight, MW.

$$\text{number of moles} = \frac{\text{wt}_g}{\text{MW}}$$

Dilute solutions are often expressed in terms of millimolarity, micromolarity, and so on, where

$$1 \text{ mmole} = 10^{-3} \text{ moles}$$

$$1 \text{ μmole} = 10^{-6} \text{ moles}$$

$$1 \text{ nmole} = 1 \text{ mμ mole} = 10^{-9} \text{ moles}$$

$$1 \text{ pmole} = 1 \text{ μμ mole} = 10^{-12} \text{ moles}$$

therefore,

$$1 \text{ mM} = 10^{-3} \text{ M} = 1 \text{ mmole/liter} = 1 \text{ μmole/ml}$$

$$1 \text{ μM} = 10^{-6} \text{ M} = 1 \text{ μmole/liter} = 1 \text{ nmole/ml}$$

$$1 \text{ nM} = 10^{-9} \text{ M} = 1 \text{ nmole/liter} = 1 \text{ pmole/ml}$$

Normality (N) = the number of equivalents of solute and its equivalent weight, EW.

$$\frac{\text{wt}_g}{\text{EW}} = \frac{\text{MW}}{n}$$

where $n$ is the number of replaceable $H^+$ or $OH^-$ per molecule (for acids and bases). The molarity and normality are related by:

$$N = n\text{M}$$

To calculate N, we need to know the weight of dissolved solute and its equivalent weight, EW.

$$\text{Equivalents} = \frac{\text{wt}_g}{\text{EW}}$$

One equivalent (i.e., the EW) of an acid or base is the weight that contains 1 gm (1 mole) of replaceable hydrogen, or 1 g– ion (1 mole) of replaceable hydroxyl. The EW of a compound involved in an oxidation-reduction reaction is the weight that provides or accepts 1 faraday (1 mole) of electrons. In general:

$$\text{EW} = \frac{\text{MW}}{N}$$

For example, 0.01 M solution of $H_2SO_4$ is 0.02 N.

Weight/Volume Percent (% w/v) = the weight in g of solute per 100 ml of solution.

Weight/volume percent is often used for routine laboratory solutions where exact concentrations are not too important.

Milligram Percent (mg%) = the weight in mg of a solute per 100 ml of solution.

***Problem 1:***   a) How many grams of solid NaOH are required to prepare 500 ml of a 0.04 M solution? b) Express the concentration of this solution in terms of N, g/liter, % w/v, mg%.

***Solution:***   a) The number of grams of NaOH equals the desired molarity (M) × the molecular weight for NaOH (40.0 g/mole). Number of grams to prepare 0.04 M solution = molecular weight of NaOH (40.0 g/mole) × 0.04 = 1.60 g.

Therefore, if you dissolve 1.60 g in deionized water and diluted to 1 liter, this will give you 0.04 M NaOH solution. If you dissolve 0.8 gram or 800 mg in deionized water and diluted to 500 ml, this also will give you 0.04 M NaOH solution.

b)

$$0.04 \text{ M NaOH} = 0.04 \text{ N NaOH}$$

$$1.60 \text{ g/liter} = 0.04 \text{ M NaOH}$$

$$1.60 \times 10^2 \text{ mg\%} = 0.04 \text{ M.}$$

Therefore, take 0.6 ml of the concentrated solution and dilute to 1.5 liters.

## *Concentration Based on Weight*

Weight/Weight Percent (% w/w) = the weight in g of a solute per 100 g of solution.

***Problem 2:***   Describe the preparation of 2 liters of 0.4 M HCl starting with a concentrated HCl solution (28% w/w HCl, SG = 1.15).

***Solution:***

$$\text{liters} \times \text{M} = \text{number of moles}$$

$$2 \times 0.4 = 0.80 \text{ mole HCl needed}$$

$$\text{wt}_g = 29.2 \text{ g pure HCl}$$

The stock solution is not pure HCl but only 28% HCl by weight.

$$\therefore \frac{29.2}{0.28} = 104.3 \text{ g of stock solution, we can calculate}$$
$$\text{the volume required.}$$

$$\text{vol}_{ml} = \frac{\text{wt}_g}{-0} = \frac{104.3}{1.15} = 97.7 \text{ ml stock solution needed}$$

Therefore, measure out 90.7 ml of stock solution and dilute to 2 liters with water.

All of the above relationships (between weight, density, and percent w/w) can be combined into a single expression.

$$\text{wt}_g = \text{vol}_{ml} \times 0_{g/ml} \times \% \text{ (as decimal)}$$

where  $\text{wt}_g$ = weight of pure substance required in g
$\text{vol}_{ml}$ = volume of stock solution needed in ml
$\%$ = fraction of total weight that is pure substance

$$\therefore \text{vol} = \frac{\text{wt}}{0 \times \%} = \frac{29.2}{1.15 \times 0.28} = 90.7 \text{ ml}$$

Molality (m) = the number of moles of solute per 1000 g of solvent.

Mole fraction (MF) = the fraction of the total number of moles per 1000 g of solvent.

Molality is used in certain physical calculations (e.g., calculations of boiling-point elevation and freezing-point depression). For dilute aqueous solutions, m and M will be quite close. In order to interconvert m and M, we need to know % w/w.

Mole fraction = the fraction of the total number of moles represented by the compound in question.

For example, in a solution containing $n_1$ moles of compound 1, $n_2$ moles of compound 2, and $n_3$ moles of compound 3, the mole fraction of compound 2, $MF_2$, is given by

$$MF_2 = \frac{n_2}{n_1 + n_2 + n_3}$$

***Problem 3:*** a) Calculate the molarity of the concentrated stock HCl solution described in Problem 2. b) Calculate the mole fraction of HCl in solution.

***Solution:***

a) The solution contains 28% w/w HCl, or 28 g HCl per 100 g total, or 28 g HCl per $(100 - 28) = 72$ g water.

$$\frac{28 \text{ g HCl}}{72 \text{ g H}_2\text{O}} \times 1000 = 3.88.9 \text{ g HCl/1000 g H}_2\text{O}$$

$$\frac{\text{wt}_g}{\text{MW}} = \text{moles} \quad \frac{388.9}{36.5} = 10.65 \text{ moles HCl/1000 g H}_2\text{O}$$

therefore, the solution is 10.65 m.

b) In 100 g of solution, for example, we have

$$\frac{28 \text{ g HCl}}{36.5 \text{ g/mole}} = 0.767 \text{ moles of HCl}$$

$$\frac{72 \text{ g H}_2\text{O}}{18 \text{ g/mole}} = 4.0 \text{ moles of H}_2\text{O}$$

$$MF_{HCl} = \frac{n_{HCl}}{n_{HCl} + n_{H_2O}} = \frac{0.767}{4.767}$$

$$MF_{HCl} = 0.161$$

## *Acids and Bases*

An understanding of acid-base chemistry is essential if we are to appreciate the properties of biological molecules. A great many of the low molecular weight metabolites and macromolecular components of living cells are acids and bases, and thus, have the potential to ionize. The charges on these molecules are important factors in the rate of enzyme-catalyzed reactions, in the stability and conformation of proteins, in the interactions of macromolecules with each other and with small ions, and in the analytical and purification techniques used in the biochemical laboratory.

**Bronsted Concept of Acids and Bases**

An *acid* is defined as a substance that donates protons (hydrogen ions), and a *base* as a substance that accepts protons. When a Bronsted acid loses a proton, a Bronsted conjugated base is produced. The original acid and resulting base are referred to as a conjugate acid-conjugate base pair. The substance that accepts the proton is a different Bronsted base; by accepting the proton, another Bronsted acid is produced. Thus, in every ionization of an acid or base, two conjugate acid-conjugate base pairs are involved.

$$\text{HA} \quad + \quad \text{B}^- \; \rightleftharpoons \quad \text{A}^- + \text{HB}$$

$$[\text{conjugate acid}]_1 \quad [\text{conjugate base}]_2 \quad [\text{conjugate acid}]_2$$

**pH and pOH**

pH is a shorthand way of designating the hydrogen ion activity of a solution. By definition, pH is the negative logarithm of the hydrogen ion activity. Similarly, pOH is the negative logarithm of the hydroxyl ion activity.

$$pH = -\log a_H^+ = \log \frac{1}{aH^+}$$

$$= -\log [H^+]$$

$$= \log \frac{1}{[H^+]}$$

$$pOH = -\log a_{OH}^- = \log \frac{1}{[OH^-]}$$

In dilute solutions of acids and bases in pure water, the activities of $H^+$ and $OH^-$ may be considered the same as their concentrations.

$$pH = -\log[H^+] = \log \frac{1}{[H^+]}$$

$$pOH = -\log[OH^-] = \log \frac{1}{[OH^-]}$$

In all aqueous solutions the equilibrium for the ionization of water must be satisfied; that is, $[H^+][OH^-] = K_w = 10^{-14}$. Thus, if $[H^+]$ is known, we can easily calculate $[OH^-]$. Furthermore, we can derive the following relationship between pH and pOH:

$$[H^+][OH^-] = K_w$$

Taking logarithms:

$$\log[H^+] + \log[OH^-] = \log K_w$$

$$-\log[H^+] = pH \quad -\log[OH^-] = pOH \quad -\log K_w = pK_w$$

$$\therefore pH + pOH = pK_w$$

$$K_w = 10^{-14} \qquad pK_w = -\log 10^{-14} = 14$$
$$pH + pOH = 14$$

Thus, if any one of the values $[H^+]$, $[OH^-]$, pH, or pOH is known, the other three can be calculated easily.

***Problem 4:*** What are the a) $H^+$ ion concentration, b) pH, c) $OH^-$ ion concentration, and d) pOH of 0.001 M solution of HCl?

**Solution:**

a) HCl is a "strong" inorganic acid; that is, it is essentially 100% ionized in dilute solution. Consequently, when 0.001 mole of HCl is introduced into one liter of $H_2O$, it immediately dissociates into 0.001 M $H^+$ and 0.001 M $Cl^-$.

Note that when we are dealing with strong acids, the $H^+$ contribution from the ionization of water is neglected.

b) $pH = -\log[H^+]$
$$= -\log 10^{-3}$$
$$= -(-3) = +3$$
$$ph = 3$$

c) $[H^+][OH^-] = K_w$
$$[OH^-] = \frac{KW}{[H^+]}$$
$$[OH^-] = \frac{1 \times 10^{-14}}{1 \times 10^{-3}}$$
$$[OH^-] = \frac{1 \times 10^{-14}}{1 \times 10^{-3}}$$
$$[OH^-] = 1 \times 10^{-11}$$

d) $pOH = -\log[OH^-]$
$$= -(-11)$$
$$pOH = 11$$

or

$$pH + pOH = 14, \quad pOH = 14 - pH$$
$$pH = 14 - 3$$
$$pOH = 11$$

***Problem 5:*** What are the a) $(H^+)$, b) $(OH^-)$, c) pH, and d) pOH of a 0.001 M solution of $HNO_3$?

**Solution:**

a) $HNO_3$ is a strong inorganic acid.
$$[H^+] = 0.001 \text{ M} = 1 \times 10^{-3} \text{ M}$$

b) $[H^+][OH^-] = 1 \times 10^{-14}$
$$[OH^-] = \frac{1 \times 10^{-14}}{1 \times 10^{-3}} = 1 \times 10^{-11}$$

c) $pH = \log \dfrac{1}{[H^+]}$
$$= \log \frac{1}{[H^+]}$$
$$= \log \frac{1}{[1 \times 10^{-3}]}$$
$$= \log 0 + \log 10^3$$
$$= 3$$
$$pH = 3$$

d) $pH^+ pOH = 14$
$$pOH = 14.00 - 3.00$$
$$= 11.00$$

## *Ionization of Weak Acids*

A weak acid ionizes to a limited extent as follows:

$$HA + H_2O \;\rightleftharpoons\; H_3O^+ \quad + \quad A^-$$
$$\text{[conjugate base]} \quad \text{[conjugate acid]}$$

The proton released from HA is accepted by water to form the hydronium ion $H_3O^+$. The reversible ionization reaction can be described by an equilibrium constant, $K_i$:

$$K_i = \frac{[H_3O^+][A^-]}{[HA][H_2O]}$$

Because $[H_2O]$ is itself a constant, we can define a new constant, Ka, that combines $K_i$ and $[H_2O]$.

$$Ka = \frac{[H^+][A^-]}{[HA]}$$

**Relationship between Ka and Kb for Weak Acids and Bases**

Kb represents the dissociation constant of a base. If we start with the conjugate base, $A^-$, and dissolve in water, it ionizes as a typical base; it accepts a proton from $H_2O$ to form $OH^-$, and the corresponding conjugated acid, HA. A Kb expression can be written for this ionization.

$$A^- + HOH \longrightarrow HA + OH$$

$$Kb = \frac{[HA][OH^-]}{[A^-]}$$

Solving for Ka and Kb expressions for $[H_3O^+]$ and $[OH^-]$:

$$[H_3O^+] \text{ and } [OH^-]:$$

$$[H_3O^+] = \frac{[HA]Ka}{[A^-]}$$

$$[OH^-] = \frac{[A^-]Kb}{[HA]}$$

Substituting into:

$$[H_3O^+]\,[OH^-] = Kw$$

or

$$Ka \times Kb = Kw$$

Taking logarithms:

$$\log Ka + \log Kb = \log Kw$$

$$-\log Ka - \log Kb = -\log Kw$$

$$\therefore\ pKa + pKb = 14$$

## *Laboratory Buffers*

A *buffer* system is a mixture of a weak acid or a weak base and its salt (conjugated base or acid respectively) that permits solutions to resist large changes in pH upon the addition of small amounts of H$^+$ or OH$^-$ ions. In other words, a buffer helps maintain a near constant pH upon the addition of small amounts of H$^+$ or OH$^-$ ions to a solution.

***Problem 6:***  A solution of 0.05 M acid, pKa 6.10, is mixed with an equal volume of a 0.1 M solution of its sodium salt. What is the pH of the final mixture?

***Solution:***  Since equal volumes of the two solutions were mixed, the final concentration of the acid is 0.025 M and the salt is 0.05 M.

The molar ratio of the concentrations of salt to acid is

$$\frac{[\text{salt}]}{[\text{acid}]} = \frac{0.05}{0.025} = 2.0$$

From a knowledge of this ratio and the pKa value we take equation

$$pH = pKa + \log \frac{[\text{salt}]}{[\text{acid}]}$$

$$= 6.10 + \log\ 2.0 = 6.10 + 0.30$$

$$= 6.40$$

***Problem 7:***  How many moles of sodium acetate and acetic acid are required to prepare 1 liter of a buffer, pH 5.0, which is 0.1 M in total available acetate (dissociated and undissociated)? Acetic acid has a pKa of 4.74.

***Solution:***  From equation

$$5.0 = 4.74 + \log \frac{[\text{salt}]}{[\text{acid}]}$$

$$= \log \frac{[\text{salt}]}{[\text{acid}]} = 0.26$$

The ratio of the concentration of sodium acetate to acetic acid is the antilog 0.26, or 1.82; there are 1.82

molecules of sodium acetate for each molecule of acetic acid. The mole fraction of salt is 1.82/2.82 and the mole fraction of acid is 1.00/2.82. It is known to us that the total molar concentration of acetate is 0.1 M (0.1 mole in one liter). Therefore, of the 0.1 moles, the salt must provide its fraction,

$$\frac{1.82}{2.82} \times 0.1\ \text{moles} = 0.065\ \text{moles}$$

and acid must provide its fraction,

$$\frac{1.00}{2.82} \times 0.1\ \text{moles} = 0.035\ \text{moles}$$

***Problem 8:***  Describe the preparation of 5 liters of 0.3 M acetate buffer, pH 4.47, starting from a 2 M solution of acetic acid and a 2.5 M solution of KOH.

***Solution:***  We first calculate the proportions of the two acetate species present.

$$pH = pKa + \log \frac{[\text{OAc}^-]}{[\text{HOAc}^-]}$$

$$4.47 = 4.77 + \log \frac{[\text{OAc}^-]}{[\text{HOAc}^-]}$$

$$-0.30 = \log \frac{[\text{OAc}^-]}{[\text{HOAc}^-]}$$

or

$$0.30 = \log \frac{[\text{OAc}^-]}{[\text{HOAc}^-]}$$

$$\text{antilog of } 0.3 = 2 = \frac{2}{1} = 2$$

therefore, 2/3 of the total acetate is present as acid and 1/3 of the total acetate is present as salt.

The final solution contains:

$$2/3 \times 0.3\ \text{M} = 0.2\ \text{M HOAc (1 mole in 5 liters)}$$

$$1/3 \times 0.3\ \text{M} = 0.1\ \text{M OAc}^-\ (0.5\ \text{mole in 5 liters})$$

In this buffer, all of the acetate must be provided by the HOAc. The buffer is prepared by converting the proper proportion of HOAc to OAc$^-$ by adding KOH. We need 5 liters $\times$ 0.3 M = 1.5 moles total acetate. Calculate how much stock 2 M HOAc is needed to obtain 1.5 moles.

$$\text{liters} \times M = \text{number of moles}$$

$$\text{liters} \times 2 = 1.5$$

$$\text{liters} = \frac{1.5}{2} = 0.75$$

therefore, 750 ml of the 2 M HOAc is required.

Next, convert 1/3 of the 1.5 moles to OAc⁻ by adding the proper amount of 2.5 M KOH.

$$1/3 \times 1.5 \text{ moles} = 0.5 \text{ mole KOH needed}$$

$$\text{liters} \times M = \text{number of moles}$$

$$\text{liters} \times 2.5 = 0.5 \text{ mole}$$

$$\text{liters} = \frac{0.5}{2.5} = 0.2 \text{ liter}$$

therefore, add 200 ml of 2.5 M KOH.

The solution now contains 1 mole of HOAc and OAc⁻.

Finally, add sufficient water to bring the volume up to 5 liters. The final solution contains 0.2 M acid and 0.1 M salt.

### *Buffer Capacity*

The ability of a buffer to resist changes in pH is referred to as the *buffer capacity*. Buffer capacity can be defined in two ways: 1) the number of moles per liter of H⁺ or OH⁻ required to cause a given change in pH (e.g., 1 unit), or 2) the pH change that occurs upon addition of a given amount of H⁺ or OH⁻ (for example 1 mole/liter). The first definition is better because it can be applied to buffers of any concentration.

***Problem 9:***   The pH of 10 ml of 1 M buffer, pH 3.9, was decreased to pH 2.9 when 1.3 ml N HCl was added. What is the buffer capacity at pH 3.9?

***Solution:***   2.1 ml of 1.3 N HCl contains $(2.1/1000) \times 1.3$ equivalent of H⁺ = 2.73 meq of H⁺.

1000 ml of buffer would require $(1000/10) \times 2.73 \times 10^{-3}$ equivalent of H⁺ = $2.73 \times 10^{-1}$ equivalent of H⁺

$$= 2.73 \times 10^{-1} \text{ equivalent of H}^+.$$

Therefore, the buffer capacity is 0.273.

## Review Questions

1. Calculate the H⁺ concentration and pH of a solution formed by adding 100 ml of 0.1 M acetic acid to 100 ml of 0.05 N sodium hydroxide.
2. What is the pH and buffer capacity of a 0.03 M lactic acid solution to which has been added two volumes of 0.01 M sodium lactate?
3. Suppose that you prepare a buffer by dissolving 0.1 mole per liter of a weak acid, HA, and 0.10 mole per liter of its sodium salt, A⁻. Assume that pKa = 3. What is the pH of the buffer?
4. The weak acid, HA, is 2.4% dissociated in a 0.22 M solution. Calculate a) the Ka and b) the pH of the solution.
5. What are the final hydrogen ion concentration and pH of a solution obtained by mixing 200 ml of 0.4 M aqueous NH³ with 300 ml of 0.2 M HCl (Kb = $1.8 \times 10^{-5}$)?

## REFERENCES


1. Siegel, I. H. 1982. *Biochemical Calculations*. New York: John Wiley and Sons.
2. Montgomery, R. and C. Swenson. 1969. *Quantitative Problems in Biochemical Sciences*. San Francisco, CA: W.H. Freeman and Company.
3. Dryer R. and G. Lata. 1989. *Experimental Biochemistry*. Oxford, England: Oxford University Press.
4. Atkinson, D and S. Clarke. 1989. *Dynamic models in Biochemistry*. Marinal Rey, CA: N. Simonson & Company.

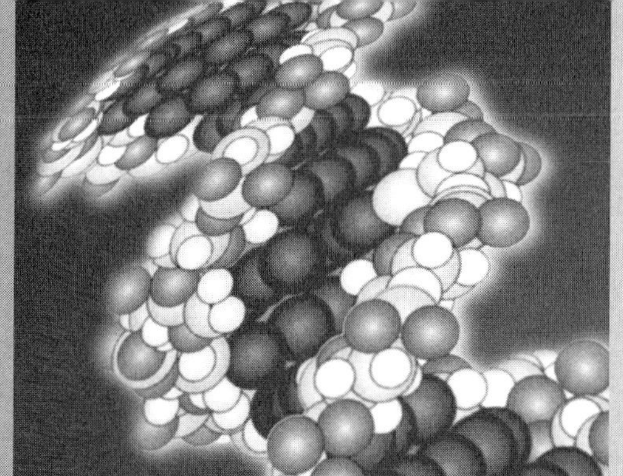



# Centrifugation—Protein Quantification

## Outline of Module

### CENTRIFUGATION

Preparations of organelles from homogenates of *Coriolus versicolor* hyphae by differential centrifugation.

> *Duration:* One laboratory period
> Combine with UV-spectroscopy assay of protein

### PROTEIN QUANTIFICATION

Quantification of protein* in organelles from homogenates of *Coriolus versicolor* hyphae.

> *Duration:* Two laboratory periods
> Team efforts
> UV spectroscopy
> Colorimetric assay
> Bradford
> Lowry
> Pierce

*Quantification procedures presented in this part of Module 2 may be applied to homogenates of either plant or animal cells and cell culture media.

# LABORATORY EXERCISES

## CENTRIFUGATION

### Supplies and Equipment

    Pasteur pipettes and amber bulbs
    10 ml pipettes
    1 ml pipettes
    Beakers for balancing centrifuge tubes
    Trip balance
    Corex tubes 8441 with adapters for fixed angle
        rotor
    Polyallomer tubes
    Ice buckets
    Screw cap tubes (culture tubes)
    Kimwipes
    Towels
    Waste box and bag for glass disposal
    Sonicator

### Chemicals

    Solution, pH 7.2, 25 mM Tris buffer containing
        250 mM sucrose and 3 mM EDTA.

### Organism

    *Coriolus versicolor* culture—3–6 day liquid
        grown.[1]

# Introduction

## Types of Centrifuge Procedures

A centrifuge is an instrument for separating particles, e.g., cells, subcellular organelles or macromolecules from solution. There are two types of centrifugation: preparative and analytical. Whereas the former is concerned with isolating specific particles, the latter centers about measuring a sedimenting particle's physical properties.

During the rotor's rotation, a relative centrifugal force (RCF) is applied to each of the sample's particles with the result that each particle will sediment at a rate proportional to the applied centrifugal force. The sedimentation rate of each particle can be influenced by its physical properties and the viscosity of the sample solution. Thus, the sedimentation rate of a particle is proportional to its molecular weight at a fixed centrifugal force and liquid viscosity.

## Theory of Sedimentation

There are many mathematical principles that govern sedimentation; refer to Table 2.1 for a review of these important principles.

## Separation Methods

The two most commonly employed separation methods in centrifugation are differential and density gradient centrifugations. The former is "pelleting" and involves filling an appropriate centrifuge tube with a uniform mixture of sample solution. As a result of an applied g force, two fractions result: a pellet consisting of sedimented sample and a supernatant solution comprised of unsedimented samples. The latter can be decanted and recentrifuged at a higher g force (Figure 2.1A) to fractionate the cell's cytoplasm into its various organelles. Differential centrifugation does not result in purification—i.e., a pelleted organellar fraction can be cross-contaminated with others. To achieve enrichment, density gradient centrifugations are beneficial. These centrifugations utilize a supporting column of fluid whose density increases toward the tube's bottom.

There are two classes of density gradient centrifugations: rate zonal and isopycnic. Figure 2.1B reveals that a sample solution containing particles to be separated is layered onto a preformed gradient column for rate zonal centrifugation. The particles sediment through the gradient into separated zones with each zone consisting of particles characterized by their sedimentation rates (Table 2.2). Rate zonal centrifugation (see Figure 2.1B) is governed by the density of sample particles. A significant aspect of this type of centrifugation is time dependency—i.e., it must be terminated prior to the separated zones reaching the bottom of the tube.

In contrast to rate zonal density centrifugation, the density gradient column (Figure 2.1C) encompasses the entire range of the sample particles' densities in isopycnic density centrifugation. Each particle sediments only to the position in a centrifuge tube at which the gradient is equal to its own density. Therefore, isopycnic density gradient centrifugation separates particles into zones exclusively on particle density differences and independent of time.

As there are many subtleties regarding centrifugation—e.g., gradient shapes, gradient materials, preparing gradients (Figure 2.1D), applying (Figure 2.1E) and recovering samples to and from gradients, and so on; refer to the manual *Techniques of Preparative, Zonal and Continuous Flow Ultracentrifugation.*[2] Portions of the previous discussion were adapted with modification from the manual, which contains many relevant conversion tables, some of which form part of the appendix to this lab manual.

# Procedures

This module is concerned, in part, with the preparation of subcellular organelles from hyphae of liquid cultured *Coriolus versicolor*.

**TABLE 2.1**   Principles That Govern Sedimentation

| | |
|---|---|
| angular velocity, $\omega$ | rate of rotation, measured in radians per second ($\omega = 2\pi\text{rpm}/60$, or $\omega = 0.10472 \times \text{rpm}$) |
| anodized coating | protective coating formed electrochemically on aluminum surfaces for corrosion resistance |
| buoyant density | the density of a particle in a specified liquid medium |
| Buna N | black rubber elastomer used for flat gaskets and O-rings in some tube caps, for the O-ring in the ultracentrifuge chamber door, and for O-rings and gaskets in rotor assemblies |
| centrifugal effect | accumulated value of: $$\int_{t_1}^{t_2} \omega^2 \mathrm{d}t$$ where $t$ is time and $\omega$ is angular velocity |
| clearing factor, $k$ | calculated for all Beckman preparative rotors as a measure of the rotor's relative pelleting efficiency: $$k = \frac{\ln\left(r_{\max}/r_{\min}\right)}{\omega^2} \times \frac{10^{13}}{3600}$$ or $$k = \frac{253303 \times \ln\left(r_{\max}/r_{\min}\right)}{\left(\text{RPM}/1000\right)^2}$$ |
| clearing time, $t$ | $t = k/s$, where $t$ is time in hours, $k$ is the clearing factor of the rotor, and $s$ is the sedimentation coefficient in Svedberg units ($S$) |
| Delrin | thermoplastic material (acetal homopolymer) used for most tube adapters, some tube spacers, the bottle cap for the Type 19 rotor, and the crown washer used on red, blue, and black aluminum caps (Delrin is a registered trademark of E. I. Du Pont de Nemours & Company) |
| derated | the condition or status of Beckman zonal, continuous flow, swinging bucket, and aluminum fixed angle rotors when limited to 90% of their original maximum rated speed after a specified time or period of usage (detailed in the rotor warranty) |
| $g$-Max™ | a system of centrifugation using a combination of short Quick-Seal® tubes and floating spacers |
| isopycnic | method of particle separation or isolation based on particle buoyant density; sedimentation equilibrium (often done with CsCl gradients); particles are centrifuged until they reach a point in the gradient where the density of the particle is the same as the density of the gradient at that point |
| konical™ tubes | thin-walled, polyallomer tubes featuring a conical tip to optimize pelleting separations; the conical tip concentrates the pellet in the narrow base of the tube. Available in both open-top and Quick-Seal® bell-top designs. |
| maximum volume | the maximum volume at which a tube should be filled for centrifugation (sometimes referred to as maximum fill volume or nominal fill volume) |
| mechanical overspeed cartridge | an assembly installed in the bases of some older rotors or swinging bucket rotor adapters as part of the mechanical overspeed protection system |
| neoprene | black elastomer used for O-rings in many tube caps and bottle cap assemblies |
| Noryl | hard thermoplastic used for some polycarbonate bottle caps, and for floating spacers used with Quick-Seal® tubes (Noryl is a registered trademark of General Electric) |
| overspeed disk | an adhesive disk, with alternating reflecting and nonreflecting sectors, attached to the bottom of rotors as part of the photoelectric overspeed protection system; the number of sectors on the disk is a function of the rotor's maximum allowable speed |
| pelleting | a centrifugal separation where particles in a sample sediment to the bottom of the tube (differential separation); differential pelleting separates particles of different sizes by successive centrifugation steps of progressively higher $g$ force and/or longer run duration |

**TABLE 2.1**   Principles That Govern Sedimentation—*Continued*

| | |
|---|---|
| polyallomer | random block copolymer of ethylene and propylene used for thinwall, thickwall, and Quick-Seal® tubes (Tenite Polyallomer is a registered trademark of Eastman Chemical Co.) |
| Quick-Seal® tubes | bell-top or dome-top thinwall tubes that are heat-sealed and require no caps |
| rate zonal | method of particle separation, based on differential rate of sedimentation, using a preformed gradient with the sample layered as a zone on top of the gradient |
| relative centrifugal field (RCF) | the ratio of the centrifugal acceleration at a specified radius and speed ($r\omega^2$) to the standard acceleration of gravity ($g$) according to the following formula: $$\mathrm{RCF} = \frac{r\omega^2}{g}$$ where $r$ is the radius in millimeters, $\omega$ is the angular velocity in radians per second ($2\pi\mathrm{RPM}/60$), and $g$ is the standard acceleration of gravity ($9807\ \mathrm{mm/s^2}$). Thus the relationship between RCF and RPM is: $$\mathrm{RCF} = 1.12r\left(\frac{\mathrm{RPM}}{1000}\right)^2$$ |
| sedimentation coefficient, $s$ | sedimentation velocity per unit of centrifugal force: $$s = \frac{\mathrm{d}r}{\mathrm{d}t} \times \frac{1}{\omega^2 r}$$ |
| Solution 555™ | Beckman concentrated rotor cleaning solution; recommended because it is a mild solution that has been tested and found effective and safe for Beckman rotors and accessories |
| Spinkote™ | Beckman lubricant for metal-to-metal contacts |
| Svedberg units, S | a unit of sedimentation velocity: $1S = 10^{-13}$ seconds |

From Beckman Instruments, Inc. 1990. *Rotors and Tubes for Preparative Ultracentrifuges: A User's Manual.* Palo Alto, CA: Beckman Instruments, Inc.

**TABLE 2.2**   Approximate Densities of Macromolecules in Sucrose Solutions

| **Macromolecules** | **Density (g/cm$^3$)** |
|---|---|
| Golgi apparatus | 1.06–1.10 |
| Plasma membranes | 1.16 |
| Smooth endoplasmic reticulum | 1.16 |
| Intact oncogenic viruses | 1.16–1.18 |
| Mitochondria | 1.19 |
| Lysosomes | 1.21 |
| Peroxisomes | 1.23 |
| Plant viruses | 1.30–1.45 |
| Soluble proteins | 1.30 |
| Rhino- and enteroviruses | 1.30–1.45 |
| Nucleic acids, ribosomes | 1.60–1.75 |
| Glycogen | 1.70 |

From Beckman Instruments, Inc. *Techniques of Preparative, Zonal, and Continuous Flow Ultracentrifugation,* 5th ed. Palo Alto, CA: Beckman Instruments, Inc.

*Data taken from *Handbook of Biochemistry.* 1968. Edited by H. A. Sober. 2nd ed. Cleveland: The Chemical Rubber Co., and *Centrifugal Separations in Molecular and Cell Biology.* 1978. Edited by G. D. Birnie and D. Rickwood. London: Butterworths.

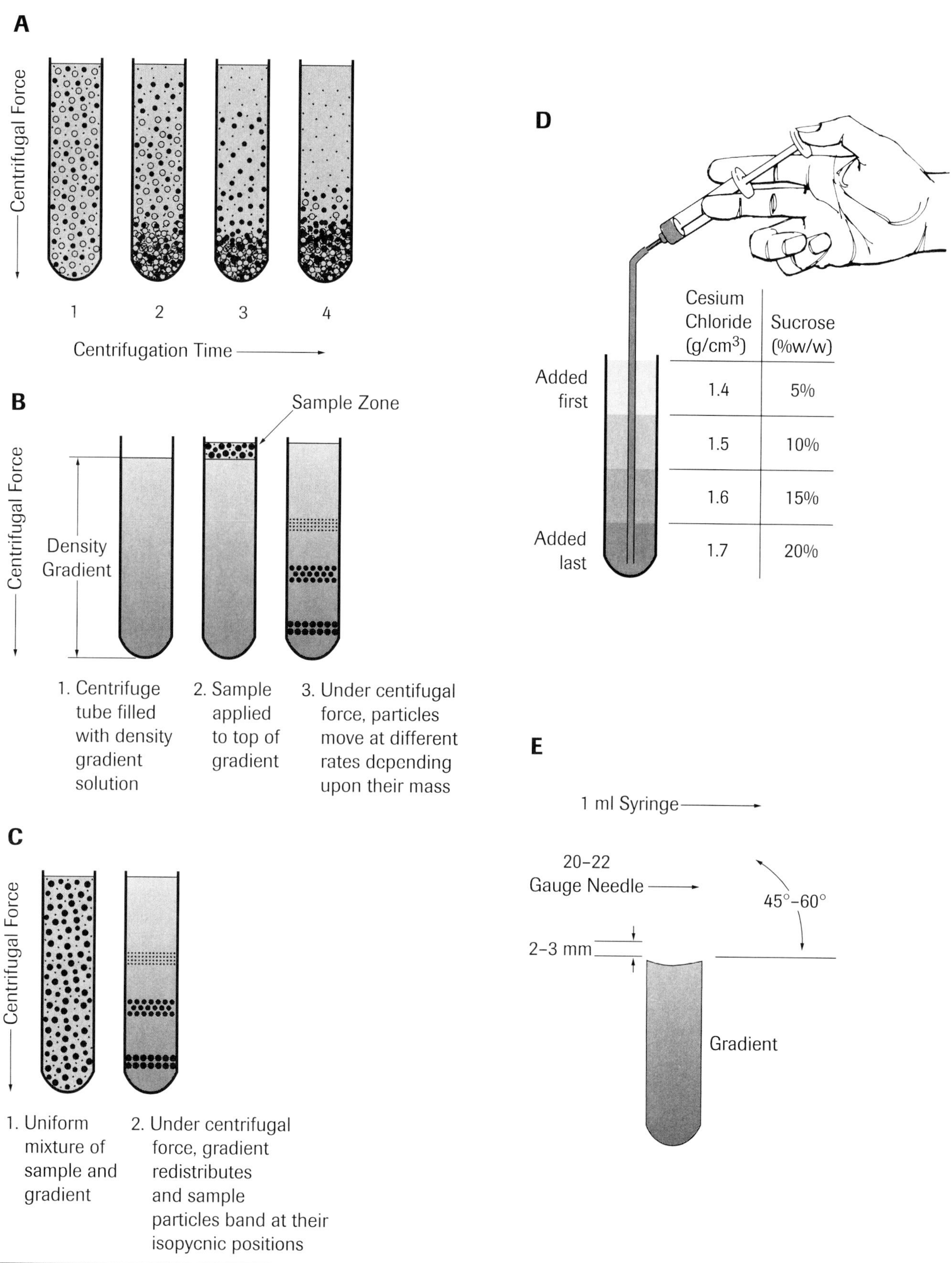


**FIGURE 2.1**   Comparison of differential (**A**), rate zonal (**B**), and isopycnic (**C**) centrifugations. Forming sucrose (**D**) solution and applying the sample (**E**) to a gradient.

From Beckman Instruments, Inc. *Techniques of Preparative, Zonal, and Continuous Flow Ultracentrifugation,* 5th ed. Palo Alto, CA: Beckman Instruments, Inc.

## Growth of *Coriolus Versicolor*

Mycelia of *Coriolus versicolor* (L. ex Fr.) Quel isolate (USDA—Northeastern Forest Exp. Stn., Durham, NH) can be maintained upon agar slants at 4° C. To obtain cultures, mycelia can be transferred to a sterile defined medium[2] containing 2% Bacto agar until a mycelial disc 30 mm in diameter is derived (5 days). Then the discs can be aspectically excised and homogenized 30s into 25 ml autoclaved distilled water within a sterile Wareing blender cup. Next, one ml aliquots of the mycelial homogenates can be transferred with a sterile manostat to autoclaved Identi plug stoppered 250 ml Erlenmeyer flasks containing 25 ml of the above sterile medium lacking agar. Cultures can be maintained at 25 ± 2° C and 150 rpm within a model 3529 Lab-line gyrotory shaker (Melrose Park, FL) for up to 16 days. Polyphenol oxidase appears in the growth medium at day 4 and increases in spc. act. to day 15.[1] The time-dependent appearances of PPO in subcellular fractions[3] are presented in Figure 2.2.

## Preparation of Subcellular Organelles

Subsequent to the separation of culture medium from hyphae by low "speed" centrifugation (500 $xg$, 10 minutes 4° C), the hyphae should be resuspended in 2 ml of 250 mM sucrose, 3 mM EDTA, 25 mM Tris, pH 7.2.[3] To disrupt the hyphae while simultaneously maintaining the structural integrity of endomembrane components, the resuspended hyphae can be sonicated on ice for 15s with a Heat Systems Ultra Sonics (Model W–220F) Sonicator (make appropriate adjustments for other model sonicators). Note: It is not advised to utilize "harsher" cell breakage techniques such as a mortar and pestle, motor-driven tissue homogenizers, and so on, as these can result in marked organellar

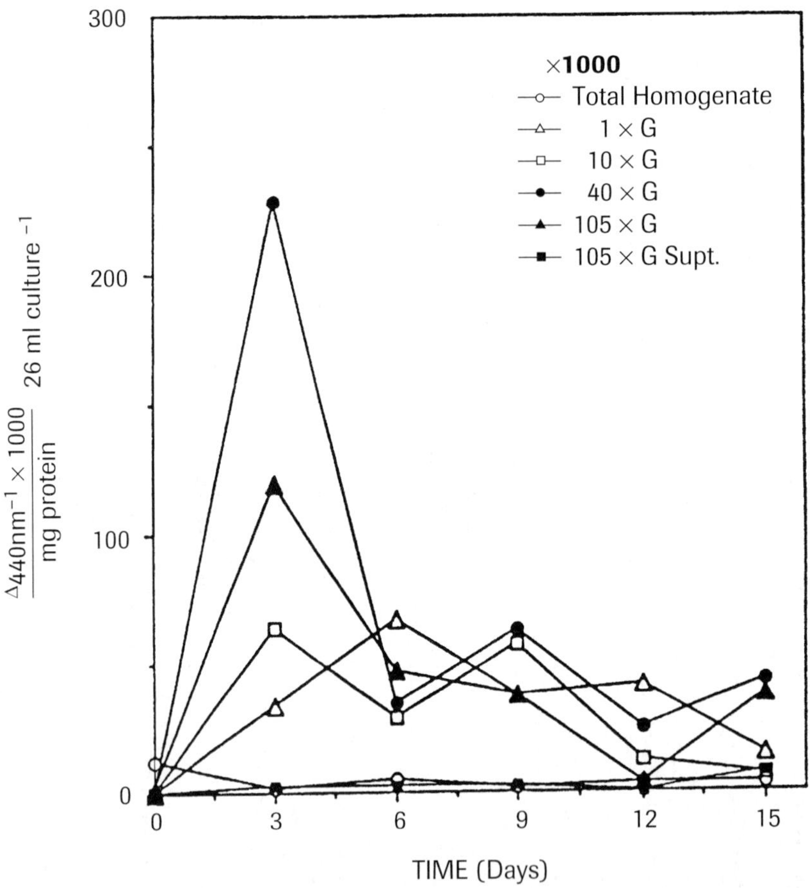


**FIGURE 2.2**  Time-dependent changes in PPO spc. act. in subcellular fractions prepared by differential centrifigation of hyphal homogenates. The RCF's are ×1,000 each.

From Moore et al. 1994. *Biotoxins, Biodegradation and Biodeterioration Research* 4. Llewellyn G.C., C.E. O'Rear and W.V. Dashek (eds.). New York: Plenum Press.

disruption. Next, the resultant sonicate can be subjected to differential centrifugation as in either Figure 2.3 or Figure 2.4 to "pellet" hyphal organelles.

Because the flow chart in Figure 2.4 is rather lengthy, yielding a number of pellets, the student is recommended to follow the flow chart in Figure 2.3 as it can be efficiently performed within a single laboratory period. In addition, the short flow chart provides a centrifugation low g force in a *preparative centrifuge* and another at high g force in an *ultracentrifuge.* In both instances, rpm "settings" to accomplish the centrifugations can be obtained by knowing both the relative centrifugal force (*xg*) and the radius (Figure 2.5) with the subsequent employment of a nomogram (Figure 2.6) and/or the formula RCF = 1.2 r (RPM/1000).[1] In this formula, RCF = relative centrifugal force or *xg* and r = radius.

When performing a centrifugation, the tubes must be positioned in the centrifuge's rotor appropriate to each other, and they must be balanced. If an ultracentrifuge is being employed, it is imperative that the tubes be balanced by weighing with a "trip" balance. Because there is a variety of preparative centrifuges, e.g., Beckman, International, Sorvall, and so on, consult the operations manual for the centrifuge that your laboratory possesses. Most of these companies provide technical booklets regarding rotors, bottles, and tubes as well as the conditions mandating their usages. Subsequent to the 1,000 *xg* centrifugation (see Figure 2.3), decant the supernatant and add it to a 5 ml polyallomar tube (see manufacturer's rotor/bottles/tubes booklet, e.g., Beckman Instruments, Inc., for the appropriate selection of tubes of the wide variety of available swinging bucket and fixed angle rotors). This exercise will employ a swinging bucket rotor (SW 65Ti) and a L8-70M ultracentrifuge (Beckman Instruments, can substitute another company's equivalent)[4] to carry out the 1 h, 105,000 *xg*

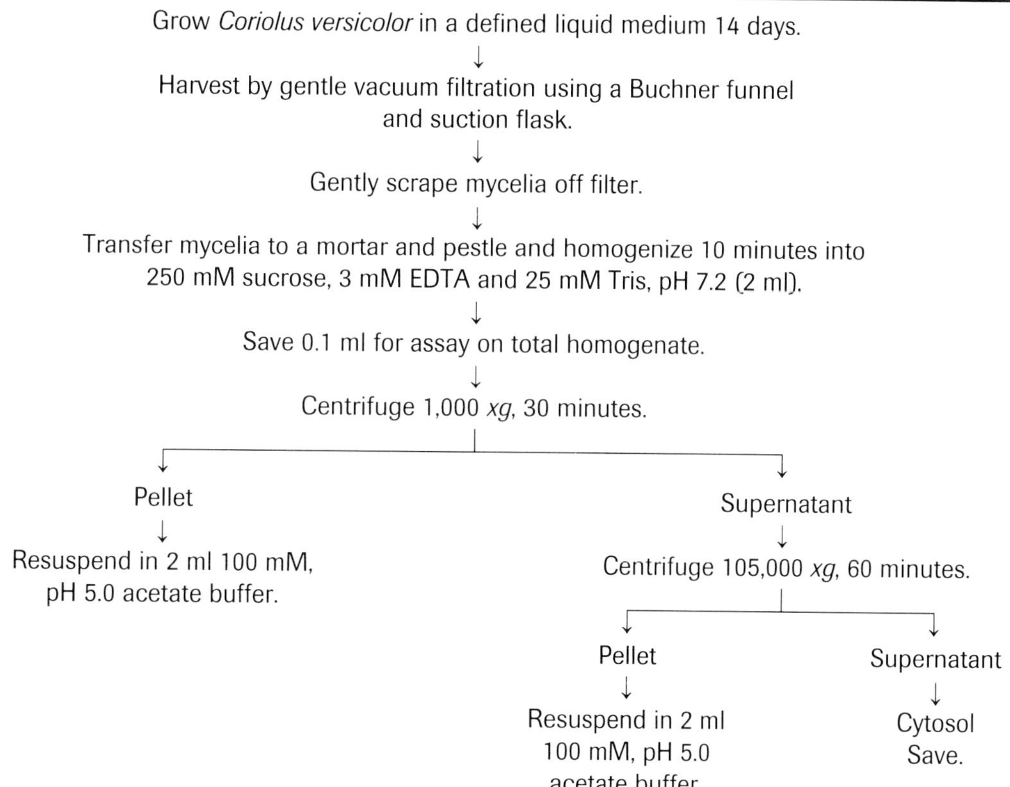



**FIGURE 2.3**  Differential centrifugation procedure for preparation of *Coriolus versicolor* cytoplasmic polyphenol oxidase.

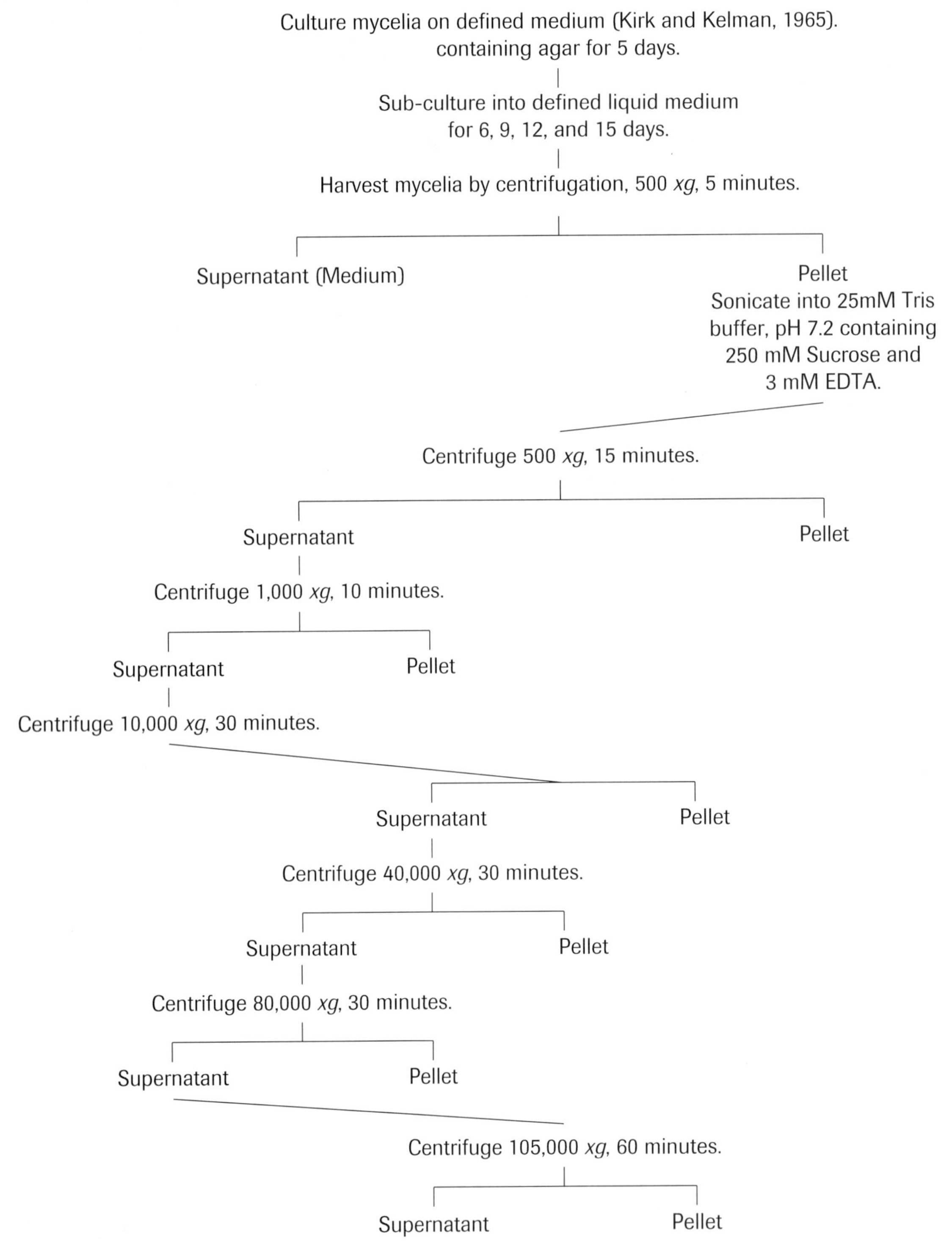


**FIGURE 2.4**  Expanded differential centrifugation procedure for preparation of *Coriolus versicolor* cytoplasmic polyhenol oxidase. Culture mycelia on defined medium (Kirk and Kelman 1965). Resuspend all pellets in 2 ml pH 5.0, 100 mM acetate buffer with a TenBroeck tissue homogenizer and assay 10 µl aliquots for PPO activity according to Evans and Palmer (1983).

Adapted from Danley et al., 1981, and Walker et al., 1984.

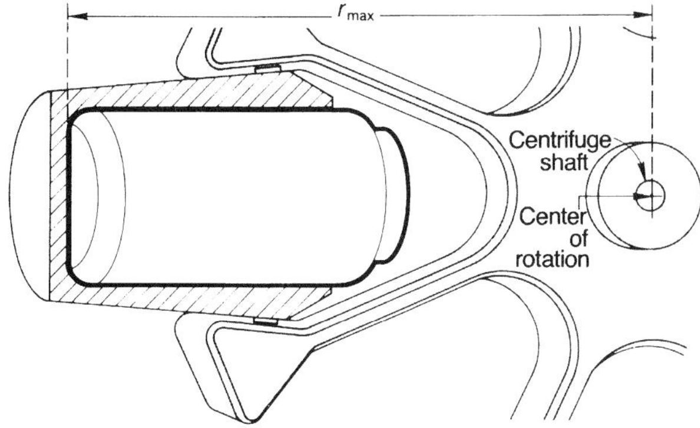



**FIGURE 2.5**  Diagrammatic representation of r max (maximum radius) of a rotor.

From Beckman Instruments, Inc. 1980. *A Centrifuge Primer.* Palo Alto, CA:  Beckman Instruments, Inc.

centrifugation. This RCF will yield the cytosol as the supernatant, and the pellet will contain all organelles not pelleted at 1,000 $xg$. It is paramount that the student recognize the costs of current day ultracentrifuges and their rotors. Furthermore, many of these centrifuges are computerized, offering advantages such as "delayed starts."

If you select Beckman's SW65 rotor, remember to install the buckets correctly, i.e., flat side inward. After the 105,000 $xg$ centrifugation has been completed, decant the supernatant and resuspend the pellet as well as that obtained from the 1,000 $xg$ in 2 ml pH 5.0, 100 mM acetate buffer. Then, store both the pellet and supernatant on ice for the protein quantification module. Note: Before the 105,000 $xg$ supernatant can be assayed for protein by UV spectroscopy, it must be dialyzed (see Module 3, Enzyme Purification) to remove free aromatic amino acids and low molecular weight peptides containing these acids.

# Review Questions

1. Compare and contrast differential and gradient centrifugations.
2. State the similarities and dissimilarities between rate zonal and isopycnic gradient centrifugations.
3. How can you identify the organellar contents of either a pellet prepared by differential centrifugation or a band obtained by combined differential/ sucrose gradient centrifugations?
4. Explain the use of a nomogram—i.e., how can one convert rpm into relative centrifugal force?
5. If one does not have access to a nomogram, what formula can be employed to convert $xg$ into rpm and vice versa?

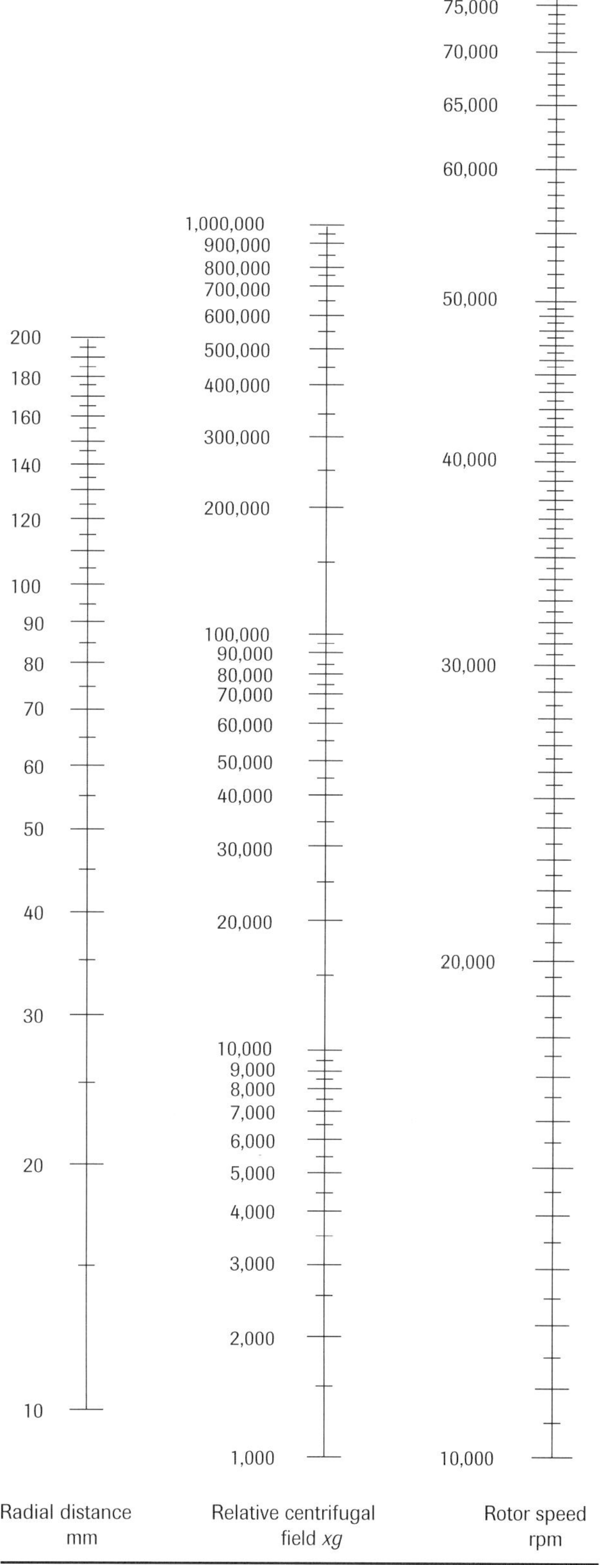



**FIGURE 2.6**  Nomogram for interconversion of rpm and g force. Align a straightedge through known values in two columns; read the desired figure where the straightedge intersects the third column.

From Beckman Instruments, Inc. 1990. *Rotors and Tubes for Preparative Ultracentrifuges: A User's Manual.* Palo Alto, CA: Beckman Instruments, Inc.

# PROTEIN QUANTIFICATION

### Supplies and Equipment
Assay tubes (disposable)
Beakers
Cuvettes, Bio-Rad plastic, quartz, and glass
Filter paper
Funnels
Graph paper
Kimwipes
Pasteur pipettes and bulbs
Pipettes and pro-pipettes
Racks for assay tubes
Wash bottles
UV-visible spectrophotometer

### Chemicals
Bio-Rad Bradford reagent
Bovine serum albumin (BSA) 140 µg
   100 µl$^{-1}$ stock
Lowry reagents
   A.   2% $Na_2CO_3$ in 0.1N NaOH
   B.   1.1% potassium tartrate
   C.   Dissolve 25 mg of $CuSO^4$ completely in
       5 ml of reagent B and dilute to 250 ml with
       reagent A
BCA reagents—Pierce

# Introduction

## Properties of Light

Light is an oscillating stream of photons or quanta with associated electric and magnetic fields. The electromagnetic radiation, which moves through space as a wave, possesses energy defined in terms of either wavelength or frequency. Wavelength designates the distance between the crests of two different waves and the number of waves transversing a given point per time is frequency (Figure 2.7).

Wavelength is currently expressed in nanometers ($10^{-9}$ meter); terms that were used in the literature in years past were ɜ or mµ. Although the light energy spectrum extends from gamma and X rays (1–10Å) to microwaves and radio frequency (25 cm), the portion of the spectrum of interest to life scientists is the near ultraviolet (200 nm–400 nm), visible (400 nm–700 nm) and the near infra-red (700 nm–900 nm). As for the relationship between photons and wavelength, a photon of short wavelength possesses greater energy than one at longer wavelength. Thus, the relationship is inverse.

## Absorption of Light

Although the interaction of light and matter can result in the absorption and possible re-emission as well as

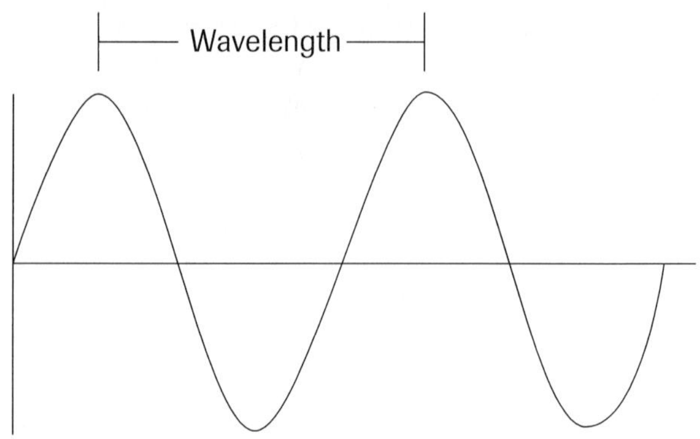


**FIGURE 2.7** Wavelength designates the distance between the crests of two different waves and the number of waves transversing a given point per time is frequency.

the refraction, reflection, dispersion, interference, diffraction or polymerization of light, the absorption of light is of most significance to biologists.

When light passes through a transparent solution of molecules, the molecules may absorb some of the light. The absorption enhances the energy content of the molecules. The amount of this enhancement is equivalent to the energy of the absorbed light and is dictated by the laws of quantum mechanics. A finite amount of energy is required to raise a molecule from the ground state (normal energy level) to the excited state (higher energy level). Furthermore, a molecule can only absorb the amount of energy required for its transition. Thus, for absorption to occur, the light energy must precisely equal the energy differential between the molecule's ground and excited states.

There are two laws that govern the absorption of light: the Lambert and Beer Laws, which are usually combined as the Beer Lambert Law. Beer Lambert's Law can be formulated mathematically, using base 10 logarithms:

$$T = \frac{P}{P_0} = 10^{-ab}$$

where  $T$ = transmittance (amount of light passing through a substance)

      $P$ = the power or intensity of the light transmitted through the substance

      $P_0$ = the power or intensity of the light hitting the substance

or

$$A = \log T = -\log \frac{P}{P_0} = \log \frac{P_0}{P} = abc$$

where  $A$ = absorbance
      $a$ = absorptivity
      $b$ = path length (in centimeters)
      $c$ = concentration (usually in grams/liter)

The Beer-Lambert Law in the form $A = abc$ is extensively used in quantitative analysis since absorbance is directly proportional to concentration. A plot of absorbance versus concentration should give a linear curve whose slope is the absorptivity when the cell length is 1.00 cm. In this connection, most modern spectrophotometers possess scales calibrated in absorbance rather than transmittance. Absorptivity is a parameter of a given substance only for precisely delineated conditions of wavelength, solvent, temperature and other parameters. If path length is provided in centimeters and concentration in g solute $1^{-1}$, the absorptivity is liters g centimeter$^{-1}$. Molar absorptivity is often useful and is the product of the absorptivity ($a$) and the molecular weight of the substance. The Beer-Lambert discussion is paraphrased from Beckman Instruments, Incorporated.[5]

The protein amount in either a pure protein sample or sample containing protein, e.g., cellular homogenates or column eluates obtained during purification of an enzyme, can be quantified by either UV spectroscopy or colorimetry. The former is dependent upon the presence of aromatic amino acids (phenylalanine, tryptophan, tyrosine) within most proteins. These amino acids absorb maximally at 280 nm. Thus, UV spectroscopy does not require the addition of an exogenous dye to both detect and quantify protein. Because UV spectroscopy does not differentiate between peptide-bound aromatic amino acids and free amino acids, the samples must be either dialyzed (e.g., Spectrofluor tubing MW cut-off 14,000) or Amicon ultrafiltrated using YM 10 filters (MW cut-off 10,000).

The most commonly employed colorimetric assay procedures are the Lowry[6] and Bradford[7] assays. Recently, the Pierce Company introduced the use of BCA protein assay reagent for the quantification of protein. Whereas the Lowry method and its subsequent modifications[8, 9, 10] involved the employment of alkaline copper salts, the Bradford technique takes advantage of a wavelength shift to 595 nm when protein binds to the dye Coomassie Brilliant Blue R–250.

The Pierce BCA protein assay reagent is especially applicable to protein samples that contain detergents. It measures proteins in the range of 10µg ml$^{-1}$ – 200 µg ml$^{-1}$. Which colorimetric assay procedure should be employed in a research project? The data of Eze and Dumbroff[11] are relevant. Figure 2.8 compares both the sensitivities and linearities of the Bradford and Lowry et al. assays. In general, the Bradford assay is more sensitive but less linear than the Lowry et al. method over the same protein amount range. In this connection, Ghosh et al.[12] modified the Bradford assay to detect ng amounts of protein. The modifications consisted of spotting ng protein amounts upon Whatman 3MM chromatography paper followed by staining of the dried samples with Coomassie Brilliant Blue R–250 for 20 minutes. Subsequent to drying, the protein amounts were quantified by either reflectance using a scanning densitometer or an ELISA plate reader. In addition to its sensitivity, the method avoids interference by chlorophyll, dithiothreitol, SDS, β–mercaptoethanol, Nonidet P–40 and phenylmethylsulfonyl fluoride.

Beside the above colorimetric assay, Naleczynska[13] reported the use of bromphenol blue for the quantitative estimation of protein in plant extracts. This method, a modification of the colorimetric assay of Flores,[14] adds gum arabic to Flores' assay mixture enabling direct quantification of the bromphenol blue protein complex at acid pH. The complex can be precipitated under denaturing conditions in acid alcoholic solutions onto paper. It does not appear that this

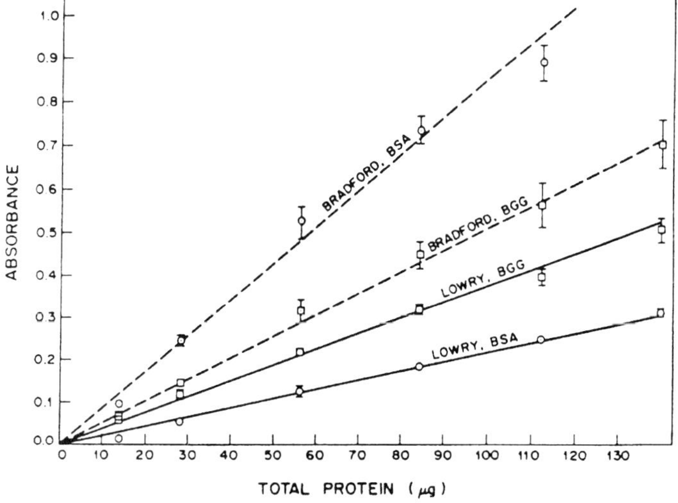


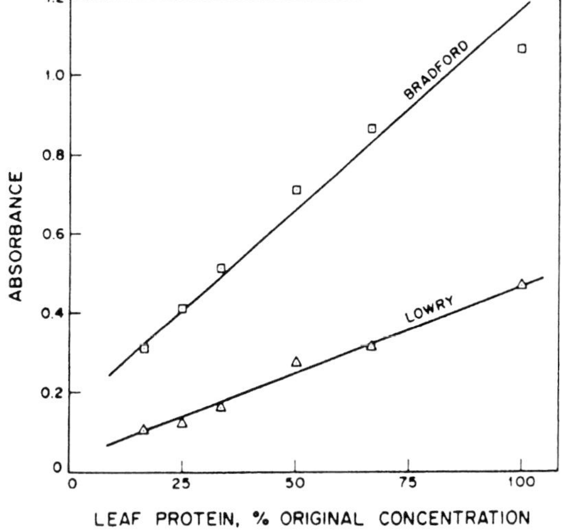


**FIGURE 2.8**  Comparison of standard curves for the Bradford and Lowry quantification of leaf proteins.
Eze, J. M. O. and E. B. Dumbroff. 1982. "A comparison of the Bradford and Lowry methods for the analysis of protein in chlorophyllous tissue." *Can. J. Bot.* 60:1046–49.

method has been widely accepted to estimate the protein content of plant extracts.

# Procedures

This portion of the module concerns quantification of protein contents of subcellular organelles.

## UV Spectroscopy

Following resuspension of subcellular organelles prepared by differential centrifugation in 2 ml 0.15 M NaCl, remove 10, 25, 50 and 100 μl aliquots and perform protein assays by UV spectroscopy (Figure 2.9) as well as Lowry et al. (Figure 2.10), Bio-Rad Bradford (Figure 2.11A,B) and Pierce BCA protein reagent (Figure 2.12) colorimetric assays. Note that you must construct a BSA standard curve, as in Table 2.3.

Perform the UV spectroscopy assay on the same day that you prepare the organelles. Complete Table 2.4 after determining the protein content in your aliquot size through the application of a standard curve (see Figure 2.13A,B for typical Bradford standard curves). If necessary refer to the Method of Least Squares section to help you in finding the line that "best" fits the data points. The colorimetric assays can be accomplished during the subsequent laboratory period. Please make sure that you maintain the organelle preparations on ice at 4° C.

## Colorimetric Assays

During the next laboratory, use the same sample aliquot sizes that you employed in the previous laboratory to perform the colorimetric assays as detailed

---

Prepare a stock solution of 140 μg bovine serum albumin (BSA) 100 μl solvent$^{-1}$.
↓
Construct a series of BSA amounts as in Table 2.3 for standard curve construction, note that the curve can be extended to 1 ml – 1400 μg.
↓
Dilute samples to 1 ml with solvent.
↓
Record absorbances at 280 nm (absorption maximum for aromatic amino acids) with the Deuterium lamp of a UV-visible spectrophotometer and using semi-micro quartz curvettes.
↓
Plot $^A$280 nm on (Y axis) and μg BSA (X-axis). (Use least squares method if necessary.)

---

**FIGURE 2.9** UV spectroscopy—flowchart for quantification of protein contents of subcellular organelles.

---

Prepare a stock solution of 140 μg BSA 100 μl solvent,$^{-1}$ note appropriate solvent for Lowry *et al.* assay is dilute NaOH.
↓
Construct a series of BSA amounts as in Table 2.3 for standard curve construction.
↓
Add 5 ml of reagent C to each assay tube; vortex and maintain at room temperature for 10 minutes.
↓
Add 0.5 ml of 1N Folin-Ciocalteu phenol agent to each tube; vortex and maintain at room temperature for 30 minutes.
↓
Record absorbances at 660 nm with the Tungsten lamp of a spectrophotometer using plastic or glass cuvettes.
↓
Plot $^A$660 nm (Y axis) vs. μg BSA (X axis). (Use least squares method if necessary.)

---

**FIGURE 2.10** Flow chart for the quantification of protein by the Lowry colorimetric assay.

Adapted from Lowry et al. 1951. "Protein measurement with the Folin phenol reagent." *J. Biol. Chem.* 193:265-75.

---

Prepare a stock solution of 140 μg bovine serum albumin (BSA) 100 μl solvent$^{-1}$.
↓
Construct a series of BSA amounts as in Table 2.3 for standard curve construction.
↓
Dilute commercially-available Bio-Rad Bradford reagent 1:5 with distilled $H_2O$ and filter through Whatman No. 1 filter paper.
↓
Add 5 ml of diluted and filtered dye to each tube.
↓
Maintain at room temperature for 5-60 minutes.
↓
Record absorbances at 595 nm with the Tungsten lamp of a spectrophotometer using plastic or glass cuvettes.
↓
Plot $^A$595 nm for (Y-axis) vs. μg BSA (X-axis). Note: See Figure 2-12 for a typical standard curve. (Use least squares method if necessary.)

---

**FIGURE 2.11A** Flow chart for the quantification of protein by the standard Bio-Rad Bradford assay.

Adapted from Bio-Rad Laboratories. *Protein Assay Instruction Manual.* Richmond, CA: Bio-Rad.

Construct several dilutions of protein standards from 1-25 µg ml$^{-1}$.

↓

Add 0.8 ml of protein standards to clean assay tubes; add 0.8 ml sample buffer to a blank assay tube.

↓

Add 0.2 ml Bio-Rad dye reagent concentrate.

↓

Vortex or mix several times by gentle inversion.

↓

Record absorbances at 595 nm after 5-60 minutes.

↓

Plot $^A$595 nm on Y-axis and µg protein on X-axis. (Use least squares method if necessary.)

**FIGURE 2.11B**  Flow chart for the quantification of protein by the Bio-Rad Bradford microassay.

Adapted from Bio-Rad Laboratories. *Protein Assay Instruction Manual.* Richmond, CA: Bio-Rad.

in Figures 2.10–2.12. Use the table format in Table 2.4 for protein calculations.

## Colorimetric Assay Sensitivity

The Pierce BCA protein assay[15] is a highly sensitive, flexible protein quantification procedure that does not require the precisely timed addition of reagents and vortexing as the Lowry procedure. In addition, other advantages include: "compatibility with ionic and non-ionic detergents, a stable working reagent, less protein-to-protein variation than with other methods, broad linear working ranges with excellent sensitivity and the ability to change the protocol which provides flexibility"[15]

The main reagent in the patented BCA protein assay is Bicinchoninic acid (BCA) as a water-soluble salt (see Figure 2.12). It is a sensitive, stable and highly specific reagent for cuprous ions. Color development appears to be dependent upon cysteine, cystine, tryptophan and tyrosine.

The sensitivity of the BCA assay is affected by both time and temperature (Table 2.5). Figure 2.14 presents the standard curves obtained with standard, room temperature, and enhanced protocols. The assay procedure is as in Figure 2.12. Refer to Pierce BCA Protein Assay Reagent Instructions Booklet for discussions regarding compatibility with selected biochemical reagents, protein-to-protein variation, and applications of a microtiter plate protocol.

## *Numerical Curve Fitting: The Method of Least Squares*

Many types of data collection result in a linear curve when the dependent variable $y$ is plotted against the independent variable $x$. The equation for a straight line that results when a set of data $(x_1y_1)$, $(x_2y_2)$, . . ., $(x_ny_n)$ is obtained is:

$$y = mx + b \qquad (1)$$

where $m$ is the slope of the line and $b$ is the $y$-intercept that results when $x = 0$.

When a set of data points lies almost, but not quite, on a straight line, the experimenter must try to find the line that "best" fits the data points. "Best" can be defined in terms of the so-called **least-squares criterion,** according to which the line must be such that the sum of the squares of the vertical departures of the points from the best line, commonly called the residuals, are a minimum.

The residual is defined as $r = y - (mx + b)$; hence the square of the residual is $r^2 = [y - (mx + b)]^{-2}$. When the sum of the squares of the residuals are a minimum, the best line for the data points will result, and the normal equations for the linear least-squares line for a set of $N$ data points can be shown to be:

$$\sum y = m \sum x + Nb \qquad (2)$$

$$\sum xy = m \sum x^2 + b \sum x \qquad (3)$$

equations 1.2 and 1.3 to give:

$$m = \frac{N \sum xy - \left(\sum x\right)\left(\sum y\right)}{N \sum x^2 - \left(\sum x\right)^2} \qquad (4)$$

$$b = \frac{\left(\sum x\right)\left(\sum x^2\right) - \left(\sum x\right)\left(\sum xy\right)}{N \sum x^2 - \left(\sum x\right)^2} \qquad (5)$$

The line that has a slope of $m$ from Equation 4 and a $y$-intercept of $b$ from Equation 5 is then the best fitting line or the linear least-squares line for the data. The normal equations (2 and 3) can be remembered by observing that Equation 2 is obtained by summing on both sides of Equation 1, and Equation 3 is obtained by multiplying both sides of Equation 1–5 by $x$ and then summing. This is not a derivation of the normal linear least-squares equations, but a means to remember them. For a derivation, see any calculus text.

Thus, to compute the slope $m$, and the $y$-intercept $b$, for a set of $n$ data points using the linear least-squares method, it is necessary to calculate $\Sigma x$, $\Sigma y$, $\Sigma x^2$, and $\Sigma xy$ for the set of data having $N$ variables and to put these values into Equations 4 and 5.

The manual determination of the values of $m$ and $b$ by the least-squares method is time-consuming, but with the use of computers, the method becomes a practical way for finding the best values for $m$ and $b$ from the set of data, which in turn will give the best-fit linear line with a slope of $m$ and a $y$-intercept of $b$ for the set of data.

## BCA Assay Procedure

Because the ratio of working reagent to sample volume is the same (i.e. 20 parts working reagent to 1 part sample) all of the recommended protocols are identical except for their incubation temperatures and times.

Prepare a set of protein standards of known concentration by diluting the stock 2 mg/ml BSA standard (bovine serum albumin), or other suitable protein, in the same diluent as your unknown samples. The set of protein standards should cover the range of concentrations suitable for the assay protocol you are following. For best results, prepare standards using known concentrations of the protein you are testing.

↓

Pipet 0.1 ml of each standard or unknown protein sample into the appropriately labeled test tube. For blanks, use 0.1 ml of diluent.

↓

Add 2.0 ml working reagent to each tube. Mix well.

↓

Incubate all tubes at the selected temperature and time:
**Standard Protocol** .........................................**37° C for 30 minutes**
**Room Temperature Protocol** ...............................**R.T. for 2 hours**
**Enhanced Protocol** ......................................**60° C for 30 minutes**

↓

After incubation, cool all tubes to room temperature.

↓

Measure the absorbance at 562 nm of each tube vs. water reference.

↓

Subtract the absorbance of the blank from the value found for the standards of unknowns.
**Note**: If you have a double beam spectrophotometer, blank correction can be made automatically. Simply zero your instrument with both the sample and reference cuvettes filled with the solutions from duplicate developed blanks.

↓

Leave the reference cell filled with blank solution and then read the absorbance of the samples/standards placed in the sample cuvette. These absorbance readings are now blank corrected.

↓

Prepare a standard curve by plotting the net (blank corrected) absorbance at 562 nm vs. protein concentration. Using this standard curve, determine the protein concentration for each unknown protein sample.

---

**FIGURE 2.12**   Flow chart for the quantification of protein by the Pierce BCA assay procedure.
From Pierce Chemical Co. *BCA Protein Assay Reagent Instructions Booklet.* Rockford, IL: Pierce Chemical Co.

**TABLE 2.3**   For Protein Standard Curve Construction

| Tube No. | µg BSA | µl BSA stock[a] | µl Solvent |
|---|---|---|---|
| 1 | 140 | 100 | 0 |
| 2 | 70 | 50 | 50 |
| 3 | 35 | 25 | 75 |
| 4 | 21 | 15 | 85 |
| 5[b] | 0 | 0 | 100 |

[a]BSA stock = 140 µg BSA 100 µl solvent⁻¹
[b]Tube 5 will be used as the blank

**TABLE 2.4**  Quantification of Protein Contents of *Coriolus Versicolor* Hyphal Subcellular Organelles

| Organelle Fraction $xg$ | Aliquot Size µl | µg Aliquot$^{-1}$ | Total Volume ml | mg Total Volume$^{-1}$ |
|---|---|---|---|---|

**TABLE 2.5**  Summary of Three Different Types of Protocols for Temperature and Time-Dependent Assays of Protein Using the Pierce BCA Protein Assay[a]

| Protocol | Temperature and Time | Sensitivity |
|---|---|---|
| Standard | 37° C 30 minutes | Applicable Protein Range = 20 µg ml$^{-1}$ – 1200 µg ml$^{-1}$; excellent sensitivity of about 0.012 absorbance units (AU) per microgram of protein; short analysis time |
| Room temperature | Room temperature 120 minutes | Applicable Protein Range = 20 µg ml$^{-1}$–1200 µg ml$^{-1}$; after 120 minutes, the typical sensitivity is 0.010 AUµg protein, after 21 hours, the sensitivity doubles |
| Enhanced | 60° C 30 minutes | Applicable Protein Range = 5 µg ml$^{-1}$ – 250 µg ml$^{-1}$; protocol suitable for situations where the sample concentration is expected to be low; sensitivity is excellent at about 0.032 AU/µg |

Adapted from Pierce Chemical Co. *BCA Protein Assay Reagent Instructions Booklet.* Rockford, IL: Pierce Chemical Co.

[a]Color stability—Following cooling, the blank corrected absorbance at 562 nm continues to increase at the following rates:

| | |
|---|---|
| Standard protocol | 2.3% 10 minutes$^{-1}$ |
| Room temperature protocol | 2.3% 10 minutes$^{-1}$ |
| Enhanced protocol | 0.1% 10 minutes$^{-1}$ |

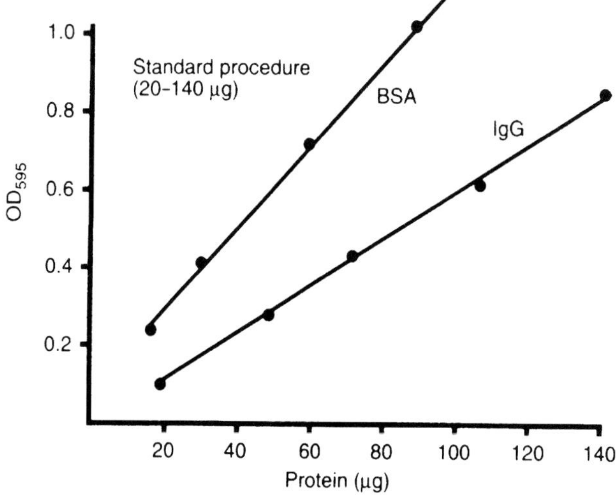


**FIGURE 2.13A**  Typical standard curve for the Bio-Rad protein assay, bovine gamma globulin (standard I), bovine serum albumin (standard II). O.D.$_{595}$ corrected for blank –200–1,400 µl/ml × 0.1 ml = 20–140 µg protein.

From *Bio-Rad Protein Assay.* Hercules, CA: Bio-Rad Life Science Group.

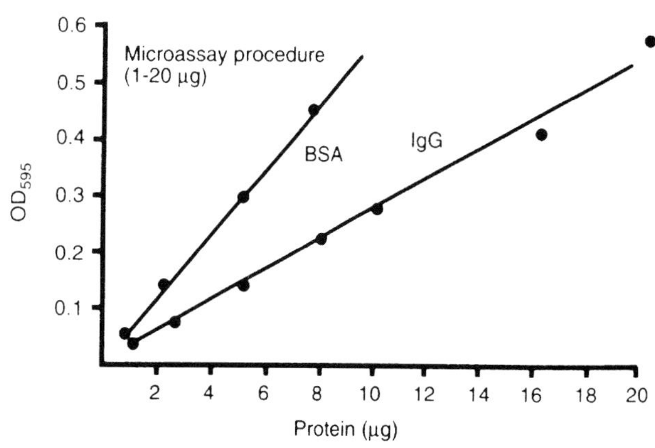


**FIGURE 2.13B**  Typical standard curve for the Bio-Rad protein microassay (1–20 mg/ml), bovine gamma globulin (standard I), bovine serum albumin (standard II). O.D.$_{595}$ corrected for blank 1.25–25 mg/ml × 0.8 ml = 1–20 mg protein.

From *Bio-Rad Protein Assay.* Hercules, CA: Bio-Rad Life Science Group.

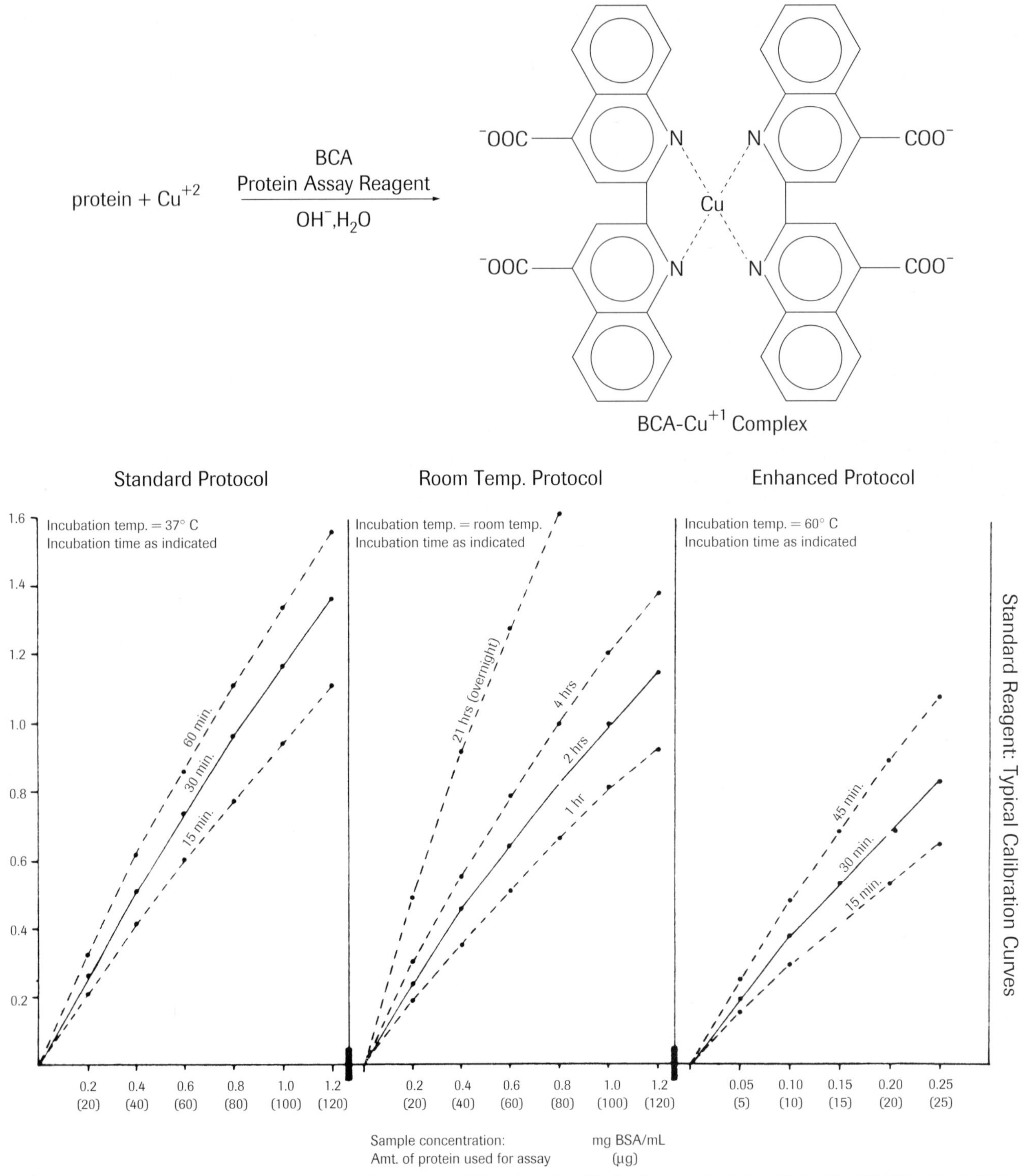

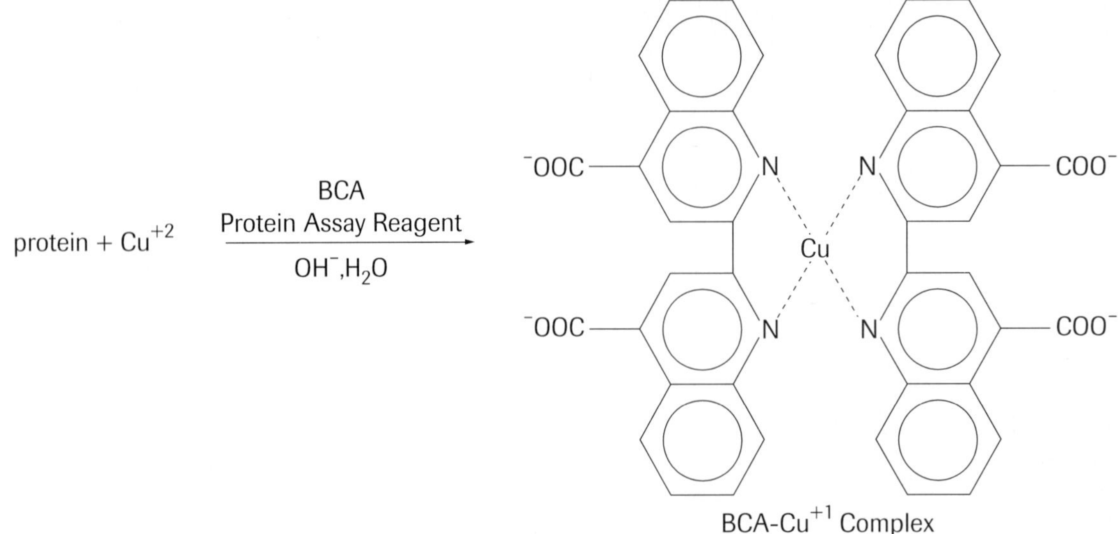



**FIGURE 2.14**   Illustrates the reaction of protein with BCA reagent in the presence of $Cu^{2+}$ ions. The water-soluble reaction product possesses an absorption maximum at 562 nm.

From Pierce Chemical Co. *BCA Protein Assay Reagent Instructions Booklet.* Rockford, IL: Pierce Chemical Co.

If a given dependent variable and independent variable produce a linear fit, the experimental points will deviate from the best-fit line only by experimental error and the residuals will be a mix of positive and negative values, with no set pattern.

In addition to inspecting the residuals, a quantity called the experimental linear correlation coefficient can be calculated. This coefficient gives information about the overall closeness of a least-squares fit to the data. If the value is zero, there is no correlation. If the data points lie exactly on the best-fitted line, the correlation coefficient will be 1 if the slope is positive and –1 if the slope is negative. In a fairly close fit, the magnitude of the linear correlation coefficient might be 0.99.

Computer programs make it relatively easy to carry out a linear least-squares fit by automatically calculating the slope, the $y$-intercept, the correlation coefficient, and the residuals for each data point. A wise experimenter, however, will make a plot of the experimental data points and draw the least-squares lines on the same graph. Any errors in the results generated by the computer should be evident, as well as any bad data points.

## Review Questions

1. Detail the components of a protein standard curve and how the curve is prepared.
2. Enumerate how the curve is used to quantify protein in biological preparations, e.g., cellular homogenates, chromatographic column eluates, and so on.
3. Compare and contrast the sensitivities and linearities of the Bradford, Lowry, and Pierce colorimetric assays for quantifying protein.
4. When is it desirable to use quartz, plastic, or glass cuvettes and either the tungsten or deuterium lamps of a UV-visible spectrophotometer?

## REFERENCES


1. Moore, N. L., D. H. Mariam, A. L. Williams, and W. V. Dashek. 1989. "Substrate specificity, *de novo* synthesis and partial purification of polyphenol oxidase from the wood decay fungus, *Coriolus versicolor.*" *J. Indust. Microbiol.* 4:349–63.
2. Beckman Instruments, Inc. *Techniques of Preparative, Zonal, and Continuous Flow Ultracentrifugation*, 5th ed. Palo Alto, CA: Beckman Instruments, Inc.
3. Moore, N. L., L. A. Brako, C. Clausen, B. R. Jones, and W. V. Dashek. 1994. "Distribution of polyphenol oxidase in organelles of hyphae of the wood-deteriorating fungus, *Coriolus versicolor.*" *Biodeterioration Res.* 4. New York: Plenum Press. 333–56.
4. Beckman Instruments, Inc. 1980. *A Centrifuge Primer.* Palo Alto, CA: Beckman Instruments, Inc.
5. Beckman Instruments, Inc. *A Brief Introduction to Ultraviolet/Visible Spectroscopy.* Palo Alto, CA: Beckman Instruments, Inc.
6. Lowry, O. H., N. J. Rosenbrough, A. L. Farr, and R. J. Randall. 1951. "Protein measurement with the Folin phenol reagent." *J. Biol. Chem.* 193:265–75.
7. Bradford, M. M. 1976. "A rapid and sensitive method for the quantitation of microgram quantities of protein utilizing the principle of protein dye binding." *Anal. Biochem.* 72:248–54.
8. Peterson, G. L. 1977. "A simplification of the protein assay method of Lowry et al. which is more generally applicable." *Anal. Biochem.* 83:346–56.
9. Peterson, G. L. 1979. "Review of the Folin phenol protein method of Lowry, Rosenbrough, Farr and Randall." *Anal Biochem.* 100:201–20.
10. Larson, E., B. Howlett, and A. Jagendorf. 1986. "Artificial reductant enhancement of the Lowry methods for protein determination." *Anal. Biochem.* 155:243–48.
11. Eze, J. M. O. and E. B. Dumbroff. 1982. "A comparison of the Bradford and Lowry methods for the analysis of protein in chlorophyllous tissue." *Can J. Bot.* 60:1046–49.
12. Ghosh, S., S. Gepstein, J. Heikkila and E. B. Dumbroff. 1988. "Use of a scanning densitometer or an ELISA plate reader for measurement of nanogram amounts of protein in crude extracts from biological tissue." *Anal. Biochem.* 169:227–33.
13. Nalezynska, A. 1981. "Use of bromphenol blue for quantitative estimation of protein in plant extracts." *Bulletin, de l' Academie Poonaise des Sciences.* 24:431–37.
14. Flores, R. 1978. "A rapid and reproducible assay for quantitative estimation of proteins using bromophenol blue." *Anal. Biochem.* 88:605–11.
15. Pierce Chemical Co. *BCA Protein Assay Reagent Instructions Booklet.* Rockford, IL: Pierce Chemical Co.


## Supplementary


Burton, M. L., L. T. Onstott, and A. S. Polars. 1989. "The use of gold reagents to quantitate antibodies eluted from nitrocellulose blots applications to electron microscopic immunocytochemistry." *Anal. Bicochem.* 183:225–30.

Cheley, S., and H. Bayley. 1991. "Assaying nanogram amounts of dilute proteins." *Biofeedback.* 10:2.

Ciesiolka, T., and H.-J. Gobruis. 1988. "An 8 to 10-fold enhancement in sensitivity for quantitation of proteins by modified application of colloidal gold." *Anal. Biochem.* 168:280–83.

Condio, J. "Ultracentrifugation. Use of k factor for estimating run times from previously established run times. Applications data." Palo Alto, CA: Beckman Instruments, Inc.

Holleman, J. M., and J. L. Key. 1967. "Inactive and protein precursor pools of amino acids in the soybean hypocotyl." *Plant Physiol.* 42:29–36.

Hunter, J. B. and S. M. Hunter. 1987. "Quantification of proteins in the low nanogram range by staining with the colloidal gold stain Auro Dye." *Anal. Biochem.* 164:430–33.

Li, K. W., W. P. M. Geraerts, R. Van Elk, and J. Joosee. 1989. "Quantification of proteins in the sub-nanogram and nanogram range. Comparison of Auro dye, Ferri Dye and India ink staining methods." *Anal. Biochem.* 182:44–47.

Schuffner, W., and C. Weissman. 1973. "A rapid, sensitive and specific method for the determination of protein in dilute solution." *Anal. Biochem.* 56:502–14.

Stoschek, C. M. 1987. "Protein assay sensitive at nanogram levels." *Anal. Biochem.* 160:301–05.

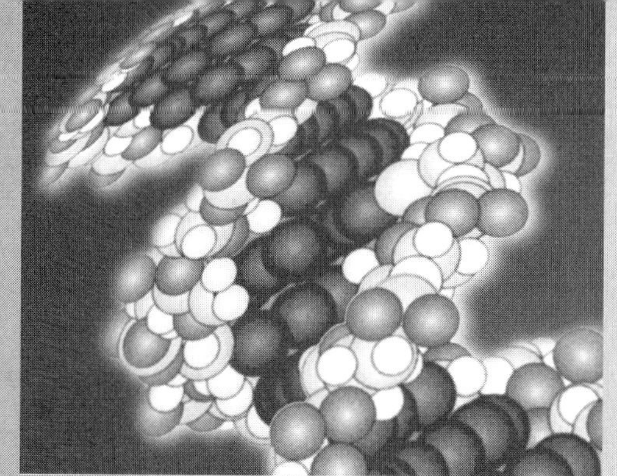



# Enzyme Purification

## Outline of Module

Theory of dialysis, ammonium sulfate fractionation, affinity, gel filtration, hydroxyapatite, hydrophobic interaction and ion exchange chromatographies as well as SDS-PAGE.

### MONAMINE OXIDASE (MAO) PURIFICATION

Preparation of mitochondria from beef liver
*Duration:*   One or two days based on the quantity. It must be prepared by either a technician or a student assistant prior to MAO purification.

Solubilization of MAO from mitochondria
*Duration:*   One 4 hour laboratory period

Purification of MAO
*Duration:*   One 3-4 hour laboratory period

Ammonium sulfate fractionation dialysis
Calcium phosphate titration
DEAE chromatography
Dialysis
Affinity chromatography
Assessment of purity—SDS-PAGE
Storage of purified MAO
*Duration:*   One 3 hour laboratory period

Under aerobic conditions in the presence of β-mercaptoethanol
Plasma amine oxidase (PAO) purification
*Duration:*   One 3 hour laboratory period

Preparation of plasma

Ammonium sulfate fractionation
Ion exchange chromatography
Affinity chromatography
Enzyme assays and protein determination
Plasma amine oxidase assay
Protein concentration

Determination of molecular weight by SDS polyacrylamide disc electrophoresis
*Duration:*   Two 3 hour laboratory periods

Preparation of samples for electrophoresis
*Duration:*   One 3 hour laboratory period

Electrophoretic run
*Duration:*   One 3 hour laboratory period

Protein solubilities
Removal and staining of gels
*Duration:*   One 3 hour laboratory period

Electrophoretic destaining
*Duration:*   One 2–3 hour laboratory period

Lowry protein assay
Evaluation of gels
*Duration:*   One 2–3 hour laboratory period

Determination of molecular weight of mono and plasma amine oxidases.
Molecular weight determination by gel filtration
*Duration:*   One 2–3 hour laboratory period

Separation of proteins based on molecular weights
Gel filtration
*Duration:*   One 2–3 hour laboratory period

Molecular weight determination by comparison of elution volumes
*Duration:*   One 3 hour laboratory period

Assessment of purity of mono and plasma amine oxidase by UV-visible spectroscopy
*Duration:*   One 3 hour laboratory period

Amino acid analysis
*Duration:*   One 2–3 hour laboratory period


## POLYPHENOL OXIDASE PURIFICATION (PPO)

Growth and harvesting of *Coriolus versicolor* and review of polyphenol oxidase assay and the concept
of fold-purified
*Duration:*   One laboratory period to prepare sterile, defined-growth medium with and without
2% agar and to inoculate agar solidified medium; grow fungus to 30 mm disc (5 days,
aseptically remove disc, homogenize and inoculate into defined liquid medium (one
laboratory period) and culture 10–13 days according to Moore et al.[17] During the latter
laboratory, review modules (polyphenol oxidase assay and the concept of fold-purified;
protein assay, calculations for spc. act. quantification).

Dialysis of *C. versicolor* growth medium and $(NH_4)_2SO_4$ fractionation of dialyzed growth medium
   *Duration:*   Harvest growth medium, and rapidly freeze in liquid nitrogen and lyophilize 2 days
            before laboratory. Resuspend lyophilized growth medium and dialyze overnight the day
            before ammonium sulfate[17] fractionation/dialysis (one laboratory); overnight dialysis.[16]
            Maintain enzyme preparation on ice for PPO assay. The next laboratory period Amicon
            ultrafiltration can be accomplished within one of the laboratory periods with YM10
            filters; dialysis can be omitted if Amicon device available.

Gel filtration of dialyzed, $(NH_4)_2SO_4$ fractionated growth medium
   *Duration:*   Two laboratory periods: one laboratory—gel filtration; one laboratory—PPO assay and
            calculations. Maintain enzyme preparation on ice.

Ion exchange chromatography of dialyzed, $(NH_4)_2SO_4$, fractionated, gel-filtrated growth medium
   *Duration:*   One laboratory—ion exchange; one laboratory—PPO assay and calculations. Maintain
            enzyme preparation on ice.

Application of other chromatographies—hydroxyapatite, hydrophobic interaction and affinity
   chromatographies
   *Duration:*   Two laboratories each: one laboratory—performance of technique; one laboratory—
            for PPO assay and calculations. Maintain enzyme preparation on ice.

Assessment of purity: SDS-PAGE
   *Duration:*   Two laboratories: one laboratory—SDS-PAGE; one laboratory—measurement and
            calculations of $m_r$, plot of calibration curve. Between laboratory one and laboratory two,
            the gels must be stained and destained.

Molecular weight determination of purified enzymes by calibrated gel filtration
   *Duration:*   One laboratory

Growth and harvesting of *Coriolus versicolor*
Review of polyphenol oxidase assay and the concept of fold-purified
Dialysis of *C. versicolor* growth medium
$(NH_4)_2SO_4$ fractionation of dialyzed *C. versicolor* growth medium
Gel filtration of dialyzed $(NH_4)_2SO_4$, fractionated *C. versicolor* growth medium
Application of other chromatographies—affinity, hydroxyapatite, hydrophobic interaction
Ion exchange chromatography of dialyzed $(NH_4)_2SO_4$ fractionated, gel filtrated *C. versicolor*
   growth medium
Assessment of PPO purity—SDS-PAGE
Molecular weight determination of purified enzymes by calibrated gel filtration resins and SDS-PAGE

# MODULE INTRODUCTION

This module is concerned with the partial and/or highly pure preparation purification of two mammalian oxidases, liver monoamine oxidase (MAO), plasma amine oxidase (PAO), and fungal polyphenol oxidase (PPO). Please refer to the enzyme assay Module 4 for a discussion of monoamine oxidase's importance in biological systems. The protocols resulting in their partial or highly pure purifications involve a variety of biochemical techniques commonly employed in protein purifications. The sequences of techniques, which result in purification of proteins to homogeneity, are protein-dependent. Usually, the purification protocol begins with dialysis and ammonium sulfate fractionation followed by a combination of diverse chromatographies—e.g., affinity,[1,2] gel filtration,[3,4] high pressure liquid,[5] hydroxyapatite, hydrophobic interaction,[6] and ion exchange.[7] Often, the progress of purification is monitored by SDS-PAGE.[8,9] Prior to providing laboratory exercises concerning the partial or highly pure purifications of MAO, PAO and PPO, the theory underlying each of the above techniques is discussed.

As mentioned, the initial stages of purification often involve dialysis and ammonium sulfate fractionation. Dialysis is one of the steps through which a protein preparation (e.g., cell/tissue homogenates) can be partially purified utilizing tubing (e.g., Spectrofluor) possessing various molecular weight "cut-offs." The value of this procedure is in its ability to remove low molecular weight peptides and free aromatic amino acids that absorb at 280 nm. This is significant as UV-spectroscopy is often employed to quantify protein (see Module 2) throughout an evolving purification protocol, since this quantification method is based upon the absorption of aromatic amino acids at 280 nm in the absence of any added dye. Thus, the quantification method is nondestructive, permitting further use of the sample. Dialysis will, of course, remove other low molecular weight compounds and ions.

Subsequent to dialysis, ammonium sulfate fractionation ("salting out") is frequently utilized to delete contaminating proteins other than the one being purified. This technique is based upon the knowledge that proteins can be precipitated by high concentrations of certain salts.[10] In this technique, "cuts" are constructed—e.g., 0–30%, 30–60%, 60–70% saturation. Different proteins will "salt-out" at various percent saturations. Subsequent to ammonium sulfate fractionation and centrifugation, precipitated proteins and/or those remaining in the ammonium sulfate supernatant after centrifugation (see following sec-

tion) must be extensively dialyzed to remove the salt prior to enzyme assay.

## Additional Procedures for Enzyme Purification

After dialysis and ammonium sulfate fractionation, proteins can be further purified by a sequence of the previously mentioned chromatographies. Affinity chromatography is a very powerful purification procedure that relies on the binding of a stereospecific ligand to a support matrix (Figure 3.1). The ligand possesses marked specificity for the compound being purified—e.g., a substrate for its enzyme, an antibody for its antigen, and so on.

Following binding of the compound to its ligand,[1] other nonbound compounds can be eluted with appropriate reagents. Finally, the bound species is selectively desorbed (see Figure 3.1).

Another chromatographic technique frequently employed in protein purification is gel filtration. This biochemical technique separates proteins and other compounds by molecular weight. In molecular-exclusion chromatography, a mixture of proteins, dissolved in a suitable buffer, is allowed to flow by grav-

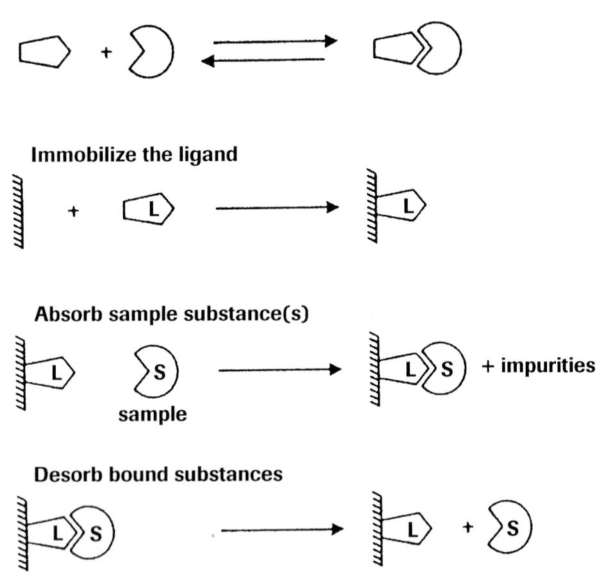


**FIGURE 3.1** Affinity chromatography mechanism for separating proteins.

From Pharmacia Laboratory Separation Division. 1967. *Affinity Chromatography: Principles and Methods.* Uppsala, Sweden: Pharmacia.

**TABLE 3.1A**  Summary of Pharmacia Sephadex Resins

| Sephadex | | Fractionation Range (MW) Peptides/Globular Proteins | Destrans | Dry Bead Diameter (μm) | Approx. Bed Volume (ml/g Dry Sephadex) |
|---|---|---|---|---|---|
| G-10 | | –700 | –700 | 40–120 | 2–3 |
| G-15 | | –1500 | –1500 | 40–120 | 25–35 |
| G-25 | Coarse | 1000–5000 | 100–5000 | 100–300 | 4–6 |
| | Medium | | | 50–150 | |
| | Fine | | | 20–80 | |
| | Superfine | | | 20–50 | |
| G-50 | Coarse | 1500–30000 | 500–10000 | 100–300 | 9–11 |
| | Medium | | | 50–150 | |
| | Fine | | | 20–80 | |
| | Superfine | | | 20–50 | |
| G-75 | | 3000–80000 | 1000–50000 | 40–120 | 12–15 |
| | Superfine | 3000–70000 | | 20–50 | |
| G-100 | | 4000–150000 | 1000–100000 | 40–120 | 15–20 |
| | Superfine | 4000–100000 | | 20–50 | |
| G-150 | | 5000–300000 | 1000–150000 | 40–120 | 20–30 |
| | Superfine | 5000–150000 | | 20–50 | 18–22 |
| G-200 | | 5000–600000 | 1000–200000 | 40–120 | 30–40 |
| | Superfine | 5000–250000 | | 20–50 | 20–25 |



ity down a column packed with beads of an inert, highly hydrated polymeric material that had previously been washed and equilibrated with only the buffer. The common column materials are Sephadex G or Bio-Gel P, commercial dextran and polyacrylamide derivatives (Table 3.1), all of which can be prepared with different degrees of internal porosity. In the column, proteins of different molecular weights penetrate the pores of the beads to different degrees and thus travel down the column at different rates (Figure 3.2 and 3.3). Very large protein molecules cannot enter the pores of the beads; they are said to be excluded and thus remain in the column's excluded volume, defined as the volume of the aqueous phase outside the beads. On the other hand, very small proteins can enter the pores of the beads freely (see Figure 3.2). Small proteins are retarded by the beads while large proteins pass through readily. Since they cannot enter the hydrated polymer particles, proteins of intermediate size will be excluded from the beads to a degree that depends upon their size; hence, the term exclusion chromatography.

The Sephadex G or Bio-Gel P gels possess upper and lower MW fractionation ranges—i.e., exclusions (see Table 3.1A, 3.1B). The upper range (void volume—Vo) can be determined through the use of Pharmacia's blue dextran 2000, a carbohydrate polymer of 2,000,000 MW. For example, if the Vo of a Sephadex superfine G-150 column was established, any protein with a molecular weight of 150,000 or more would elute within the void volume coincident with the blue dextran. In contrast, proteins of molecular weights less than 150,000 would be retarded with the protein possessing the least MW being retarded the most.

Molecular-exclusion chromatography can also separate mixtures of other macromolecules, as well as very large biostructures—e.g., viruses, ribosomes, cell nuclei, or even different degrees of internal porosity. The resolving power of molecular-exclusion chromatography is so great that this simple method is now widely used as a way of determining the molecular weight of proteins (see Figure 3.2 and 3.3).

Another chromatographic technique that is often valuable to protein purification is ion exchange chromatography. This technique capitalizes upon the net charge on molecules to achieve their separations. Both the Pharmacia (Table 3.2 and 3.3) and Bio-Rad companies manufacture a variety of strong and weak ion exchange resins. The chemistries and applications of these resins to the life sciences have been thoroughly explained (Pharmacia Ion Exchange Chromatography and Bio-Rad Ion Exchange Chromatography). The CM (carboxymethyl) and DEAE (diethylaminoethyl) Sephadexes available from Pharmacia are two commonly employed ion exchange resins, and their

**TABLE 3.1B**  Summary of Bio-Rad Bio-Gels

| Product Polysacrylamide Gels | | Applications | U.S. Standard Wel Mesh Designation (Hydrated) | Diameter of Hydrated Beads (Microns) | Fractionation Range (Daltons) | Hydrated Bed Volume ml/g Dry Gel |
|---|---|---|---|---|---|---|
| Bio-Gel P-6DG | Desalting Gel | Desalting, buffer exchange | 80–170 | 90–180 | 1,000–6,000 | 7 |
| Bio-Gel P-2 | Fine | Analytical fractionation | 200–400 | 40–80 | 100–1,800 | 3.5 |
| | Extra fine | Medium pressure fractionations | –400 | <40 | | |
| Bio-Gel P-4 | Medium | Group separations | 100–200 | 80–150 | 800–4,000 | 5 |
| | Fine | Analytical fractionation | 200–400 | 40–80 | | |
| | Extra fine | Medium pressure fractionations | –400 | <40 | | |
| Bio-Gel P-6 | Course | Preparative industrial use | 50–100 | 150–300 | 1,000–6,000 | 7 |
| | Medium | Group separations | 100–200 | 80–150 | | |
| | Fine | Analytical fractionation | 200–400 | 40–80 | | |
| | Extra fine | Medium pressure fractionations | –400 | <40 | | |
| Bio-Gel P-10 | Course | Preparative industrial use | 50–100 | 150–300 | 1,500–20,000 | 9 |
| | Medium | Group separations | 100–200 | 80–150 | | |
| | Fine | Analytical fractionation | 200–400 | 40–80 | | |
| | Extra fine | Medium pressure fractionations | –400 | <40 | | |
| Bio-Gel P-30 | Course | Group separations | 50–100 | 150–300 | 2,500–40,000 | 11 |
| | Fine | Analytical fractionation | 100–200 | 80–150 | | |
| | Extra fine | Medium pressure fractionations | –400 | <80 | | |
| Bio-Gel P-60 | Course | Group separations | 50–100 | 150–300 | 3,000–60,000 | 14 |
| | Fine | Analytical fractionation | 100–200 | 80–150 | | |
| | Extra fine | Medium pressure fractionations | –400 | <80 | | |
| Bio-Gel P-100 | Course | Group separations | 50–100 | 150–300 | 5,000–100,000 | 15 |
| | Fine | Analytical fractionation | 100–200 | 80–150 | | |
| | Extra fine | Special applications | –400 | <80 | | |
| Bio-Gel P-150 | Course | Group separations | 50–100 | 150–300 | 15,000–150,000 | 18 |
| | Fine | Analytical fractionation | 100–200 | 80–150 | | |
| | Extra fine | Special applications | –400 | <80 | | |
| Bio-Gel P-200 | Course | Group separations | 50–100 | 150–300 | 30,000–200,000 | 25 |
| | Fine | Analytical fractionation | 100–200 | 80–150 | | |
| | Extra fine | Special applications | –400 | <80 | | |

From Bio-Rad Laboratories. Hercules, CA.

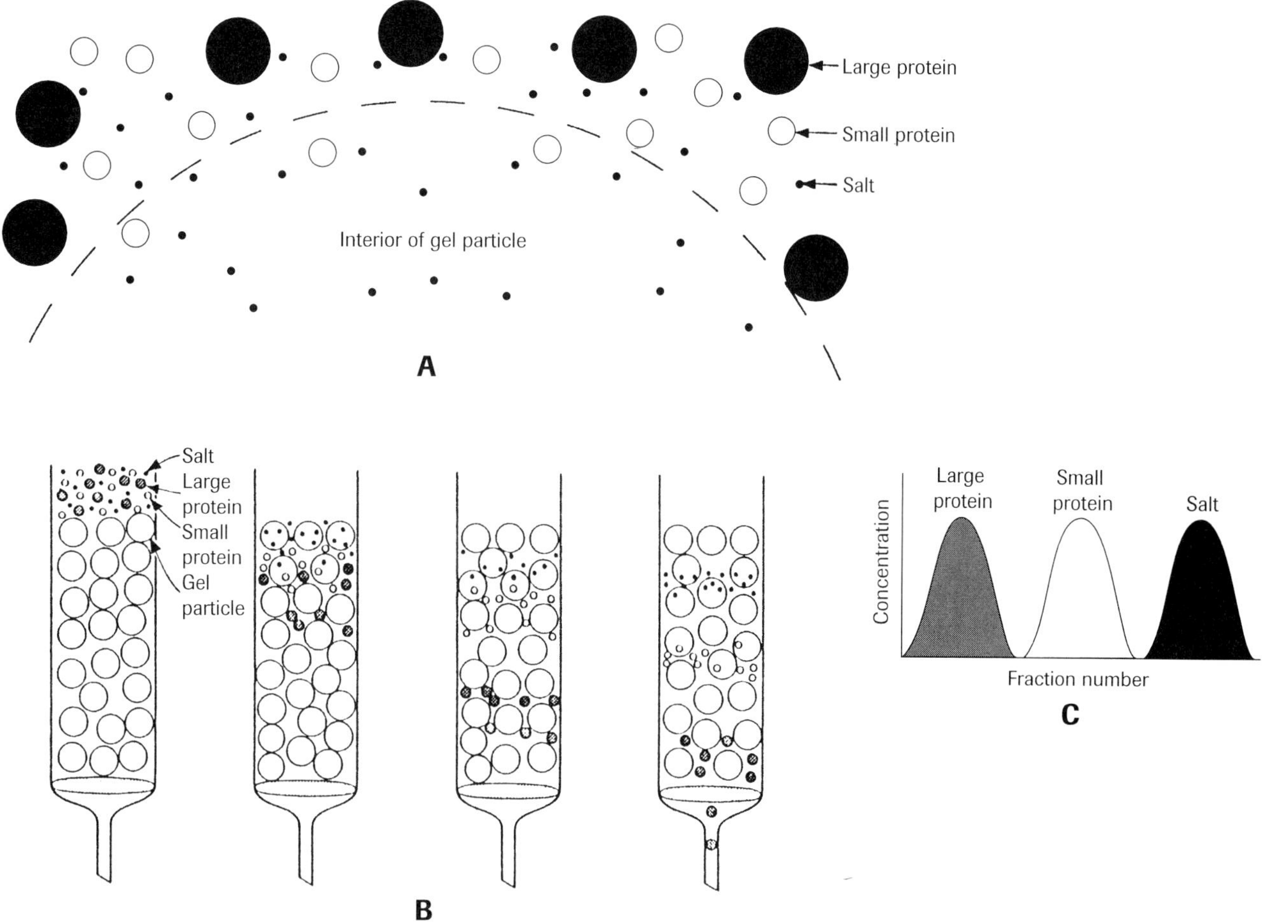



**FIGURE 3.2**   Gel filtration. **(A)** Diagrammatic representation of a gel particle showing "pores" of a given size range. **(B)** Progressive separation of three kinds of molecules of different sizes. **(C)** Elution pattern.
From Segel, Irwin H. 1976. *Biochemical Calculations,* 2nd ed. New York: John Wiley and Sons.

mechanisms of action illustrate the theory of ion exchange chromatography (Figure 3.4). Whereas CM Sephadex is a cation ion exchange resin, DEAE is a weak anion exchange resin. Each Sephadex consists of a backbone derivatized with a charged functional group to which an appropriately charged counterion is attached. In the case of CM Sephadex, the functional group is $OCH_2COO^-$, and the counterion is $Na^+$ (see Table 3.2). In contrast, the functional group and counterion of DEAE are $C_2H_5$ and $Cl^-$, respectively. If a mixture of positively and negatively charged as well as neutral molecules is layered onto Sephadex CM C-50, the negatively charged and neutral molecules will "wash through." In contrast, the positively charged molecules will exchange with the counterions, to varying degrees—i.e., those molecules may be either tightly or weakly bound to the functional group depending upon the affinity of that group (see Figure 3.4). The exchanged molecules can be desorbed by application of either salt (low to high ionic strength) or pH (neutralized charges) gradients. Desorption by

salt gradients is dependent upon an order of ion affinities (selectivity) for the functional group (Table 3.4). For example, an ion with a greater affinity for the functional group than the exchanged species will replace the latter. Tightly bound molecules can be desorbed by increasing the ionic strength of the eluting salt. Finally, for ion exchange chromatography, it is important to realize that cations exchange with cations and anions with anions. In practice, this means that the counterions of CM Sephadex will only exchange with cations, and vice versa, for a DEAE column.

## Hydroxyapatite and Hydrophobic Interaction Chromatographies

Hydrophobic compounds are nonpolar compounds that participate in nonenergetically favorable interactions with water molecules and interfere with the hydrogen binding among water molecules. Hydrophobic interactions are forces that hold nonpolar regions

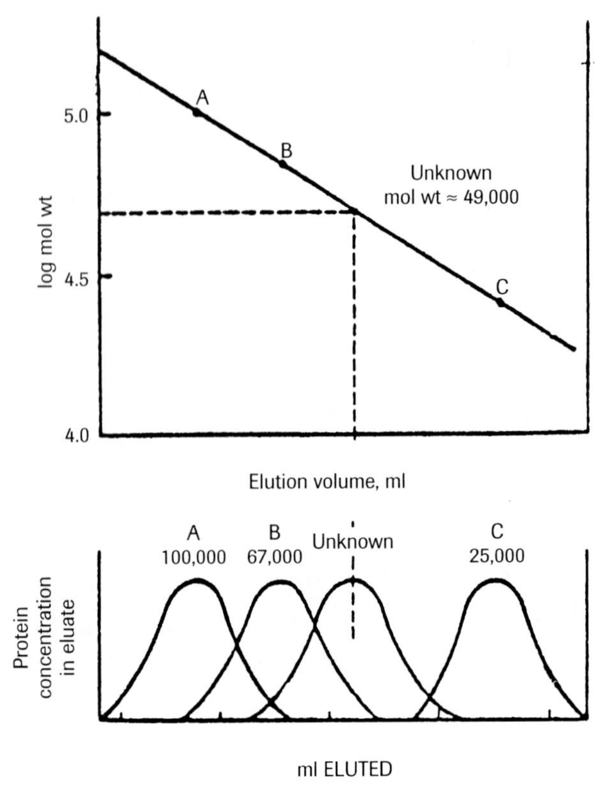

**FIGURE 3.3** Sephadex calibration curve for determining molecular weight.

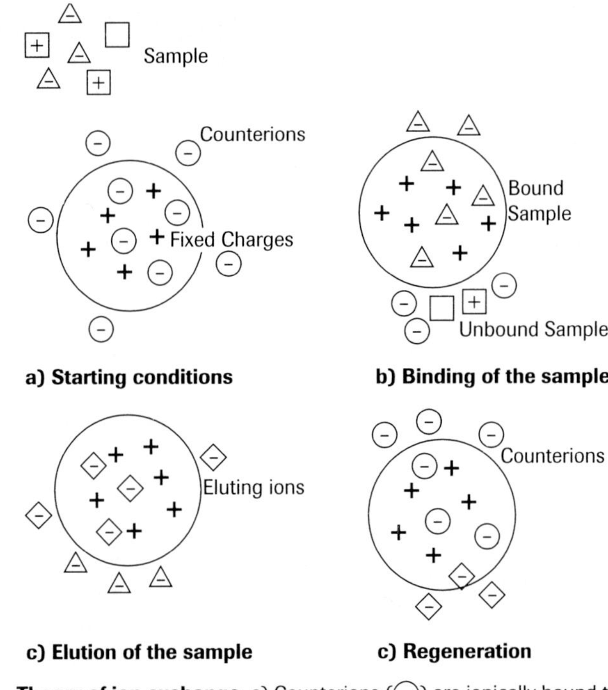


**Theory of ion exchange.** a) Counterions ($\ominus$) are ionically bound to the fixed charge ($+$) on the matrix. b) Sample molecules ($\triangle$) of the opposite charge of the matrix bind displacing the counterions while neutral molecules and molecules of the same charge ($\square$ $+$) as the matrix do not bind (pass straight through). c) The sample is displaced by eluting ions ($\diamondsuit$). d) The resin is regenerated by exchanging the eluting ions for the original counterions. The ion exchange resin is ready for re-use.

**FIGURE 3.4** Mechanism by which proteins can be separated via ion exchange chromatography.

From Bio-Rad Laboratories. *Guide to Ion Exchange Resins.* Richmond, CA: Bio-Rad Laboratories.

**TABLE 3.2**  Composition of Pharmacia Ion Exchange Resins

| Ion ExchangeResin | Functional Group | Counterion |
|---|---|---|
| Anion exchangers<br>DEAE<br>(diethylaminoethyl) | $O-CH_2CH_2N^+-H$ with $C_2H_5$ groups | $Cl^-$ |
| QAE<br>(quaternaryaminoethyl) | $OCH_2CH_2N^+-CH_2CHCHCH_3$ with $C_2H_5$ groups | $Cl^-$ |
| Q<br>(quarternary amine)<br>Cation exchangers | $CH_2-N^+(CH_2)_3$ | $SO_4^{2-}$ |
| CM<br>(carboxymethyl) | $OCH_2COO$ | $Na^+$ |
| SP<br>(sulphopropyl) | $OCH_2CH_2CH_2SO_3$ | $Na^+$ |
| S<br>(sulphonate) | $CH_3-SO_3$ | $Na^+$ |

From Pharmacia Laboratory Separation Division. 1986. *Ion Exchange Product Profile.* Uppsala, Sweden: Pharmacia.

**TABLE 3.3**   Ion Exchange Selection Guide

| Solute | Total Charge | Molecular Weight | Chemical Stability (pH) | Sorption | Fractionation | Purification |
|---|---|---|---|---|---|---|
| Inorganic | Positive ($\leq$4) | <1000 | 1–14 | AG 50W resin<br>AG MP-50 resin<br>Chelex 100 resin<br>AMP-1 resin | AG 50W resin<br><br>Chelex 100 resin | AG 1 resin<br>AG 2 resin<br>Chelex 100 resin |
| | Negative ($\leq$6) | <1000 | 1–14 | AG 1 resin<br>AG 2 resin<br>AG MP-1 resin<br>Chelex 100 resin | AG 1 resin<br>AG 2 resin<br>AG MP-1 resin<br>AG 11 A8 resin<br>Chelex 100 resin | AG 50W resin<br>AG MP-50 resin<br>Chelex 100 resin |
| Inorganic or Organic | Neutral (0) | >100 | 1–14 | — | — | AG 501-X8 resin<br>AG 11A8 resin |
| | | | 1–7 | — | — | AG3-X4 resin<br>AG 2 resin<br>AG 11 A8 resin |
| | | | 7–11 | — | — | Bio-Rex 70 resin<br>AG 11 A8 resin |
| Organic | Positive ($\geq$1) | <1000 | 1–14 | AG 50W resin<br><br>Chelex 100 resin | AG 50W resin | AG 1 resin<br>AG 2 resin<br>Chelex 100 resin<br>AG 11 A8 resin |
| | | | 3–10 | Bio-Rex 70 resin | Bio-Rex 70 resin | AG 1 resin<br>AG3-X4 resin<br>AG 11 A8 resin |
| Organic | Positive ($\geq$3) | >500 | 1–14 | AG MP-50 resin | AG MP-50 resin | AG 1 resin<br>AG 2 resin<br>AG MP-1 resin<br>AG 11 A8 |
| | | | 2–9 | CM Bio-Gel® A gel<br>Bio-Rex 70 resin | CM Bio-Gel A gel<br>Bio-Rex 70 resin | AG 11 A8 resin<br>AG3-X4 resin |
| Organic | Negative ($\geq$1) | <1000 | 1–14 | AG 1 resin<br>AG 2 resin<br>AG MP-1 resin | AG 1 resin<br>AG 2 resin<br><br>AG 11 A8 resin | AG 50W resin<br><br>AG 11 A8 resin |
| | | | 4–11 | Bio-Rex 5 resin<br>AG 1 resin | Bio-Rex 5 resin<br>AG 1 resin | Bio-Rex 70 resin<br>Chelex 100 resin<br>Ag 11 A8 resin |
| Organic | Negative ($\geq$3) | >100 | 1–14 | AG MP-1 resin | AG MP-1 resin | AG 50W resin<br>AG 501-X8 resin<br>Chelex 100 resin<br>AG 11 A8 resin |
| | | | 3–10 | Bio-Rex 5 resin<br>DEAE Bio-Gel A gel | Bio-Rex 5 resin<br>DEAE Bio-Gel A gel | AG 50W resin<br>AG 501-X8 resin<br>AG 11 A8 resin |

From Bio-Rad Laboratories. *Guide to Ion Exchange Resins.* Richmond, CA: Bio-Rad Laboratories.

**TABLE 3.4**  Bio-Rad Ion Exchange Resins and Order of Ion Affinities

| Resin Type | Active Group*† | Order of Selectivity for Monovalent Ions / for Multivalent Ions | Thermal | Stability: [1]Solvent (Alcohols, Hydrocarbons, etc.) [2]Oxidation | Reduction |
|---|---|---|---|---|---|
| AG 1<br>Bio-Rex MSZ 1<br>Aminex anion resins<br>strongly basic<br>anion exchangers | $R\text{-}CH_2N^+(CH_3)_3$ | 1>phenolate>$HSO_4$>$ClO_3$><br>$NO_3$>Br>CN>$HSO_3$>$NO_2$><br>Cl>$HCO_3$>$IO_3$>$H_2COO$>Ac><br>OH>F | $OH^-$ form-fair to<br>50°C<br>$Cl^-$ and other forms<br>good to 150° C | [1]Very good<br>[2]Slow solution in<br>hot 15% $HNO_3$ or<br>conc. $H_2O_2$ | Good[‡] |
| AG 2<br>strongly basic<br>anion exchanger | $R\text{-}CH_2N^+(CH_3)_2$<br>$C_2H_4OH$ | phenolate>i>$HSO_4$>$ClO_3$>$NO_3$><br><br>Br>CN>$HSO_3$>$NO_2$>Cl><br>OH>$IO_3$>$H_2COO$>Ac>F | $OH^-$ form-fair to<br>30° C<br>$Cl^-$ form<br>Good to 150° C | [1]Very good<br>[2]Slow solution in<br>hot 15% $HNO_3$ or<br>conc. $H_2O_2$ | Good[‡] |
| AG 50W<br>Bio-Rex MSZ 50<br>Aminex cation resins<br>strongly acidic<br>cation exchangers | $R\text{-}SO_3^-$ | Ag>Rb>Cs>K>$NH_4$>Na><br>H>Li<br>Zn>Cu>Ni>Co | Good to 150° C | [1]Very good<br>[2]Slow solution in<br>hot 15% $HNO_3$ | Very good |
| Bio-Rex 5<br>intermediate base<br>anion exchanger | $R\text{-}N^+(CH_3)_2HCl$<br>&<br>$R\text{-}N^+(CH_3)_2$<br>$C_2H_4OH$ | — | Good to 60° C | [1]Good<br>[2]Good | Excellent |
| Bio-Rex 70<br>weakly acidic<br>carboxylic cation<br>exchanger | $R\text{-}COO^-$ | H>Ag>K>Na>Li<br>H>Fe>Ba>Sr>Ca>Mg | Good to 100° C | [1]Good<br>[2]Good | Good |

| Resin | Functional group | Selectivity | Thermal stability | Chemical stability | Radiation stability |
|---|---|---|---|---|---|
| AG 4-X4<br>  AG 3-X4<br>  weakly basic<br>  anion exchangers | CH$_2$-COO-<br>R-CH$_2$N CH$_2$COO$^-$ | $\underline{H>Li>Na>K}$<br>Cu>Pb>Fe$^{-3}$>Al>Ni><br>Zn>Ag>Co>Cd>Fe>Mn><br>Ba>Ca>Na> | Good to 75° C | [1]Good<br>[2]Breakdown in<br>  strong oxidizing<br>  agents | Unknown |
| Chelex 20<br>  Chelex 100<br>  weakly acidic cation<br>  chelating resins | R-CH$_2$N$^+$(CH$_3$)$_3$ &<br>R-CH$_2$COO$^-$ | — | Tentatively good to<br>100° C | [1]Good<br>[2]Slow solution in<br>  hot 15% HNO$_3$ or<br>  conc. H$_2$O$_2$ | Good |
| AG 11 A8<br>  strongly basic anion<br>  plus weakly acidic<br>  cation<br>  (Zwitterion)<br>  exchanger | R-SO$_3^-$<br>R-CH$_2$N$^+$-(CH$_3$)$_3$ | — | Good to 50° C | [1]Very good<br>[2]Slow solution in<br>  hot 15% HNO$_3$ or<br>  conc. H$_2$O$_2$ | Good† |
| AG 501-X8<br>  Bio-Rex MSZ 501<br>  strongly basic anion<br>  plus strongly acidic<br>  cation exchangers | R-CH$_2$N$^+$(CH$_3$)$_2$ | SO$_3$H>HCit>CrO$_3$>HCO$_3$<br>tartaric>oxalic>H$_3$PO$_4$><br>H$_3$AsO$_4$>HNO$_3$>Hl>HBr><br>HCl>HF>HCOO>HAc>H$_2$CO$_3$* | Good to 60° C | [1]Good<br>[2]Good | Unknown |

From Bio-Rad Laboratories. *Guide to Ion Exchange Resins.* Richmond, CA: Bio-Rad Laboratories.

*Apply in acid solution only.

†AG 1 and AG 2 resins will slowly break down in the presence of sulfur-containing reducing agents.

*†"R" represents the matrix of the resin.

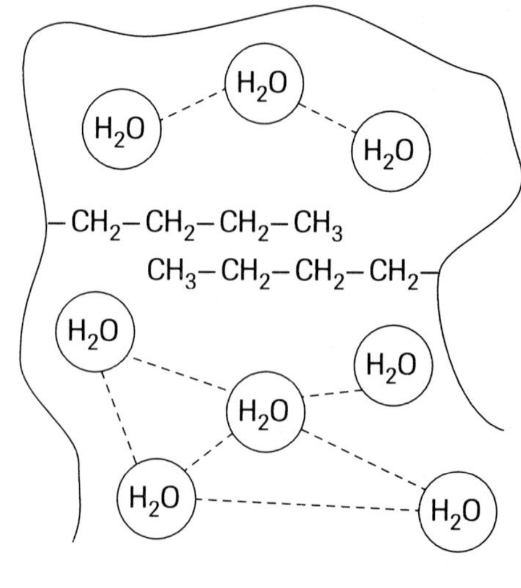



(Dashed lines represent hydrogen bonds)

No hydrophobic interaction between hydrocarbon sidechains: increased non-polar surface area.

(II)  Hydrophobic clustering of lipid-like sidechains; water "sees" less lipid surface.

**FIGURE 3.5**  Hydrophobic interactions.

Bhagavan, N.V. *Biochemistry: A Comprehensive Review.*  Philadelphia, PA: J. P. Lippincott Family.

of the molecules together. These interactions are similarly weak and are continually formed and broken. Hydrophobic interaction chromatography, which separates proteins via their relative hydrophobicities employs insoluble support such as either agarose or polyacrylamide (Figure 3.5). Hydrophobic alkyl or aryl groups are covalently attached to the supports. Proteins that have been bound to the resin through hydrophobic interactions can be eluted according to the strength of this interaction. Changes in hydrophobicity, ionic strength, pH or temperature of the eluting buffer promote elution.[6,11] In contrast to hydrophobic interaction chromatography, hydroxyapatite chromatography involves the use of calcium phosphate gels.

# LABORATORY EXERCISES

## MONOAMINE OXIDASE (MAO) AND PLASMA AMINE OXIDASE (PAO) PURIFICATION

### *Supplies and Equipment*
Cheesecloth
Con A—sepharose column
DEAE—cellulose column
Electrophoretic cell
Hydroxyapatite column
International centrifuge
Potter-Elvehim homogenizer
Sephadex G-100
Sorvall RC2B preparation centrifuge
Spectrophotometer
Spectroscope
Waring blender

### *Chemicals*
Sigma Chemical Company
  Benzylamine sulfate
  DEAE cellulose
  Calcium phosphate
  Potassium phosphate
  Tris (hydroxymethyl) amino
    methane
  Con A sepharose
  Sucrose
  Hydroxyapatite
Aldrich Chemical Company, Inc.
  Acrylamide

Bio-Rad Laboratories
    Bio-gel A 1.5 beads 100–200 mesh
    Con-A sepharose
Nutritional Biochemicals Corporation
    Bovine serum albumin (2X crystallized)
J. T. Baker Chemical Company
    Sucrose
Eastman Organic Chemicals
    Benzylamine

# Introduction

## Reagents and Buffers

### *Buffers and Solutions for Preparation of Mitochondria*

a. Mitochondria washing buffer, 0.1 M potassium phosphate buffer (pH 8.2) (0.5% Triton X-100). Add 50 ml Triton X-100 20% to 2 l buffer.
b. 0.1 M potassium phosphate, pH 8.2 (1% Triton X-100, 0.5% potassium cholate). Add 100 ml of 20% Triton X-100, 100 ml of 10% potassium cholate.

### *Solutions for Protein Determination*

a. Lowry A    2% Sodium Carbonate in 0.1 N Sodium hydroxide solution.
b. Lowry B    0.5% $CuSO_4 \cdot 5\ H_2O$ in 1% Sodium Citrate.
c. Lowry C    Has to be prepared fresh 5:0.1 ratio of a and b respectively.

### *Determination of Dry Weight of Calcium Phosphate*

a. Mix the commercially available calcium phosphate very well, then place 1 ml of the gel into 50 ml beaker and dry it in the oven at 120° C overnight.
b. Weigh out the dried 1 ml gel.
c. Calculate how many ml of wet gel is needed to give you one gram of gel.

## Purification of Amine Oxidases

### *Monamine Oxidase (MAO)*

The authors highly recommend that students should be trained in all aspects of the purification of this enzyme. The enzyme is located in the outer mitochondrial membrane, and thus preparation of the mitochondria is essential to achieve enzyme purification. Students will gain valuable experience in performing all the steps, including differential centrifugation, to obtain the organelle where the enzyme is located in a highly pure form.

### **Preparation of the Mitochondria**

The mitochondria can be both isolated and purified by the methods of Schneider and Hogeboom.[12] How-

ever, an additional step has been added to improve the final specific activity of the enzyme. The mitochondria were washed a second time with 1.15% potassium chloride after freezing and before purification of the enzyme. The preparation can be accomplished in either one or two days, depending upon the amount of the mitochondria you are interested in isolating and purifying. For example, if you started with 5.0 kg of liver, the recommendation is to complete the isolation in two days. For 3.0 kg or less, the complete isolation can be accomplished in one day. It is almost essential to freeze the mitochondria after you complete the isolation of the mitochondria. Freezing helps to liberate the enzyme from the phospholipid domain where it resides.

To save time, the mitochondria can be prepared by a technician.

a. Briefly rinse superficial blood off beef liver, cut off fat and large blood vessels, and cut into about 2.5–5.0 cm cubes.
b. Homogenize liver in a Waring blender into 0.25 M sucrose within 0.01 M, pH 7.4 potassium phosphate buffer, 200 ml beaker filled with liver to 1800 ml sucrose/phosphate.
c. Centrifuge the homogenate in an International Centrifuge for 15 minutes at 2500 RPM.
d. Strain the supernatant through a cheesecloth and discard the bottom sediment.
e. Centrifuge the supernatant in a Sorvall RC2B preparative centrifuge with a rotar at 5,000 RPM for 20 minutes.
f. Discard the supernatant and resuspend the pellet in sucrose with one-third the original volume of sucrose used for the homogenization.
g. Centrifuge the above suspension in the above Sorvall (large head) at 10,000 RPM for 10 minutes.
h. Discard the supernatant and resuspend the pellet in 1.5% potassium chloride in 0.01 M potassium phosphate, pH 7.4 buffer, one-third the original sucrose volume.
i. Centrifuge in the Sorvall at 5,000 RPM for 10 minutes.
j. Discard the supernatant and suspend the pellet in 0.01 M potassium phosphate buffer at a ratio 1:2, liver:buffer, then freeze at –20° C.
*k. Wash the mitochondria again with 1.15% potassium chloride solution after freezing and immediately before attempted purification of the enzyme.
l. Assay for monoamine oxidase following the procedure described in Module 4.

## **Purification of Monoamine Oxidase**

1. Wash the purified mitochondria first with 0.1 M potassium phosphate buffer containing 0.5% Triton

---

*Highly recommended

X-100, pH 8.2, and centrifuge in the above Sorvall at 10,000 RPM for 20 minutes.

2. Suspend and homogenize the washed mitochondria with 0.1 M potassium phosphate buffer containing 1% Triton X-100, 0.5% potassium cholate, pH 8.2, and 0.002 M β-mercaptoethanol by using a Potter-Elvehem homogenizer. Note, use an ice bucket.

3. Maintain the extract from step 2 with gentle stirring in the cold for 4 hours.

4. Assay for monoamine oxidase as described in Module 4.

5. *Ammonium sulfate fractionation.*

   a. Add crystalline ammonium sulfate to the extract between 0%–25% saturation (use the ammonium sulfate percent saturation chart presented in Table 3.8 on page 51). Make sure that all the ammonium sulfate is completely dissolved, then centrifuge at 10,000 RPM for 25 minutes. Assay for monoamine-oxidase in the supernatant. LET THE CENTRIFUGE GO TO A COMPLETE STOP. DO NOT TRY TO STOP IT BY HAND.

   b. Discard the pellet and add crystalline ammonium sulfate to the extract between 25%–45% saturation (again use the ammonium sulfate chart). Centrifuge at 10,000 RPM for 25 minutes. After centrifugation, observe that the enzyme forms a film floating on top of the ammonium sulfate. Be very careful in the removing the enzyme. It is highly recommended to use suction to eliminate the ammonium sulfate and leave the enzyme in the container.

   c. Suspend the enzyme in 100 ml of 0.05 M, pH 7.4 potassium phosphate buffer containing 0.1% Triton X-100 and 0.002 M β-mercaptoethanol.

   d. Assay for monoamine oxidase using the procedure in Module 4.

6. *Dialysis.* Dialyze the partially purified enzyme against 0.05 M potassium phosphate, pH 7.4 buffer containing 0.1% Triton X-100, 0.002 M β-mercaptoethanol, pH 7.4 with a change of buffer twice over 24 hours. Assay for the enzyme prior to calcium phosphate titration.

7. *Calcium phosphate titration.* First set up a small titration scale as follows:

   a. Determine the protein concentration using the Biuret, Bradford, Lowry or UV-spectroscopic methods as outlined in Module 2.

   b. Pipette 1 ml of enzyme into five separate small test tubes.

   c. Add 0.2 ml, 0.4 ml, 0.6 ml, 0.8 ml, and 1 ml of calcium phosphate (either commercially prepared or prepared after you determine the dry weight first). Mix and then centrifuge using a clinical centrifuge at 4° C for five minutes, and then assay for the enzyme using the assay procedure in Module 4. Calculate the specific activity as described in Module 4.

   d. Select the optimum titration range that gives you the maximum specific activity from the titration scale.

   e. Measure the total volume of the enzyme and use the optimum range you selected in d (normally 1:4 ratio). Mix well and make sure that all the calcium phosphate is dissolved, then centrifuge at 10,000 RPM for 10 minutes.

   f. Make sure that the specific activity of the partially purified enzyme is 1500 U/mg–1600 U/mg as calculated using the procedure in Module 4. If the specific activity is below 1500, use a 1:5 ratio. A student may select to use calcium phosphate to run Module 4. However, to obtain a pure enzyme, it is recommended to go on to DEO.

8. *Ion exchange chromatography.* It is highly recommended that students very carefully read the introductory material on the General Methods section of this module and understand the complete theory for using ion exchange chromatography prior to starting this phase of the purification.

   a. Packing the column. The column should contain some water in it before the slurry of gel is poured into it (a volume of 1/3 the bed height is satisfactory). The slurry of gel is then carefully poured down the wall of the column (a stirring rod may be helpful in accomplishing this). If the column is not filled with slurry, add eluant to fill the column (a stirring rod may be helpful in accomplishing this). Start the flow (1 ml/min–2 ml/min is sufficient).

   b. Dilute the enzyme to about one-and-a-half volume with 0.01 M potassium phosphate buffer containing 0.1% Triton X-100 and 0.002 M β-mercaptoethanol, pH 7.6.

   c. Apply the diluted enzyme to a 30.0 DEAE-cellulose column (4.5 × 30 cm) that had been equilibrated with the starting buffer (use about 500 ml of 0.01 M potassium phosphate, pH 7.4 buffer containing 0.1% Triton X-100 and 0.002 M β-mercaptoethanol for equilibration.

   d. Elute the enzyme by employing a sodium chloride gradient (0.0 M–0.05 M) in 0.02 M potassium phosphate buffer, pH 7.6 containing 0.1% Triton X-100 and 0.002 M β-mercaptoethanol.

   e. Collect the eluted enzyme in separate 10 ml test tubes, either manually or using a fraction collector in which you set the flow rate as 1 ml/min.

   f. Determine enzyme activity and protein concentration for each fraction, then select the fractions that have a specific activity of 2500 U/mg or higher and pool these fractions together.

   g. Measure the total volume of the combined fractions and assay for the enzyme and amount of protein.

h. Concentrate the enzyme employing 45% ammonium sulfate saturation using Table 3.8 presented in this module.

i. Centrifuge at 10,000 RPM for 25 minutes, and then dissolve the enzyme in 20 ml of 0.02 M potassium phosphate buffer containing 0.1% Triton X-100, 0.002 M β-mercaptoethanol, pH 7.4.

j. Dialyze. the enzyme at 4° C against 500 ml 0.02 M potassium-phosphate buffer containing 0.1% Triton X-100, 0.002 M β-mercaptoethanol, pH 7.4 with at least two changes of buffer.

**A student may select to use the DEAE eluted enzyme to run Module 4. However, to obtain a highly purified enzyme that shows only a single band on SDS polyacrylamide gel electrophoresis, the student must apply the enzyme to an affinity chromatography column.**

9. *Affinity chromatography.* The eluted enzyme can be further purified on a hydroxyapatite column as follows:

a. Packing the column. Use the same packing procedure listed in step 7a.

b. Apply the enzyme to a 5.2 × 4.5 cm hydroxyapatite column.

c. Wash the column with 100 ml 0.02 M potassium phosphate buffer containing 0.1% Triton X-100, 0.002 M β-mercaptoethanol, pH 7.4.

d. Apply 100 ml 0.2 M potassium phosphate buffer containing 0.2% potassium cholate, 0.002 M β-mercaptoethanol, pH 7.4 to the column. Collect enzyme fractions in 10 ml test tubes. With this elution buffer system, you should obtain a minor fraction with a specific activity about 3200 U/mg protein.

e. Apply 100 ml 0.2 M potassium phosphate buffer containing 0.5% potassium cholate and 0.2% Triton X-100, 0.002 M β-mercaptoethanol pH 7.4 to the column. Collect enzyme fractions in 10 ml test tubes. With this elution buffer system, you should obtain the major highly purified enzyme with a specific activity greater than 7,000 U/mg protein.

f. Collect the major and minor fractions separately and concentrate the enzyme using 40% saturation. Dissolve the pellet in 25 ml–30 ml 0.02 M potassium phosphate buffer, 0.1% Triton X-100, 0.005 M β-mercaptoethanol, pH 7.4. Complete Table 3.5.

**This is considered a highly purified enzyme that can be stored at 4° C and covered with aluminum foil. If the student wishes to obtain the highest purification step, he/she can proceed through gel-filtration step, but this step is not normally needed.**

## Purification of Plasma Amine Oxidase

The purification procedure for plasma amine oxidase (PAO) is much simpler and faster compared with monoamine oxidase. PAO is a very stable enzyme and can be kept frozen for years. The enzyme can be purified as follows:

1. *Preparation of the plasma.*

a. To prevent blood coagulation, immediately after you obtain fresh bovine blood from a local abattoir, mix it with a 20% citrate solution.

b. Centrifuge the citrated blood at 10,000 RPM for 30 minutes to remove erythrocytes.

2. *Ammonium sulfate fractionation.*

a. Add crystalline ammonium sulfate to a saturation level of 35% utilizing Table 3.8. Mix and make sure that all the ammonium sulfate is

---

**TABLE 3.5**  Purification Steps of Bovine Liver Monoamine Oxidase from 5.0 kg of Liver

| Step | Volume (ml) | Activity (Units/ml) | Protein (mg/ml) | Spec. act. m/mg | Total Protein (mg) | Yield (%) | Fold-purified |
|---|---|---|---|---|---|---|---|
| 1. Washing with 0.5% Triton X-100 | | | | | | | |
| 2. Homogenate | | | | | | | |
| 3. 0.25–0.45 $(NH_4)_2SO_4$ precipitate | | | | | | | |
| 4. $Ca_3PO_4$ gel | | | | | | | |
| 5. DEAE eluate | | | | | | | |
| 6. HA eluates<br>a. 0.2 M buffer-0.2% cholate<br>b. 0.5 M buffer-0.5% cholate | | | | | | | |

dissolved completely. Centrifuge at 10,000 RPM for 30 minutes. Discard the pellet and assay for PAO in the supernatant as described in the following assay section.

  b. Add ammonium sulfate to the supernatant to a saturant level of 55% utilizing Table 3.8. Again make sure all the ammonium sulfate is dissolved, then centrifuge at 10,000 RPM for 30 minutes. Most if not all of the enzyme is in the pellet. However, to make sure, assay for PAO in the supernatant.

  c. Dissolve the 35%–55% ammonium sulfate precipitate in 250 ml 0.01 M potassium phosphate buffer, pH 7.0.

3. *Dialysis.* Dialyze the enzyme against 2 l 0.01 M potassium phosphate, pH 7.0 for 3 hours, then change the buffer and dialyze overnight at 4° C. Assay for PAO after dialysis (see PAO).

4. *Ion exchange chromatography.*

  a. Set up a DEAE-cellulose column ($4.5 \times 30$ cm), then wash the column with 2 l 0.01 M potassium phosphate buffer, pH 7.0.

  b. Apply the enzyme to the column, then wash the column with 0.01 M potassium phosphate buffer, pH 7.0.

  c. Elute the enzyme with 2 l of 0.0–0.5 M pH 7.0 potassium phosphate gradient.

  d. Concentrate the enzyme by adding ammonium sulfate to a saturation level of 45%, then dissolve the enzyme in 200 ml of 0.01 M potassium phosphate buffer, pH 7.0.

5. *Affinity chromatography.*

  a. Set up a Con A-Sepharose 4B column ($4.5 \times 30.0$ cm), then wash the column with 0.01 M potassium phosphate buffer, pH 7.0.

  b. Elute the enzyme by passing 0.1 M potassium phosphate buffer containing 0.1 M methyl-D-glucoside, pH 7.0, through the column.

  c. Collect the fractions containing the enzyme and assay for PAO.

  d. You can apply the enzyme to a Bio-Gel A 1.5 m column directly to eliminate high molecular weight protein impurities.

  e. Assay for PAO obtained from step d. Normally a highly purified enzyme has a specific activity of 1,000–1,200 units per mg protein.

  f. Concentrate the enzyme by using ammonium sulfate to a saturation level of 45%. Save the enzyme in the freezer.

### *Enzyme Assays and Protein Determination*

***Plasma amine oxidase assay.***  The enzyme activity can be measured spectrophotometrically by using a common spectrophotometer. The specific activity can be calculated as units of enzyme per mg of protein. One unit is defined as the amount of enzyme that catalyzes a change of 0.001 absorbance units under the standard conditions. A typical assay can be performed as follows:

1. Add 0.2 ml of PAO in a total volume of 2.9 ml of 0.1 M potassium phosphate buffer, pH 7.2.
2. Add 0.1 ml of 0.1 M benzylamine sulfate. Mix the cuvette very well. DO NOT USE MORE THAN THIS CONCENTRATION. The final benzylamine concentration of $3.33 \times 10^{-3}$ M is a saturating but not inhibiting level.
3. Measure the absorbance increase at 250 nm with time using a spectrophotometer.

**At this time the student should compare the assay procedure for MAO with PAO. Is it different?**

***Protein concentration.***  Measure the protein concentration spectrophotometrically at 280 nm using an extinction coefficient $E^{1\%}_{CM}$ of 20.8.

**At this time the student should compare the procedure used for protein concentration for MAO with PAO. Is it different?**

## Purification and Determination of Molecular Weight by SDS Polyacrylamide Gel Electrophoresis

The basic and acidic amino acids in proteins render them highly ionized; the net charge is then specific for a given protein at a given pH. Proteins can, therefore, be separated on the basis of their charges by allowing them to migrate through a medium in an electric field.

The most popular electrophoresis medium is the polyacrylamide gel, which is transparent, chemically inert, and offers good resolution. Polyacrylamide gels are prepared by the free radical polymerization of acrylamide:

$$CH_2 \!=\! CH - CONH_2 \longrightarrow CH_2 - \overset{\displaystyle \overset{CONH_2}{\|}}{CH} \rightarrow$$

For gelation it is necessary that the long chains formed by the above reaction be linked together, so a "crosslinker" (bis-acrylamide) is also present in the reaction mixture:

$$CH_2\!=\!CH-\overset{\displaystyle\overset{O}{\|}}{C}-NH-CH_2-NH-\overset{\displaystyle\overset{O}{\|}}{C}-CH\!=\!CH_2$$

The amount of crosslinker added specifies the gel pore size. Because proteins must travel through holes in the gel, they are separated on the basis of their size and conformation run; however, their mobilities should not be related to any conformational differences. The detergent SDS not only denatures proteins but also binds to each amino acid, leaving an integral negative charge for each peptide bond. The charge of the protein is then directly proportional to its size, and a protein's molecular weight can be estimated within 10%, by comparing its mobility with that of proteins whose molecular weights are known.[8,25]

For this experiment, work in pairs. Each pair will run two gels, one that contains standards of known molecular weight, and one with an unknown protein whose molecular weight is to be determined.

## *Gel Preparation*

To expedite the experiment your gels will be prepared for you by a technician or student assistant. For 20 gels, 15 ml gel buffer is added to 13.5 ml acrylamide solution; 1.5 ml of freshly made ammonium persulfate solution 0.045 ml of N, N, N', N' tetramethylethylene diamine (TEMED) are added as catalysts. After mixing, each tube is quickly filled with 1.3 ml of gel solution. Before the gels harden, a few drops of water are added to the top of the gel to ensure a flat surface is formed. Just before use, the water layer is sucked off the electrophoresis apparatus.

## *Preparation of Samples*

Into two small test tubes, place five drops of tracking dye, two drops of glycerine, and five drops of β-mercaptoethanol. In one tube add 10 μl of the following standards: serum albumin, fumarase, trypsin, and ribonuclease. To the other tube add 10 μl of your preparation of highly purified monoamine and plasma amine oxidases and 40 μl dialysis buffer. Flick each tube to mix contents and layer each sample on top of a gel. Gel buffer diluted 1:1 with water is very carefully layered on top of each sample so that a meniscus bulges up over each tube. This is done very slowly so that the underlying sample is not disturbed. The gel tubes are next attached to the electrophoresis apparatus, and the gel buffer is added to both compartments of the electrophoresis apparatus. The run is conducted under a consistent current of 8-10 MA per gel. After the tracking dye have moved to the bottom of the gels they are removed from the apparatus. and taken from their tubes by squirting water from syringe between the gels and the wall of the tubes and forcing them out with a pipette bulb. The length of the gels and the distance moved by the dye are to be measured and carefully recorded.

## *Staining and Destaining*

To stain the protein bands, the gels are placed in tubes filled with the dye Coomassie blue. The student will partially destain the gels in acetic acid, and in the second period destaining will be completed electrophoretically. Since the gels swell during the staining and destaining, the length must be measured and

recorded after destaining. The relative mobility of a protein is then given as:

$$\text{Mobility} = \frac{\text{distance of protein migration} \times \text{length before staining}}{\text{length after staining} \times \text{distance of dye migration}}$$

**Preparation of Solutions**

Solution for SDS protein incubation:

0.1 M sodium phosphate pH 7.0 in 1% SDS and 1% in β-mercaptoethanol.

To prepare 100 ml of 0.1 M sodium phosphate, pH 7.0:

0.1 M $NaH_2PO_4$ $H_2O$          0.55 g/40 ml
0.1 M $Na_2HPO_4$              0.99 g/70 ml

Then mix 39 ml of $Na_2HPO_4$ with 61 ml of $Na_2HPO_4$. Dilute 1:10 for 0.01 M protein incubation solution (50 ml).

49 ml buffer
0.5 ml β-mercaptoethanol
0.5 g SDS

Dialysis solution:

0.01 M sodium phosphate buffer, pH 7.0

0.1% SDS and 0.1% β-mercaptoethanol similar to above, only cut SDS and β-mercaptoethanol concentrations to 1/10 of above.

Gel buffer:

7.8 g $NaH_2PO_4H_2O$
20.4 g $Na_2HPO_4$
2.0 g SDS
$H_2O$ to liter

10% acrylamide solution:

22.2 g acrylamide
0.3 g methylenebisacrylamide
$H_2O$ to 100 ml

Protein solutions:

1 mg/ml so that a load of 10μl gives a final concentration of 0.01 mg

Staining solution, 100 ml:

0.25 g Coomassie brilliant blue
91 ml of 50% methanol
9 ml glacial acetic acid

Tracking dye:

0.05% Bromophenol blue in $H_2O$

Ammonium persulfate solution (fresh):

15 mg/ml

Destaining solution:

> 75 ml acetic acid
> 50 ml methanol
> 875 ml $H_2O$

Gel storing solution:

> 7.5% acetic acid solution

# Molecular Weight Determination of Mono and Plasma Amine Oxidases by Gel Filtration

One of the most useful and powerful tools for separating proteins from each other on the basis of size is molecular-exclusion chromatography[13,14] also known as gel filtration or molecular-sieve chromatography. It differs from ion-exchange chromatography, which separates solutes on the basis of electric charge and acid-base properties. In molecular-exclusion chromatography the mixture of proteins, dissolved in a suitable buffer, is allowed to flow by gravity down a column packed with beads of an inert, highly hydrated polymeric material that has previously been washed and equilibrated with the buffer alone. The common column materials are Sephadex, Bio-Gel, a commercial polyacrylamide derivative and agrose— all of which can be prepared with different degrees of internal porosity. In the column, proteins of different molecular size penetrate into the internal pores of the beads to different degrees and thus travel down the column at different rates (see Figure 3.2). Very large protein molecules cannot enter the pores of the beads; they are said to be excluded and thus remain in the excluded volume of the column, defined as the volume of the aqueous phase outside the beads. On the other hand, very small proteins can enter the pores of the beads freely (see Figure 3.2). Small proteins are retarded by the column while large proteins pass through rapidly, since they cannot enter the hydrated polymer particles. Proteins of intermediate size will be excluded from the beads to a degree that depends on their size; hence, the term exclusion chromatography. From measurements of the protein concentration and perhaps enzyme activity in a small fraction of the elute, an elution curve can be constructed (see Figure 3.3).

Molecular-exclusion chromatography can also be used to separate mixtures of other kinds of macromolecules, as well as very large biostructures— e.g., viruses, ribosomes, cell nuclei, or even bacteria— simply by using beads or gels with different degrees of internal porosity. The resolving power of molecular-exclusion chromagraphy is so great that this simple method is now widely used as a way of determining the molecular weight of proteins (see Figure 3.3— dashed line inset). The main objective of the gel filtration experiment is to separate proteins in a sample on the basis of their molecular weights. Theoretically, the smaller proteins should be temporarily trapped inside the Sephadex bead and therefore take more time to come off the column. Conversely, the larger protein should be eluted first.

## Gel Filtration

### Gel Preparation

3.6 g of Sephadex G-150 are placed in an Erlenmeyer flask and 200 ml of water added. The flask will be placed in a boiling water bath for four hours. Swirling of the flask may be necessary to remove any lumps that may form. At the end of the period, the swollen Sephadex is ready for packing into the column. (The swollen gel can be stored in the refrigerator until the column is to be packed.)

### Packing the Column

The column should contain some water in it before the slurry of gel is poured in (a volume of 1/3 the bed height is satisfactory). The slurry of gel is then poured carefully down the wall of the column (a stirring rod may be helpful in accomplishing this). If the column is not filled with slurry, add eluant to fill the column. Start the flow (1 ml/min–2 ml/min is sufficient).

### Sample Application

A 2 ml aliquot of the protein mixture will be pipetted onto the column by letting the sample solution run down the side of the column near the top of the Sephadex bed (1" above bed is adequate). Care must be taken so as not to disturb the packed Sephadex. Once the sample has been applied, allow the sample to flow into the Sephadex by starting the flow. Once the sample solution has penetrated the Sephadex bed, stop the flow and wash the sides of the column with the eluting solvent (ml vol) and allow this volume to penetrate into the bed. Stop the flow and then carefully layer more eluting solvent onto the Sephadex bed. Continue until the desired amount of solvent head is achieved, then start the flow at the desired rate (1 ml/min–2 ml/min).

### Analysis of Fractions

Each fraction will be assayed for absorbance at 280 nm using the Beckman Spectrophotometer. A plot of the absorbance versus volume of eluting solution will be made.

The main objective of the gel filtration experiment is to determine the molecular weights of mono and plasma amine oxidases by comparing the sample's elution volume to the elution volumes of standard proteins of known molecular weights.

# Procedure

## *Monoamine Oxidase*

In the first part of the experiment, the student will run two mixtures of two protein standards through a column of Sephadex G-150 or BioGel P-100 (for MAO) at a flow rate of approximately 0.5 ml per minute. Analysis of the protein contents of the aliquots will provide a plot of the elution volume versus absorption curve, which can be used to determine the elution volumes for each of the standard protein solutions, Bovine serum albumin, ribonuclease, ovalbumin and chymotrypsinogen. You will then run an aliquot of monoamine oxidase in 0.02 M potassium phosphate buffer containing 0.002 M β-mercaptoethanol and 0.1% Triton X-100 through the column with the same flow rate to determine its elution volume. Proteins with similar elution volumes can be assumed to have similar molecular weights, and therefore, molecular weight determination can be made by use of a standard curve of log molecular weight versus elution volume (see Figure 3.3).

| Protein | MW |
| --- | --- |
| ribonuclease | 13,700 |
| chymotrypsin | 25,000 |
| ovalbumin | 43,000 |
| serum albumin | 67,000 |

## *Plasma Amine Oxidase*

Repeat the above procedure except use chymotrypsin, ovalbumin, serum albumin, phosphorylase and paramyocine as a standard, and Sephadex superfine G-150.

| Protein | MW |
| --- | --- |
| phosphorylase | 94,000 |
| paramyocine | 100,000 |

## Assessment of Purity of Mono and Plasma Amine Oxidases by UV-Visible Spectroscopy

This section presents to students two highly interesting enzymes.[15,16] In this experiment, both mono and plasma amine oxidase have a distinctive UV-visible spectra.

## *Monoamine Oxidase*

Because of the flavin moiety, the UV-visible spectrum is distinguished by a peak at a wavelength of 475. A hypo- (to a lower wave length) or hyper- (to a higher wave length) chromic shift can be an indication of impurities. Absorption spectrum can be taken between wavelength of 300 nm–600 nm using spectrophotometer. Figure 3.6 shows a typical UV-visible spectrum for MAO. See Module 5 on physical characterization of mono and plasma amine oxidase.

## *Plasma Amine Oxidase*

A typical UV-visible spectrum for PPO is presented in Figure 3.7. PAO is a chromophoric enzyme. Upon addition of substrate—for example, benzylamine or an inhibitor like phenylhydrazine—the enzyme turns

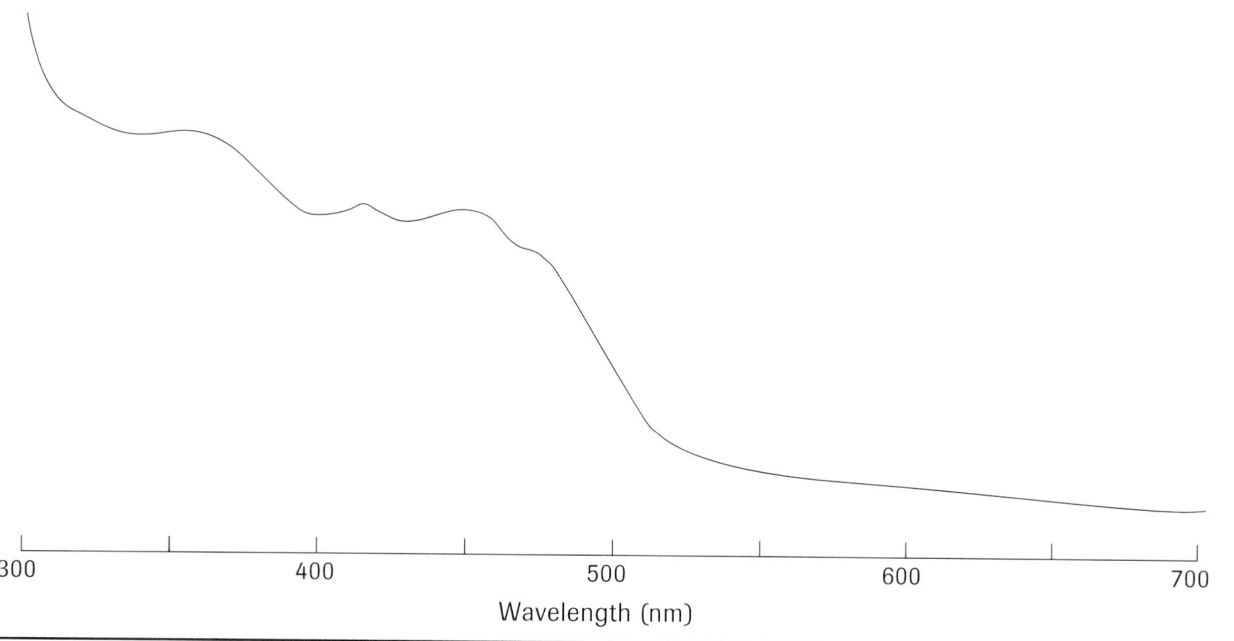


**FIGURE 3.6**  UV-visible spectrum for monoamine oxidase.

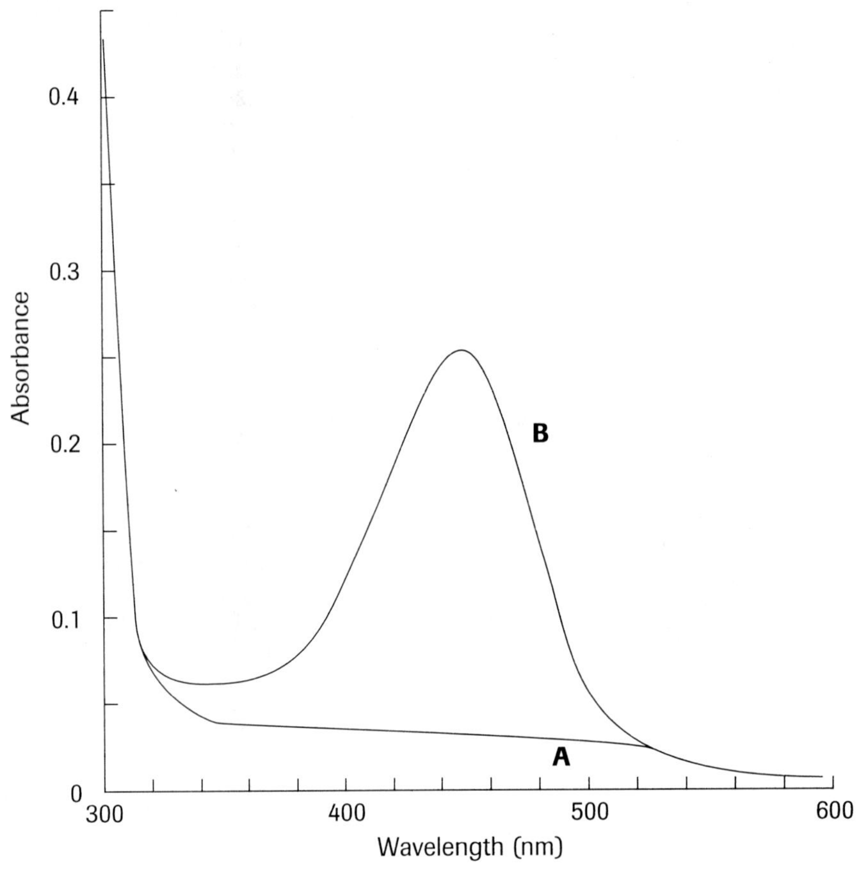



**FIGURE 3.7**   UV-visible spectrum for plasma amine oxidase. **(A)** Plasma amine oxidase at a concentration of 1 mg/ml. **(B)** Plasma amine oxidase after addition of 1:1 moles of phenylhydrazine.

to colorless, or yellow derivatives respectively. See Module 5 on physical characterization of mono and plasma amine oxidases. Students are encouraged to study and compare the effect of addition of 0.1 ml of 0.1 M substrate and or phenyl hydrazine with 1:1 molar ratio of PAO to phenylhydrazine.

## Assessment of Molecular Weight and Purity of Mono and Plasma Amine Oxidases by Determination of Amino Acid Analysis

For departments and or institutions that have access to amino acid analyzers, students can determine amino acid compositions. In addition, there are several departments where the service can be offered to institutions at a minimal fee. It is the intention of the authors to complete this section and give a wide spectrum of methods for assessment of molecular weight and purity of mono and plasma amine oxidases. Perform the procedure as follows:

a. Dissolve an aliquot of either mono and or plasma amine oxidases in 0.5 ml of 5.7 N hydrochloric acid, which had previously been redistilled three times.

b. Transfer an aliquot of dissolved enzyme to 10 ml test tube and then seal evacuate the test tube after you evacuate all the air. Make sure that the sample is in a highly evacuated, sealed tube after it was frozen in an acetone-dry ice bath before you hydrolyze it.

c. Hydrolyze the sample at 110° C in an oven for 24 hours.

d. After the hydrolysis, evaporate the excess acid by flushing with nitrogen gas and dissolve the sample in 0.2 M sodium citrate buffer, pH 2.2.

e. Standardize the automated amino acid analyzer with a Beckman amino acid calibration mixture.

f. Apply your sample to the automatic amino acid analyzer.

g. Calculate the number of amino acids on a mole to mole basis and compare and calculate the molecular weight of the enzyme.

# Review Questions

1. The subunit molecular weights of the standard proteins are:

   | | |
   |---|---|
   | serum albumin | 68,000 |
   | glyceraldelyde phosphate dehdrogenase | 36,000 |
   | fumarase | 49,000 |
   | catalase | 60,000 |
   | aldolase | 40,000 |

   Construct a standard curve of mobility versus log molecular weight from your gel containing these standards. Determine the molecular weight of monoamine oxidase you purified from the curve and the table of molecular weight supplied.

2. The subunit molecular weights of the standard proteins for plasma amine oxidase determination:

   | | |
   |---|---|
   | β-galactosidase | 130,000 |
   | phosphorylase | 94,000 |
   | serum albumin | 68,000 |
   | L-amine oxidase | 63,000 |

   Again construct a standard curve of mobility versus the log of molecular weight from your gel containing these standards. Determine the molecular weight of plasma amine oxidase from the standard curve and the table of molecular weights supplied.

2. What is the reasoning behind choosing the 280 nm wavelength for assay?
3. Is it permissible to use the 260 nm?
4. If it is permissible to use the 260 nm wavelength when would you be least apt to use this wavlength?
5. Is your monoamine oxidase preparation pure?
6. Is the plasma amine oxidase preparation pure?
7. What is the clinical significance of monoamine oxidase?
8. Why is the addition of Triton X-100 crucial in the purification of monoamine oxidase?
9. What is the role of β-mercaptoethanol? Show by equation the function of this reducing agent.
10. What is the physiological reducing agent in biological systems?
11. Explain the role of calcium phosphate in monoamine oxidase purification.
12. Name all the natural substrates for monoamine oxidase.
13. Name ten different inhibitors for monoamine oxidase. Classify these inhibitors as competitive, noncompetitive, uncompetitive or unknown (use the references on page 55).
14. What is the molecular weight of monoamine oxidase?
15. How many bands did you get by SDS-Electrophoresis? Why? Explain you results.

# POLYPHENOL OXIDASE PURIFICATION

### *Supplies and Equipment*

Aluminum foil
Beaker 4,000 ml
Borosilicate tubes
Cold room (4° C)
Cuvettes plastic (semi-micro)
Electrophoresis cell
Erlenmeyers
Eppendorf pipettes
Fraction collector—automated
Ice buckets
Hot plate
Magnetic stirrer and bars
Metric rulers
Pasteur pipettes
Petri dishes
pH meter
Pipettes—1 ml, 5 ml, and 10 ml
Power pack
Screw cap culture tubes
Serum caps
Shaker gyrotory type with controlled temperature
Spectrofluor tubing
Spectrophotometer (UV-visible)
Syringes and 20-gauge needles
Thermometer
Water bath
UV cuvettes (quartz)

### *Chemicals*

100mM acetate buffer, pH 5.0
Acetic acid-glacial
Acrylamide
Ammonium persulfate
Ammonium sulfate
Bis-acrylamide
Blue dextran—Pharmacia 2000
Bovine serum albumin
Bromphenol blue
Catechol
CM-Sephadex C-50
Coomassie blue
DEAE-Sephadex A-50
Glycerol
Glycine
Hydroxyapatite—Pharmacia
Kirk and Kelman's growth medium
β-mercaptoethanol
Methanol
Phenyl sepharose
Phosphate buffer

Sigma's tyrosinase
Sodium dodecyl sulfate
TEMED
Triton X-100

# Introduction

## Growth of *Coriolus versicolor*

Mycelia of a *Coriolus versicolor* (L. ex Fr.) Quel isolate (USDA—Northeastern Forest Exp. Stn., Durham, NH) can be maintained upon agar slants at 4° C. To obtain cultures, mycelia can be transferred to a sterile defined medium[18] containing 2% Bacto agar until a mycelial disc 30 mm in diameter is derived (5 days). Then the discs can be asceptically excised and homogenized 30 s into 25 ml autoclaved distilled water within a sterile Waring blender cup. Next 1 ml aliquots of the mycelial homogenates can be transferred with a sterile manostat to autoclaved Identi plug-stoppered 250 ml Erlenmeyer flasks containing 25 ml of the above sterile medium lacking agar. Cultures can be maintained at 25 ± 2° C and 150 RPM within a model 3529 Lab-line gyrotory shaker (Melrose Park, FL) for up to 16 days. Polyphenol oxidase appears in the growth medium at day 4 and increases in spc. act to day 15.[17]

## Dialysis

Following the assay of PPO (see Module 4) in either mycelial extracts or growth medium,[17] subject either the extracts or medium to dialysis by placing the extracts or medium into Spectrofluor tubing possessing a molecular weight cut-off of 14,000. Dialyze either the extracts or medium against 41 of pH 5.0, acetate buffer at 4° C with constant stirring—i.e., employ a magnetic stirrer and a stirring bar. Subsequent to an 18 hour dialysis, remove the dialysis bags from the buffer and "blot" them with paper towels. Then assay the contents of the bags for 280 nm absorbing substances (see Module 2) and PPO (see Module 4). Next, express PPO activity as spc. act. (Table 3.6)

$$\frac{^{\Delta}A \ min^{-1} \times 1,000}{mg \ protein}$$

and construct a balance sheet (Table 3.7) by comparing the spc. act. of PPO within crude mycelial extracts and growth medium with that following dialysis of the extracts and growth medium.

After dialysis, the mycelial extracts and growth medium can be subjected to ammonium sulfate fractionation. To accomplish this, record the volumes of the extract and growth medium and then add solid ammonium sulfate to achieve 0%–30% saturation (Table 3.8). This is performed by placing a beaker containing either

**TABLE 3.6**    Worksheet for the Calculation of PPO Units

| Time (min) | $^{A}$440 nm | $^{\Delta A}$440 nm | $^{\Delta A}$440 nm min$^{-1}$ | $^{\Delta A}$440 nm min$^{-1}$ × 1,000 | mg Protein | $\dfrac{^{\Delta A}440nm^{-1} \times 1,000}{mg\ Protein}$ |
|---|---|---|---|---|---|---|
| 0 | | | | | | |
| 2 | | | | | | |
| 4 | | | | | | |
| 6 | | | | | | |
| 8 | | | | | | |
| 10 | | | | | | |

**TABLE 3.7**    Polyphenol Oxidase Specific Activity Following Sequential Purification of Extracellular Polyphenol Oxidase

| Medium Purification Step | Polyphenol Oxidase Spc. Act. (Units) | Fold-purified |
|---|---|---|
| Crude medium | | |
| Dialyzed medium | | |
| Dialyzed plus $(NH_4)_2SO_4$ | | |
| Fractionated medium | | |
|     Pellet | | |
|     Supernatant | | |
| Gel filtration Sephadex G-150 | | |

From Moore et al. 1989. *J. Indust. Microbiol.* 4:349–364.

**TABLE 3.8**   Ammonium Sulfate Fractionation Table

Percent  Saturation Final Concentration of Ammonium Sulfate

|  | 10 | 15 | 20 | 25 | 30 | 33 | 35 | 40 | 45 | 50 | 55 | 60 | 65 | 70 | 75 | 80 | 85 | 90 | 95 | 100 |
|---|---|---|---|---|---|---|---|---|---|---|---|---|---|---|---|---|---|---|---|---|
| 0 | 56 | 84 | 114 | 144 | 176 | 196 | 209 | 243 | 277 | 313 | 351 | 390 | 430 | 472 | 516 | 561 | 610 | 662 | 713 | 767 |
| 10 |  | 28 | 57 | 86 | 118 | 137 | 150 | 183 | 216 | 251 | 288 | 326 | 365 | 406 | 449 | 494 | 540 | 592 | 640 | 694 |
| 15 |  |  | 28 | 57 | 88 | 107 | 120 | 153 | 185 | 220 | 256 | 294 | 333 | 373 | 415 | 459 | 506 | 556 | 605 | 657 |
| 20 |  |  |  | 29 | 59 | 78 | 91 | 123 | 155 | 189 | 225 | 262 | 300 | 340 | 382 | 424 | 471 | 520 | 569 | 619 |
| 25 |  |  |  |  | 30 | 49 | 61 | 93 | 125 | 168 | 193 | 230 | 267 | 307 | 348 | 390 | 436 | 485 | 533 | 583 |
| 30 |  |  |  |  |  | 19 | 30 | 62 | 94 | 127 | 162 | 198 | 235 | 273 | 314 | 356 | 401 | 449 | 496 | 546 |
| 33 |  |  |  |  |  |  | 12 | 43 | 74 | 107 | 142 | 177 | 214 | 252 | 292 | 333 | 378 | 426 | 472 | 522 |
| 35 |  |  |  |  |  |  |  | 31 | 63 | 94 | 129 | 164 | 200 | 238 | 278 | 319 | 364 | 411 | 457 | 506 |
| 40 |  |  |  |  |  |  |  |  | 31 | 63 | 97 | 132 | 168 | 205 | 245 | 285 | 328 | 375 | 420 | 469 |
| 45 |  |  |  |  |  |  |  |  |  | 32 | 65 | 99 | 134 | 171 | 210 | 250 | 293 | 339 | 383 | 431 |
| 50 |  |  |  |  |  |  |  |  |  |  | 33 | 66 | 101 | 137 | 176 | 214 | 256 | 302 | 345 | 392 |
| 50 |  |  |  |  |  |  |  |  |  |  |  | 33 | 67 | 103 | 141 | 179 | 220 | 264 | 307 | 353 |
| 55 |  |  |  |  |  |  |  |  |  |  |  |  | 34 | 69 | 105 | 143 | 183 | 227 | 269 | 314 |
| 60 |  |  |  |  |  |  |  |  |  |  |  |  |  | 34 | 70 | 107 | 147 | 190 | 232 | 275 |
| 65 |  |  |  |  |  |  |  |  |  |  |  |  |  |  | 35 | 72 | 111 | 153 | 194 | 237 |
| 70 |  |  |  |  |  |  |  |  |  |  |  |  |  |  |  | 36 | 74 | 115 | 155 | 198 |
| 75 |  |  |  |  |  |  |  |  |  |  |  |  |  |  |  |  | 38 | 77 | 117 | 157 |
| 80 |  |  |  |  |  |  |  |  |  |  |  |  |  |  |  |  |  | 39 | 77 | 118 |
| 85 |  |  |  |  |  |  |  |  |  |  |  |  |  |  |  |  |  |  | 38 | 77 |
| 90 |  |  |  |  |  |  |  |  |  |  |  |  |  |  |  |  |  |  |  | 39 |
| 95 |  |  |  |  |  |  |  |  |  |  |  |  |  |  |  |  |  |  |  |  |

*Percent  Saturation Initial Concentration of Ammonium Sulfate*

the extract or medium in a "dishpan" partially filled with ice over a magnetic stirrer. The required amount of ammonium sulfate is added slowly, with constant stirring. After the salt has been completely dissolved, centrifuge either the extracts of growth medium by centrifugation at 12,000 $xg$ (remember to use a nonogram) for 30 minutes. Then resuspend the pellet in 2 ml acetate buffer and dialyze for 18 hours. It is critical that you change the buffer at least once. Following dialysis, assay the bags' contents for 280 nm absorbing substances and PPO. Calculate the spc. act, and compare it to those for crude serum and dialyzed extracts and medium, respectively, by completing Table 3.7. How many fold has PPO been purified?

Dialysis and ammonium sulfate fractionation are among the early stages of protein purification. Later procedures, which are often employed, are chromatographies such as affinity, gel filtration, hydrophobic interaction, hydroxyapatite, and ion exchange.

The mechanisms of these were discussed in the introduction to this module. Here, we will concentrate on gel filtration employing Pharmacia Sephadex resins (although Bio-Rad resins are also effective). The dialyzed, ammonium sulfate 12,000 $xg$ supernatant in 100 mM acetate buffer, pH 5.0 can be applied to a $0.5 \times 17.0$ cm Bio-Rad column (note: increasing column length enhances resolution) of Sephadex superfine G-150 precalibrated with 100 µl of 0.1% blue dextran. The blue dextran is eluted with pH 5.0 acetate buffer collecting 0.5 ml fractions with an automated fraction collector. The blue dextran within column elutes is quantified at 620 nm, and absorbance is plotted on the $y$-axis and ml eluted on the $x$-axis. Then, the above supernatant is layered and eluted with the protein contents in column eluates being quantified at 280 nm. Plot $^A$280 nm on the $y$-axis and ml eluted on the $x$-axis on the same graph as that for blue dextran (Figure 3.8A). In addition to assaying eluates at 280 nm, also

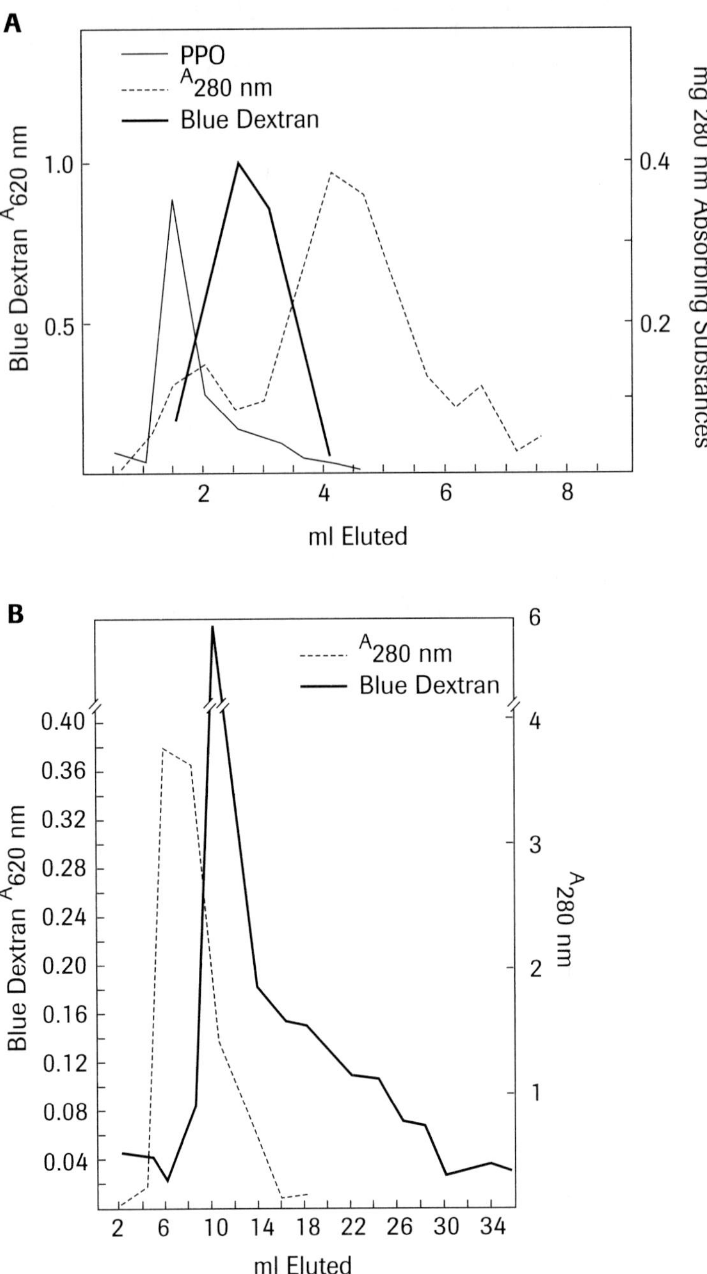



**FIGURE 3.8**   Sephadex G-150 elution profiles of 280 nm-absorbing substances and PPO within crude growth medium **(A)** and authentic PPO **(B)**. Data are xs and σs where N = 3 **(A)** with complete recovery of the layered protein being achieved. The $^A$280 nm data for **(B)** represents absorbances corrected to fraction total volume. From Moore et al. 1989. *J. Indust. Microbiol.* 4:349–364.

assay PPO activity at 440 nm (see Module 4) and plot as in Figure 3.8B. Compare the spc. act. of the PPO with those for dialyzed, dialyzed plus ammonium sulfate fractionated PPO, and dialyzed plus ammonium sulfate fractionated PPO plus gel filtration, and complete the fold purified in Table 3.7.

## Electrophoresis

To assess the effectiveness of an evolving protein purification protocol, SDS-PAGE[8,19-22] is often employed.

See Purification and Determination of Molecular Weight by SDS Polyacrylamide Gel Electrophoresis on page 44.[23,24]

If either authentic PPO or that present in *Coriolus versicolor's* growth medium are subjected to DEAE Sephadex or CM Sephadex ion exchange, PPO exchanges with the counterion of DEAE with a loss of PPO spc. act. Therefore, ion exchange chromatography on these resins is not useful in the purification of *Coriolus versicolor's* PPO. Other techniques that may be useful are affinity, hydroxyapatite and hydrophobic interaction chromatographies. Students are encouraged to develop a sequential protocol utilizing these techniques resulting in a single band on SDS-PAGE.

## Assessment of Purification—SDS-PAGE

To determine whether an evolving purification protocol does indeed result in a significant fold purification, withhold sample aliquots equivalent to 30 µg of protein during each of the above purification steps. Then subject these aliquots to SDS-PAGE[8] employing *marker* proteins of known molecular weights—e.g., Bio-Rad's markers (Figure 3.9A). The SDS-PAGE tube gel procedures are summarized in Table 3.9. The results of these procedures yield a molecular weight calibration curve (Figure 3.9B) relating log MW of the markers ($y$-axis) to relative mobility ($x$-axis).

The $m_r$ can be calculated from the following equation:[25]

$$m_r = \frac{\text{distance of protein migration}}{\text{length of gel after destaining}} \times \frac{\text{length of gel before staining}}{\text{distance of dye migration}}$$

**A**

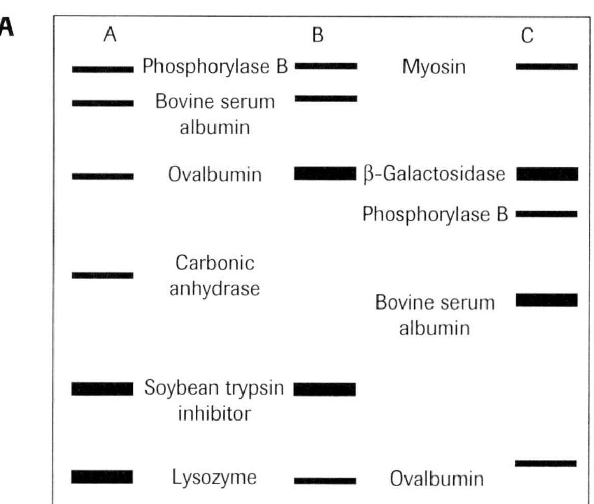


SDS polyacrylamide gels run in the Mini-PROTEAN™ II cell according to the method of Laemmli.[1] (A) Low molecular weight standards run on a 12% SDS polyacrylamide gel stained with Coomassie R-250. (B) Low molecular weight standards run on a 12% SDS polyacrylamide gel stained with Bio-Rad's Silver Stain. (C) High molecular weight standards run on a 7.5% SDS polyacrylamide gel. stained with Coomassie R-250.

**B**

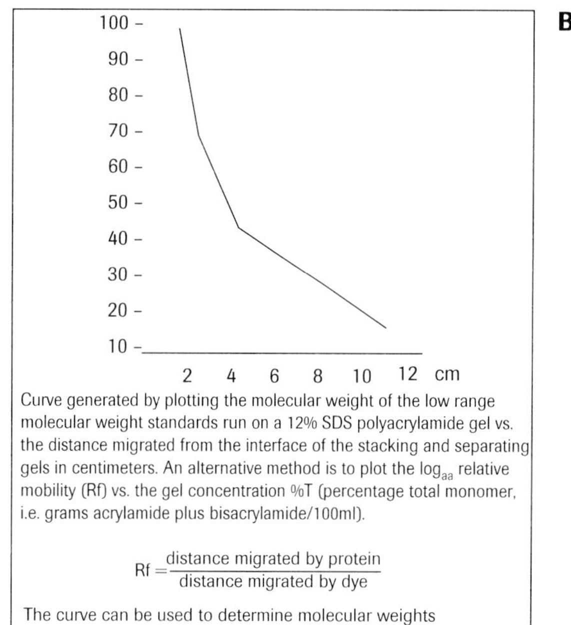


Curve generated by plotting the molecular weight of the low range molecular weight standards run on a 12% SDS polyacrylamide gel vs. the distance migrated from the interface of the stacking and separating gels in centimeters. An alternative method is to plot the $\log_{aa}$ relative mobility (Rf) vs. the gel concentration %T (percentage total monomer, i.e. grams acrylamide plus bisacrylamide/100ml).

$$Rf = \frac{\text{distance migrated by protein}}{\text{distance migrated by dye}}$$

The curve can be used to determine molecular weights of unknown proteins.

| Protein | Molecular weight | Reference |
|---|---|---|
| Myosin | 200,000 | Woods. E. F., Himmartian S. and Harrington, W. F., *J. Biol. Chem.*, 238. 2374 (1983). |
| *E. coli* β-Galactosidase | 116,250 | Fowler, A. V., and Zabin. I.*Proc. Natl. Acod. Sci. USA*. 74, 1507 (1977). |
| Rabbit muscle phosphorylase B | 97,000 | Titani. K., et al., *Proc. Natl. Acad. Sci. USA*. 74. No. 11. p. 4762 (1997). |
| Bovine serum albumin (BSA) | 86,200 | Brown. J. R., *Fed Proc.*, 34, 591 (1975). |
| Hen egg white ovalbumin | 42,600 | Niebet. A. D., *Eur. J. Biochem.*, 115. 335 (1961). |
| Bovine carbonic anhydrase | 31,000 | Davis. R. P., "Carbonic Anhydrase." *The Enzymes.*Vol. V. p. 545. (Boyer, P.O. ad.) Academic Press, New York (1977). |
| Soybean trypsin inhibitor | 21,500 | Wu. Y. V., and Scherage H. A., *Biochemistry*, 1, 698 (1962). |
| Hen egg white Lysozyme | 14,400 | Jolien P. Angeur. *Chem. Inst. Edit.* 8. 227 (1989). |

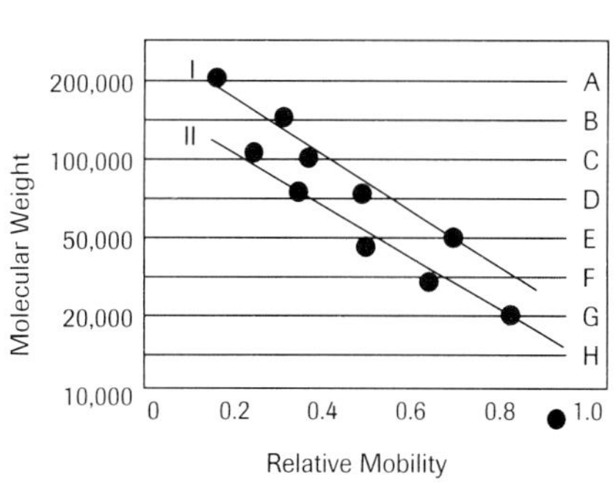


Standard curve: Molecular weight vs. relative mobility. I. HMW standards on 5% gel. II. LMW standards on 10% gel. (A) Myosin, (B) β-Galactiosidase, (C) Phosphorylase B. (D) BSA, (E) Ovalbumin, (F) Carbonic anhydrase, (G) Soybean trypsin inhibitor, (H) Lysozyme.

High Molecular Weight (HMW) Protein Standards. MW Operating Range 40,000-250,000 Dallons used for 4%-10% gels

Low Molecular Weight (LMW) Protein Standards. MW Operating Range 10,000-100,000 Dallons used for 10%-20% gels

Electrophoresis Conditions A 25µl sample (1:20 dilution) of the molecular weight standards was electrophoresed on either a 5% SDS Bio-Phore gel (for HMW Standards) or a 10% SDS Biophore gel (for LMW Standards) using preformulated SDS BioPhore buffer at pH 8.3 Electrophoresis was at 1.5 mA/lube for 2.5 hours.

**FIGURE 3.9** Molecular weights of Bio-Rad SDS-Page marker proteins **(A)** and calibration curve **(B)** using markers.

From Bio-Rad Laboratories. Hercules, CA.

**TABLE 3.9**  SDS-PAGE-Electrophoresis Protocol

**Preparation of Gel Tubes**

a) Clean gel tubes in a chromic acid solution (extreme caution)—rinse thoroughly with distilled $H_2O$.
b) Dry tubes in an oven.
c) Mark desired length of gel on borosilicate tubes.
d) Seal one end of tubes with a serum cap (some people use paraffin).
e) Place sealed tubes perfectly upright in a rack (see demonstration by instructor).

**Preparation of Gel Mix**

a) Prepare a 100 ml solution that contains 30 g of arcylamide and 0.8 g bisacrylamide (careful neuropoison).
b) Pipet 2.7 ml of arcylamide mix into an Erlenmeyer, 3.75 ml of Tris (pH 8.8), 1.01 ml of 10% SDS and 2.55 ml of $H_2O$; mix gently by swirling; add 6μl of TEMED and 0.06 ml of fresh 10% ammonium persulfate; swirl.

**Pouring of Gels**

a) Using a long-stem Pasteur pipet, add gel mix to tubes slowly until meniscus is on length marker.
b) Tap tubes gently to release trapped bubbles.
c) Carefully layer distilled water on top of gel mix.
d) Allow gels to polymerize (20 minutes).

**Preparation of Sample**

a) Prepare each protein by pipeting 0.3 mg of protein into a small test tube, add 0.01 ml of 10% SDS, one drop of glycerol, 3 μl of β-mercaptoethanol, and 2μl of tracking dye (0.1% bromphenol blue).
b) Heat samples at 100°C for 2 minutes and cool to room temperature.

**Electrophoresis of Samples**

a) Pour 750 ml of electrophoresis buffer (0.025 M Tris-HCl, pH 8.8 containing 0.2 M glycine) into the lower chamber of the electrophoresis cell.
b) Remove seal from bottom of gel tubes and carefully insert them into washers and slots in the upper chamber of the electrophoresis cell.
c) With sample applicator (usually a Hamilton syringe) layer sample onto gel.
d) Add buffer to upper chamber the same as in step a, but with 0.1% SDS.
e) Place top on gel apparatus and plug electrodes into the power supply.
f) Electrophorese samples at 3 mA/gel for 30 minutes, then increase voltage to 8mA gel and continue electrophoresis until tracking dye approaches end of gel.
g) After electrophoresis, record the pH of the buffer in the upper and lower chambers.

**Staining and Destaining**

a) Remove gel from tube (instructor will give demonstration).
b) Record length of gel and distance tracking dye traveled.
c) Place gel in staining container (test tube containing 50% TCA solution) overnight.
d) Remove gels from TCA, rinse in $H_2O$, stain in 0.1% Coomassie blue in 50% TCA at 37° C for 1 hour.
e) Remove gels from stain, rinse, and destain by stirring in several changes of solution (50 ml methanol, 75 ml glacial acetic acid brought to 1,000 ml with $H_2O$).

**Molecular Weight Calculation**

a) Record length of gel after staining.
b) Record distance each protein band traveled.
c) Calculate relative mobility ($m_r$) of each band:

$$\text{relative mobility } (m_r) = \frac{\text{Distance to band}}{\text{Distance of tracking}} \times \frac{\text{length of gel before staining}}{\text{length of gel after staining}}$$

d) Make the same calculations for gel that had the molecular weight standards on it.
e) Construct a standard curve-logs of molecular weight standards as a function of relative mobility.

Celis, J. E., and R. Bravo. 1984. *Two Dimensional Gel Electrophoresis: Methods and Application.* New York: Academic Press.

**TABLE 3.10**  Relative Mobilities of Bio-Rad Markers after Laemmli, SDS-PAGE

| Marker | Log $m_w$ | $M_r$ |
|---|---|---|
| Myosin | 5.301 | 0.097* |
|  |  | 0.117 |
| β-Galactosidase | 5.065 | 0.389 |
|  |  | 0.369 |
| Phosphorylase B | 4.966 | 0.447 |
|  |  | 0.437 |
| Bovine serum albumin | 4.821 | 0.554 |
|  |  | 0.530 |
| Ovalbumin | 4.653 | 0.593 |
|  |  | 0.573 |
| Carbonic anhydrase | 4.491 | 0.768 |
|  |  | 0.756 |
| Soybean trypsin inhibitor | 4.332 | 0.836 |
|  |  | 0.826 |
| Lysozyme | 4.158 | 0.866 |
|  |  | 0.865 |

*Top number, run 1; bottom number run 2 for each marker

Table 3.10 presents the $m_r$ for Bio-Rad markers. This calibration curve is useful in establishing the MW of a purified protein. To accomplish this, the $m_r$ of the protein being purified is established and then employed to ascertain the protein's $m_r$ via the following relationship.

Once the log $m_r$ of the protein is determined, its MW can be derived by taking the anti-log of the log MW. As an alternative to tube gel electrophoresis, slab gel electrophoresis can be employed. This technique permits the use of silver stains to detect proteins at less than μg levels.[26,27]

The use of Bio-Rad's Mini Protean II Dual electrophoresis cell has become quite popular as it together with a mini trans-blot electrophoresis transfer cell and an electroeluter permit 2D-SDS-PAGE, blotting, and electroelution. When employing 2D-SDS-PAGE, the second direction usually is isoelectric focusing.[28–30]

## Review Questions

Each of the following should be written in essay form.

1. Explain how ammonium sulfate fractionation can be used to purify proteins.
2. Describe the mechanism by which gel filtration separates proteins.
3. Suppose that you wish to separate a mixture of proteins possessing the following molecular weights: 121,000 (MW of Sigma Chemical Company's PPO), 60,000 and 12,000; design a gel filtration experiment to accomplish this.
4. Detail how ion exchange chromatography separates proteins.
5. Provide an ion exchange chromatography experiment to determine if it would be more advantageous to utilize a cation or anion exchange resin in an evolving enzyme purification protocol.
6. Compare and contrast the mechanisms by which affinity, hydroxyapatite, and hydrophobic interaction chromatographies separate proteins.
7. Suppose that you have prepared an extract of rat liver, fungal hyphae, or a higher plant tissue (e.g., tomato or potato); design an enzyme purification protocol that would result in MAO, PAO, or fungal/higher plant PPO being purified to homogeneity. Employ the range of chromatographic and electrophoretic techniques described herein. How would you assess whether the enzyme's spc. act. was enhanced?
8. Explain how gel filtration and SDS-PAGE molecular weight calibration curves are constructed and employed.

## REFERENCES

1. Pharmacia Laboratory Separation Division. 1967. *Affinity Chromatography: Principles and Methods.* Uppsala, Sweden: Pharmacia.
2. Lowe, C. R. 1979. *An Introduction to Affinity Chromatography.* New York: Elsevier. 256.
3. Pharmacia Laboratory Separation Division. 1967. *Gel Filtration: Theory and Practice.* Uppsala, Sweden: Pharmacia.
4. Fischer, L. 1980. *Gel Filtration Chromatography.* New York: Elsevier. 270.
5. Henschev, A. 1985. *Higher Performance Liquid Chromatography in Biochemistry.* New York: VCH Publishers.
6. Zubay, G. 1993. *Biochemistry.* Dubuque, IA: Wm. C. Brown Communications, Inc.
7. Pharmacia Laboratory Separation Division. 1967. *Ion Exchange Chromatography: Principles and Methods.* Uppsala, Sweden: Pharmacia.
8. Laemmli, U. K. 1970. "Cleavage of structural proteins during the assembly of the head of bacteriophage T4." *Nature.* 227:680–85.
9. Chrambach, A. 1989. *The Practice of Quantitative Gel Electrophoresis.* New York: VCH Publishers.
10. Lehninger, A. L., D. L. Nelson, and M. M. Cox. 1993. *Principles of Biochemistry.* New York: Worth Publishers.
11. Beckman Instruments, Inc. *Hydrophobic Interaction Chromatography.* Berkeley, CA: Beckman Instruments, Inc.
12. Schneider, W. C., and Hogeboom, G. H. 1950. *J. Biol. Chem.* 183:1236.

13. Weber, K., and M. Osborn. "The reliability of molecular weight determinations by dodecyl sulfate polyacrylamide gel electrophoresis." *J. Biol. Chem.* 1969:244, 4406.

14. Loening, U. E. "The fractionation of high molecular weight RNA by polyacrylamide-gel electrophoresis." *J. Biochem.* 1967:202, 251.

15. Oreland, L. 1971. *Arch. Biochem. Biophys.* 146:410–21.

16. Kearney, E. B., J. I. Salach, W. H. Walker, R. L. Seng, W. Kenney, E. Zeszotech, and T. P. Singer. 1971. *Eur. J. Biochem.* 24:321–27.

17. Moore, N. L., D. H. Mariam, A. L. Williams, and W. V. Dashek. 1989. *J. Indust. Microbiol.* 4:349–64.

18. Kirk, T. T., and A. Kelman. 1965. "Lignin degradation as related to polyphenoloxidases of selected wood decaying basidiomycetes." *Phytopathol.* 55:739–45.

19. Gordon, A. H. 1975. *Electrophoresis of Proteins in Polyacrylamide and Starch Gels.* New York: Elsevier. 216.

20. Haines, B. D., and D. Richwood. 1981. *Gel Electrophoresis of Proteins: A Practical Approach.* London: IRL Press Ltd.

21. Anderson, B. L., R. W. Berry, and A. Telser. 1983. "A sodium dodecyl sulfate-polyacrylamide gel electrophoresis system that separates peptides and proteins in the molecular weight range of 2,500 to 90,000." *Anal. Biochem.* 132:365–75.

22. Ochs, D. 1983. "Protein contaminants of sodium dodecyl sulfate-polyacrylamide gels." *Anal. Biochem.* 135:470–74.

23. Bio-Rad Laboratories. *Bio-Rad Acrylamide Polymerization: A Practical Approach.* Bulletin 1156. Richmond, CA: Bio-Rad Laboratories.

24. Radola, J. 1989. *Electrophoresis.* New York: VCH Publishers.

25. Weber, K., and M. Osborn. 1969. "The reliability of molecular weight determinations by dodecyl sulfate-polyacrylamide gel electrophoresis." *J. Biol. Chem.* 244:4406–12.

26. Switzer, R. C., C. R. Merril, and S. Shifrin. 1979. "A high sensitive silver stain for detecting proteins and peptides in polyacrylamide gels. *Anal. Biochem.* 98:231–37.

27. Ochs, D. C., E. H. McConkey, and D. W. Sammons. 1981. "Silver stains for proteins in polyacrylamide gels—a comparison of 6 methods." *Electrophoresis.* 2:304.

28. Righetti, P. G. 1983. *Isolectric Focusing Theory, Methodology and Applications.* New York: Elsevier. 386.

29. Celis, J. E., and R. Bravo. 1984. *Two Dimensional Gel Electrophoresis: Methods and Application.* New York: Academic Press.

30. Jitsukawa, S., H. Sakuai, and T. Hashiro. 1984. "Electrostaining of two-dimensional polyacrylamide gel electrophoresis." *Electrophoresis 83.* Berlin: Walter de Gruyter and Co. 257–61.

## Supplementary—MAO/PAO

Fowler, C. J. and S. B. Ross. 1984 "Selective inhibitors of monoamine oxidase A and B, biochemical, pharmacological and chemical properties." *Med. Res. Rev.* 4:423–358.

Huang, H. K. 1980. "Lipid-protein interactions in the multiple forms of monoamine oxidase enzymatic and ESR studies with purified intact rat brain mitochondria." *Mol. Pharmacol.* 17:192–98.

Jain, M. 1977. "Monoamine oxidase." *Life Sci.* 1925–33.

Minamiura, N. and K. T. Yasaunobu. 1978. "Bovine liver monoamine oxidase-modified purification procedure and preliminary evidence for two subunits and one FAD." *Arch. Biochem. Biophys.* 189:481–89.

Murphy, D. L., N. A. Garrlick, and R. M. Cohen. 1982. "Monoamine Oxidase." *Biochemical and Physiological Aspects Relevant to Human Psychopharmacology in Antidepressants.* Burrows, N., and D. Davies (eds.). New York: Elsevier. 209.

Sandler, M., S. Bonham, B. Goodwin, C. Ruthrven, M. Youdim, E. Hanington, M. Cuthbert, and C. Pare. 1973. "Multiple forms of monoamine oxidase: Some *in vivo* correlations." *Neuropsychopharmacology of Monoamines and their Regulatory Enzymes.* Usdin, E. (ed.). New York: Raven Press. 3–28.

Schact, V. and M. Levien. 1977. *Depressive Disorders Symposium.* Rowe, Publisher F. K. Schattaner Verlag, Stutgart, NY. 21–40.

Singer, T. P., R. W. Von Korff, and D. L. Murphy. 1979. *Monoamine Oxidase, Structure, Function and Altered Functions.* New York: Academic Press. 1–57.

Singer, T. P. 1986. "Mitochondrial monoamine oxidase." *Advances in Biochemical Pharmacology,* 1. Vessey, D. A. and D. Zakim (eds.). New York: John Wiley and Sons.

Singer, T. P. 1985. *Demise and resurrection of misconceptions about monoamine oxidase in Neuropharmacology.* Kelemen, K., K. Magyar, and E. S. Vizi (eds.). Budapest: Akademiai Kiado. 29.

Tabor, C. W., H. Tabor, and S. M. Rosenthal. 1954. "Spectrophotometric determination of monoamine oxidase." *J. Biol. Chem.* 208:645.

Tabor, G. E., H. Tabor, and S. M. Rosenthal. 1951. *J. Biol. Chem.* 193:265–71.

Tiselius, A. S., S. Hjerten, and D. Levin. 1956. *Arch. Biochem. Biophys.* 65:132.

Tipton, K. F., C. J. Flower, and M. D. Houslay. 1981. "Specificities of the two forms of monoamine oxidase." *Monoamine Oxidase, Basic and Clinical Frontiers.* Kamijo, K. E. Usidin, and T. Nagatsu, (eds.). Amsterdam-Oxford-Princeton, NJ: Excerpta Medica. 87–99.

Tipton, K. F., M. D. Housley, and T. J. Mantle. 1981. *Monoamine Oxidase and Its Inhibition.* Ciba Foundation Symp. 39:5–31, Amsterdam: Elsevier.

Von Korff, R. W. 1979. "Monoamone oxidase: Unanswered questions." *Monoamine Oxidase: Structure, Function and Altered Functions.* Singer, T. P., R. W. Von Korff, and D. L. Murphy (eds.). New York: Academic Press. 1–6.

White, H. L. and R. L. Tansik. 1979. "Characterization of multiple substrate." *Monoamine Oxidase: Structure, Function and Altered Functions.* Singer, T. P., R. W. Von Korff, and D. L. Murphy (eds.). New York: Academic Press. 129–44.

White, H. T. and D. K. Stine. 1981. "Characterization of MAO-A and MAO-B sites by various biochemical techniques." *Monoamine Oxidase, Basic and Clinical Frontiers.* Kamijo, K., E. Usdin, and T. Nagatsu (eds.). Amsterdam: Excerpta-Medica. 62–73.

White, H. L. and D. Stine. 1982. "Monoamine oxidase A and B as components of a membrane complex." *J. Neurochem.* 38:1429–36.

White, H. L. and D. K. Stine. 1982. *J. Neurochem.* 38:1429–36.

White, H. T. and J. C. Wu. 1975. *Neurochemistry.* 25:21–26.

Watanabe, K., N. Minamuria, and K. T. Yasunobo. 1980. "Thiols liberated covalently bound flavin from monoamine oxidase." *Biochem. Biophys. Res. Commun.* 94:579–85.

Yasunobu, K. T., H. Ishizaki, and N. Minamiura. 1976. "Amine oxidases." *Mol Cell Biochem.* 13:3.

Yagi, K. and M. Naoi. 1982. *Biochem. Int.* 4:457.

Youdim, M. B. H. and J. P. M. Finberg. 1982. *Monoamine Oxidase Inhibitor Antidepressants in Psychopharmacology I,* Graham-Smith, D. G., H. Hippius, and G. Winoker (eds.). Amsterdam: Elsevier. 38.

Zeidan, H., P. Han, and J. Johnson. 1985. "Spin label studies of the essential sulfhydryl group environment in chicken liver fructose 1,6-bisphosphatase." *FEBS.* 192:294–97.

Zeidan, H., K. L. H. Watanabe, Piette, and K. T. Yasunobu. 1979. "ESR studies of bovine liver monoamine oxidase B." *Frontiers in Protein Chemistry.* Mayammia, L. L., T-Y Liu, and K. Yasunobu (eds.). Amsterdam: Elsevier. 133–47.

Zeidan, H. 1988. Interaction of spin-labeled tryptamine with monoamine oxidase: Probing the microenvironment of the active site by spin probe-spin label techniques." *Biochem. Biophys. Acta.* 955:111–118.

Zeidan, H., K. Watanabe, L. H. Piette, and K. T. Yasunobu. 1980. "ESR-spin labelling of plasma amine oxidase: Probing the substrate liberated essential sufhydryl group." *J. Biol. Chem.* 255:7621–26.

Zeidan, H. 1988. "Interaction of spin-labeled tryptamine with monoamine oxidase: Probing the microenvironment of the active site by spin probe-spin label techniques." *Biochem. Biophys. Acta.* 957:266–71.

Zeidan, H. 1990. "The active site of monoamine oxidase: The key enzyme of biogenic amines degradation" as seen by magnetic resonance spectroscopy. *Biodeterioration Research.* 3:555–64, Llewellyn G. C. and C. E. O'Rear (eds.). New York: Plenum Press.

Zeidan, H. and S. Buchanan. 1990. "Spectral studies on monoamine oxidase modified with N-(1-pyrene) melemide fluorescent probe. *Biodeterioration.* 577–85, Llewellyn G. C. and C. E. O'Rear (eds.). New York: Plenum Press.

## Supplementary—PPO

Bergneyer, H. V. 1983. *Methods of Enzymatic Analysis. Volume 3 Enzymes 1: Oxidoresductases: Transferases.* New York: VCH Publishers.

Brooks, K. A., and E. G. Sander. 1980. "Preparative polyacrylamide gel electrophoresis: removal of polyacrylate from proteins." *Anal. Biochem.* 107:182–86.

Bio-Rad Laboratories. *Bio-Rad Ion Exchange Manual.* Richmond, CA: Bio-Rad Laboratories.

Dodson, P. A., P. J. Harvey, C. S. Evans, and J. M. Palmer. 1989. "Properties of an extracellular ligninase from *Coriolus versicolor.*" *Biotechnology in the Pulp and Paper Industry.* 3rd International Conference, Stockholm. 185.

Fahraeus, G., and B. Reinhammer. 1967. "Large scale production and purification of laccase from culture of the fungus *Polypous versicolor* and some properties of laccase." *Acta Chem. Scand.* 21:2367–78.

Farrell, P. H. 1975. "High resolution two-dimensional electrophoretic analysis of soybean proteins." *J. Agric. Food Chem.* 31:963–68.

Flurkey, W. H. 1985. "*In vitro* biosynthesis of *Vicia faba* polyphenol oxidase." *Plant Physiol.* 79:564–67.

Flurkey, W. H. 1986. "Polyphenol oxidase in higher plants: Immunological detection and analysis of *in vitro* translation products." *Plant Physiol.* 86:614–18.

King, R. S. and W. H. Flurkey. 1987. "Effects of limited proteolysis on broad bean polyphenol oxidase." *J. Sci. Food Agric.*

Kowalski, S. P., J. Bramberg, W. M. Tingey, and J. C. Stefens. 1990. "Insect resistance in the wild potato *Solanum berthaultii:* Inheritance of glandular trichomes polyphenol oxidase." *J. Heredity.* 81: 475–78.

Lanker, T., T. G. King, S. W. Arnold, and W. H. Flurkey. 1987. "Active, inactive and *in vitro* synthesized forms of polyphenol oxidase during leaf development." *Physiol. Plant.* 69:323–29.

Lei, M-G., D. Tyrell, R. Bassette, and G. R. Reeck. 1983. "Two-dimensional electrophoretic analysis of soybean proteins." *J. Agric. Food Chem.* 31:963–68.

Leisola, M., U. Thanei-Wyss, and A. Fletcher. 1985. "Strategies for production of high ligninase activities by *Phanerochaete chrysosoporium.*" *Journal Biotechnology.* 3:97–197.

Liu, Shu-Yen, R. D. Minard, and J. M. Bollag. 1981. "Oligomerization of syringic acid, a lignin derivative, by phenol oxidase." *Soil Sci. Am. J.* 45: 1100–105.

Mayer, A. M., and E. Harel. 1990. "Polyphenol oxidases and their significance in fruits and vegetables." *Enzymes in Foods.* New York: Elsevier.

Montgomery, R., and C. A. Swenson. 1969. *Quantitative Problems in the Biochemical Sciences.* San Francisco: W. H. Freeman and Co.

Pharmacia. 1986. *Ion Exchange Product Profile.* Uppsala, Sweden: Pharmacia.

Schmidt, H. 1988. *Phenol oxidase (EC 1.14.18.1)—A Molecular Enzyme for Defense Cells.* New York: VCH Publishers.

Segel, I. H. 1976. *Biochemical Calculations*, 2d ed. New York: John Wiley and Sons.

Snedecor, G. W., and W. C. Cockran. 1979. *Statistical Methods.* Ames: Iowa State University Press.

Sonderhall, D., I. Carlberg, and T. Ericksson. 1985. "Isolation and partial purification of prophenoloxidase from *Daucus carota* L. Cell cultures." *Plant Physiol.* 78:730–33.

Winnacker, E. L. 1987. *From Genes to Clones.* New York: VCH Publishers.

Wong, T. C., B. S. Luh, and J. R. Whitaker. 1971. "Isolation and characterization of polyphenol oxidase isozyme of clingstone peach." *Plant Physiol.* 48:19.

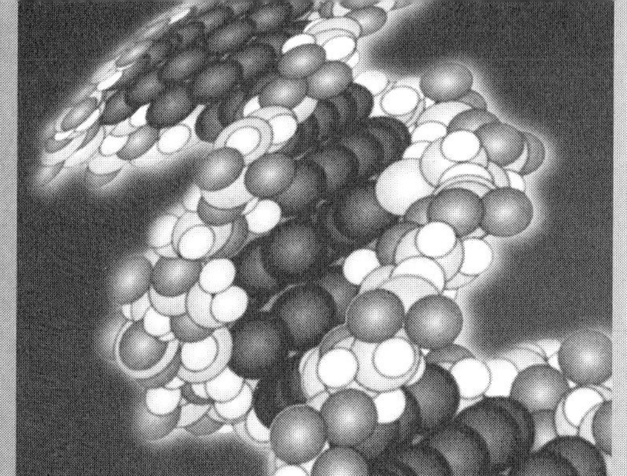

# MODULE
# 4

# Enzyme Assay, Kinetics, Inhibition, and Activation Energy

## Outline of Module

### MONOAMINE OXIDASE B

Laboratory exercises based upon three 4 hour laboratory periods for which prior laboratory preparation was accomplished. Each of the following exercises is one laboratory period.

Assessment of enzyme purity
Assay of enzyme using benzylamine as artificial substrate
Quantification of the specific activity of monoamine oxidase
Investigation of the effects of pH and temperature upon the rate of enzyme catalysis
Evaluation of substrate concentration on enzyme activity
Observation of effects of inhibitors upon the rate of enzyme catalysis
Inhibitor of monoamine oxidase

### POLYPHENOL OXIDASE ASSAY

Laboratory exercises based upon two 2–3 hour laboratory periods for which prior laboratory preparation was accomplished.

Sources of PPO[M]
*Duration:*  One laboratory period to prepare sterile defined growth medium with and without 2% agar and inoculate agar solidified medium; grow fungus to 30 mm disc (5 days); asceptically remove disc; homogenize and inoculate into defined liquid medium (one laboratory period) and culture 10–13 days according to Moore et al. (1994)

# MODULE INTRODUCTION

Enzymes are mostly proteins that catalyze a variety of chemical reactions that occur within living organisms. Enzymes generally catalyze one particular kind of reaction. The actions of these biological catalysts are sensitive to a number of factors, including temperature, pH, substrate, regulators, and product concentrations as well as inhibitors. In this module, we will study the effects of some of these factors upon an enzyme-catalyzed reaction and measure the rate of the reaction. The enzymes that will be studied are monoamine oxidase and polyphenol oxidase. These enzymes are found in humans and in other organisms as well.

# LABORATORY EXERCISES

## MONOAMINE OXIDASE B

### Supplies and Equipment
Electrophoretic cell
Spectrophotometer
Glass tubes
Water bath
1.5 ml semi-micro plastic cuvettes or 1.0 ml
   glass cuvette

### Chemicals
Potassium phosphate buffer
Triton X-100
Potassium cholate
β-mercaptoethanol
Benzylamine
P-hydroxyamphetamine

## Introduction

The amine oxidases are found in all forms of life and they all catalyze the general reaction shown in Equation 1.

$$RCH_2NH_2 + O_2 + H_2O \rightarrow RCHO + NH_3 + H_2O_2 \quad (1)$$

Monamine oxidase (MAO) is important in the control of the levels of biogenic amines in the brain and the central nervous system.[1,2] Therefore, there has been considerable interest in the physiochemical and enzymatic properties of the enzyme.

The biological role of the amine oxidases is clearly not restricted to the inactivation of toxic amines that enter the blood stream from the gastrointestinal tract. In the course of the enzymatic deamination of amines, aldehydes are formed that possess very different biological activities from those of the amines.[3,4] See Module 5 for the clinical and pathological significances of the enzyme.

In this module, we have selected the materials that are relevant to chemical kinetics, inhibition, and activation energy. Other aspects of MAO physiochemical properties—namely, chemical aspects, multiple forms of the enzyme, and physiological and clinical aspects—are presented throughout the laboratory manual as independent modules. As a consequence, this laboratory manual offers a comprehensive review of significant areas relevant to MAO research.

MAO has been characterized by amine substrate specificities and response to inhibitors as having two independent catalytic sites[3,4] or as two distinct enzyme forms (A, B) embedded in the same mitochondrial membrane.[5–7] The A-type enzyme, which deaminates serotonin, is sensitive to clorgyline inhibition. MAO-B, which deaminates benzylamine, is sensitive to deprenyl inhibition. Tyramine and tryptamine are substrates for both the A and the B enzyme.[8,9]

The broad specificity of MAO will be useful in this study. It means that the enzyme can be assayed with an artificial substrate, benzylamine, which gives no absorption at a wavelength of 250 nm, but its product, benzaldehyde, can yield a significant absorption at this wavelength (Equation 1). By following the production of benzylaldehyde, the activity of the enzyme can be measured.

You have 3 four-hour laboratory periods in which to complete the experiment. In the first period, assess the enzyme purity by methods reported in Module 3, assay the enzyme using benzylamine as an artificial substrate,[10] and investigate the effects of pH and temperature upon the rate of enzyme catalysis. During the second period, evaluate the effects of substrate concentration and storage upon enzyme activity. Devote the third period to observing the effects of the following inhibitor upon the rate of enzyme catalysis: a competitive inhibitor, parahydroxyampethamine (a common antidepressant agent). Try to understand the principles and do the calculations for each laboratory period before beginning the next. Students are encouraged to work in pairs. Students are encouraged to be familiar and review all monoamine oxidose inhibitors.[1–11]

# Procedures

## Assessment of Enzyme Purity

Prior to assaying the enzyme, the purity of the enzyme should be assessed by one or by both methods presented in Modules 2 and 3: a) the UV-visible absorption spectrum of the enzyme should be similar to the UV-visible absorption spectrum presented in Module 3. This can be done easily and in a very short time prior to assaying the enzyme; b) the SDS-polyacrylamide gel electrophoresis should clearly show two identical bands, each of which has a molecular weight of 52,000 daltons.[9,10] If the instructor and or students selected this method, the purity of the enzyme can be checked in a separate experiment (see Module 3), and thus the three weeks allowed for this experiment can be changed to four weeks.

## *Week One:*
## *Monoamine Oxidase Activity*

### Assay of Monoamine Oxidase Activity
To assay the activity of MAO, the following procedure must be followed:

Each variable investigated will require two tubes; one is a reagent blank in which the enzyme is replaced by phosphate buffer. (All buffers should include 0.01% Triton X-100 and 0.2% potassium cholate to maintain the enzyme in a soluable form; 0.002 M β-mercaptoethanol.) See reagents preparations on page 69.

The enzyme reactions are performed as follows:

a. *Buffer:* Add 1.00 ml of 0.01 M potassium phosphate buffer, pH 7.6 containing 0.01 M Triton X-100 and 0.2% potassium cholate and 0.002 M β-mercaptoethanol (page 69).

b. *Substrate:* Add 0.05 ml (50 µl) of 0.1 M benzylamine, pH 8.2 to each tube (see reagent preparation of benzylamine).

c. *Temperature equilibration:* Place tubes in appropriate water bath at room temperature for 5 minutes.

d. *Enzyme:* Transfer the substrate buffer solution to a 1 ml cuvette and add 0.05 ml (50 µl of the enzyme) directly to the cuvette. Mix well and begin the reaction by recording time (to the second) or monitoring the absorbance change of the spectrophotometer at 250 nm versus time.

   Students can run the assay on three different enzyme dilutions so they can gain familarity with the assay before analyzing the properties of the enzyme. Students can take 0.1 ml (100 µl), 0.2 ml (200 µl) and 0.4 ml (400 µl) of the purified enzyme and dilute this up to 1 ml with the buffer prepared in step a. 0.1 ml (100 µl) of the diluted enzyme can be added directly to the cuvette as mentioned above.

e. *Replication:* You can repeat the experiment by changing the enzyme concentration. For example, you may use 25 or 75 µl of the enzyme and record the activity as mentioned in step d.

f. *Absolute enzyme activity:* Determine the amount of benzylaldhyde (RCHO) produced per minute either directly from the slope of your absorbance spectrum or calculating the change in absorbance per minute (manually). Assuming the molecular weight of the enzyme is 52,000, express your absolute enzyme activity as µmoles benzylaldhyde formed per ml/min. This is enzyme *absolute activity*. Remember, this procedure as described will produce only the relative activity change in absorbance units/ml not the absolute activity expressed in µmoles of benzylaldhyde formed/ml. The latter is used in assaying for polyphenol oxidase as described later in this module.

g. *Protein concentration:* Determine the protein concentration of the enzyme following Lowry or Biuret method reported in Module 2. Express protein concentration as mg/ml.

h. *Specific activity of MAO:* Determine the specific activity of the enzyme from the following equation:

$$\text{Specific activity} = \frac{\text{absolute activity}}{\text{protein concentration}}$$

$$= \frac{\text{activity/ml (step e)}}{\text{mg protein/ml (step f)}}$$

The units for specific activity are expressed as units of enzyme/mg protein.

i. *Storage of the enzyme:* Now add one ml of MAO to each of three test tubes and leave one at room temperature, keep the second in the freezer, and put the third in the cold for a week. Assay the enzyme after one week as explained in Week Two of this module, and compare the specific activity with the previously reported activity. Discuss your results utilizing your background knowledge about the enzyme from the information mentioned in this module and references 7–11.

### The Effect of pH on the Rate of the Reaction Catalyzed by MAO

Number ten tubes 1 through 10. After adding 50 µl substrate to each, add the following to each tube:

Tubes 1 and 2—1 ml of pH 6.0, 0.01 M buffer (see reagents preparation) containing 0.01% Triton X-100, 0.2% potassium cholate.

Tubes 3 and 4—1 ml of pH 7.0, 0.01 M buffer containing 0.01% Triton X-100, 0.2% potassium cholate.

Tubes 5 and 6—1 ml of pH 8.0, 0.01 M buffer containing 0.01% Triton X-100, 0.2% potassium cholate.

Tubes 7 and 8—1 ml of pH 9.0, 0.01 M buffer containing 0.01% Triton X-100, 0.2% potassium cholate.

Tubes 9 and 10—1 ml of pH 10.0, 0.01 M buffer containing 0.01% Triton X-100, 0.2% potassium cholate.

Carry out the reaction as indicated in steps c–h. Complete Table 4.1.

## Review Questions

1. Plot the rate of deamination (v) against pH. What is the optimum pH for MAO? Why does the pH have such a strong influence on enzyme activity? Can the effect of pH on enzyme activity provide you with any information about groups essential for catalysis? Explain.

2. Repeat the above experiment, but add 0.01M phosphate buffer containing (0.1% Triton X-100 and 0.2% potassium cholate). Plot your data and compare them with your plot obtained in the previous exercise. From your results, what is the effect of Triton X-100 on MAO? Explain your results.

### Effect of Temperature on MAO

Most biological reactions proceed very slowly at room temperature in the absence of a catalyst because the reacting molecules do not collide with each other with sufficient energy to form an "activated complex." The reaction cannot go to completion unless this activation energy is supplied. Raising the temperature will supply the requisite energy, but in biological systems reactions occur at relatively low temperatures, indicating that the energy barrier is surmounted by some other means. Enzyme catalyzed reactions proceed very rapidly at room temperature because the enzyme is able to combine with its substrate to form an intermediate complex, the formation of which requires less energy, and which therefore bypasses part of the activation barrier. The activation energy may be defined as the minimal, or extra, energy that must be imparted to the reacting molecules for them to react.

Different amounts of activation energy are required by different kinds of reactions. The activation energy for any given reaction may be calculated from the rates of the reaction at two different temperatures according to the Arrhenius equation:

$$\log v_2/v_1 = E_a/2.303R\,(1/T_1 - 1/T_2)$$

**TABLE 4.1**  Effect of pH on Monoamine Oxide Activity

| Tube Number | pH | $A_{250}$ nm/min/ml | Absolute Activity µmoles RCHO/ml | Protein mg/ml MAO | Specific Activity µmoles/mg MAO |
|---|---|---|---|---|---|
| 1 | 6.0 | | | | |
| 2 | | | | | |
| 3 | 7.0 | | | | |
| 4 | | | | | |
| 5 | 8.0 | | | | |
| 6 | | | | | |
| 7 | 9.0 | | | | |
| 8 | | | | | |
| 9 | 10.0 | | | | |
| 10 | | | | | |

in which $v_1$ = rate of reaction at $T_1$
$v_2$ = rate of reaction at $T_2$
$R$ = a constant, 1.987 cal/degree/mole
$E_a$ = activation energy in calories
$T_1$ and $T_2$ = temperatures in degrees absolute, which is equal to °C + 273.

# Procedure

Set up 7 test tubes. Pipette 1 ml of pure or partially purified enzyme as prepared in Module 3 into each tube. Place the first tube in the ice bath at 0° C, the second on the bench at room temperature (record this temperature), the third, fourth, fifth, sixth, and seventh in water baths at 23° C, 37° C, 55° C, 65° C, 80° C, and 90° C, respectively. Number 14 test tubes. Pipette 1 ml 0.01 M potassium phosphate buffer, pH 7.6 containing 0.01% Triton X-100, 0.2% potassium cholate and 50 μl of 0. 1 M benzylamine, pH 8.2 into each tube. After the tubes have been equilibrated with the environment, add 50 μl of MAO which has been incubated at 0° C to tubes one and two, then 50 μl of MAO which has been incubated at 23° C to test tubes three and four, continue on with MAO that has been incubated at 37° C, 55° C, 65° C, 80° C, and 90° C respectively. Assay the enzyme as reported on page 61. Record the data in Table 4.2.

Calculate the rate of deamination at each temperature and plot these rates against temperature. What is the optimum temperature for MAO? Is this the optimum temperature you would have predicted? Explain. Describe the effect of higher temperatures (80° C and 90° C) on MAO activity.

From your data (only data obtained from the first 10 tubes), plot the logarithms of the rates of hydrolysis (log $v$) versus $1/T$ (use absolute $T$). What kind of a curve did you obtain?

At any two points on the curve, read the corresponding reaction rates (log $v_1$ and log $v_2$) and temperatures ($1/T_1$ and $1/T_2$) and calculate the activation energy ($E$) using the Arrhenius equation.

$E_0$ can also be calculated from the slope of the line on your graph, which is equal to $E_a/2.303R$.

How do these two values for $E_a$ compare?

Since heat supplies energy to the reacting molecules, more of them will be able to surmount the activation energy barrier in a warm solution than in a cooler one, and the reaction will proceed at a faster rate. In general, increasing the temperature 10° C will approximately double the reaction rate. The temperature coefficient, conventionally designated $Q$, expresses how many times a reaction is accelerated by an increase in temperature. The temperature coefficient for a 10° C rise in temperature ($Q_{10}0$) can be calculated from the Van't Hoff equation:

$$\log Q_{10} = \frac{10}{T_2 - T_1} \log\ v_2/v_1$$

in which $T_1$ and $T_2$ are the two different temperatures and $v_2$ and $v_1$ the corresponding reaction rates at those temperatures. From your data, calculate $Q_{10}$ for MAO. Explain the significance of these thermodynamic parameters.

## *Week Two:*
## *Monoamine Oxidase Activity II*

### Effect of Storage on MAO Activity

Complete your experiment on Week One (enzyme reaction, step 10) and calculate the **specific activity of the enzyme** after one week of storage at room temperature, in the freezer, and in the cold and record the data in Table 4.3. What is the optimum temperature for storage? Discuss your results.

**TABLE 4.2**   Effect of Temperature on Monoamine Oxide Activity

| Tube Number | Temperature (°C) | $A_{250}$nm/min | μmoles/ml RCHO | mg/ml MAO | μmoles/mg MAO |
|---|---|---|---|---|---|
| 1 | 0 | | | | |
| 2 | | | | | |
| 3 | 23 | | | | |
| 4 | | | | | |
| 5 | 37 | | | | |
| 6 | | | | | |
| 7 | 55 | | | | |
| 8 | | | | | |
| 9 | 65 | | | | |
| 10 | | | | | |
| 11 | 80 | | | | |
| 12 | | | | | |
| 13 | 90 | | | | |
| 14 | | | | | |

**TABLE 4.3**  Effect of Storage on Monoamine Oxidase Activity

| Tube Number | Temperature (°C) | $A_{250}$nm/min | µmoles/ml RCHO | mg/ml MAO | µmoles/mg MAO/min |
|---|---|---|---|---|---|
| 1 | 25° C | | | | |
| 2 | (Room temp.) | | | | |
| 3 | 5° C | | | | |
| 4 | (Cold) | | | | |
| 5 | –15° C | | | | |
| 6 | (Freezer) | | | | |

**TABLE 4.4**  Effects of Substrate Concentration on Monoamine Oxidase Activity

| Tube Number | Concentration of Substrate Added | Final Concentration of Substrate | Rate at 250 nm/(µ/min/ml) |
|---|---|---|---|
| 1 | $1.0 \times 10^{-2}$ M | $4.4 \times 10^{-4}$ M | |
| 2 | $1.0 \times 10^{-2}$ M | | |
| 3 | $2.0 \times 10^{-2}$ M | $9.1 \times 10^{-4}$ M | |
| 4 | $2.0 \times 10^{-2}$ M | | |
| 5 | $4.0 \times 10^{-2}$ M | $1.7 \times 10^{-3}$ M | |
| 6 | $4.0 \times 10^{-2}$ M | | |
| 7 | $8.0 \times 10^{-2}$ M | $4.5 \times 10^{-3}$ M | |
| 8 | $8.0 \times 10^{-2}$ M | | |
| 9 | $15.0 \times 10^{-2}$ M | $6.5 \times 10^{-3}$ M | |
| 10 | $15.0 \times 10^{-2}$ M | | |

### Effect of Substrate Concentration [S] on MAO Activity

To study the effect of varying the substrate concentration on the enzymatic activity, prepare 10 tubes, each containing 1 ml of 0.01 M potassium phosphate buffer containing 0.01% Triton X-100, 0.2% potassium cholate, pH 7.6. Add 0.05 ml of substrate at the concentration indicated in Table 4.4.

Warm the tubes at 37° C in water and then prepare to add the enzyme (MAO). Add 50 µl or 100 µl (depends on the level of enzyme purity as discussed in Module 3) to test tubes one and two that has final concentration of $4.4 \times 10^{-4}$ M benzylamine. Monitor the increase in absorbance (Abs) at 250 nm for 4–5 minutes. Plot the Abs versus time. You should be able to get a linear plot. The slope of this plot will give you the initial velocity ($v_0$) at this concentration ($4.4 \times 10^{-4}$). This is the rate at 250 nm/min. Follow the same procedure for all test tubes at different final concentrations. You should be able to complete Table 4.4. Calculate the µmoles benzylaldehyde produced and the rate of deamination in each.

Plot the rate of deamination ($v_0$) against the substrate concentration [S].

From the data you have collected, what can you conclude about the effect of substrate concentrations on the rates of enzymatic reactions? Plot ($v$) versus [S] and estimate $V_{max}$, the maximum velocity that can be obtained by this enzyme at any substrate concentration.

### Calculation of the Michaelis-Menten Constant, $K_m$

The Michaelis-Menten constant, $K_m$, is one of the most fundamental numbers used in characterizing an enzyme. The units in which it is expressed are those of molar concentration, and the $K_m$ corresponds to that concentration of substrate at which the enzyme catalyzed reaction proceeds at one half its maximum rate. That is, $K_m = [S]$ when $v_0 = 1/2\ V_{max}$. Thus, $K_m$ is useful in estimating the substrate concentrations necessary to obtain the optimal rates of the reaction. A high $K_m$ is indicative of a low enzyme-substrate affinity, and a low $K_m$ of a high affinity.

The form of the Michaelis-Menten equation is such that about 10 and 90% of $V_{max}$ are achieved when the substrate concentrations, [S], correspond to $K_m \times 10^{-1}$ and $K_m \times 10$, respectively.

In its usual form, the Michaelis-Menten equation appears as follows:

$$v_o = \frac{V_{max} \times [S]}{K_m + [S]}$$

in which  $v_o$ = the initial velocity of the reaction at any substrate concentration
$V_{max}$ = the maximum velocity
$[S]$ = the substrate concentration in moles per liter
$K_m$ = the Michaelis-Menten constant

When the reaction rate $v_o$ is plotted against $[S]$, a curved line is obtained according to this equation. It is difficult to estimate $K_m$ from this curve. To simplify the determination of $K_m$, Lineweaver and Burk pointed out that rearranging this equation by taking the reciprocal of both sides would yield an equation of a straight line of the form $y = mx + b$. Thus,

$$\frac{1}{v_o} = \frac{K_m}{V_{max}} \frac{1}{[S]} + \frac{1}{V_{max}}$$

Since $V_{max}$ and $K_m$ are constants for a given enzyme under conditions, while $V_o$ and $[S]$ are variables, this form of the equation is equivalent to the straight line equation above. The advantages of plotting the data in this fashion are several. The slope of the line obtained is $K_m/V_{max}$ and at the intercept where $1/[S] = 0$, obtained by extrapolating the straight line to the $1/v_o$ axis, $1/v_o = 1/V_{max}$.

If the line is extended to its intercept on the $1/[S]$ axis, the negative reciprocal of $K_m - 1/K_m$, is obtained. Thus, this way of plotting the data permits the determination of $K_m$ and $V_{max}$ with great ease.

Using the data in Table 4.4, fill out Table 4.5.

Plot $1/v_o$ against $1/[S]$ and determine $V_{max}$ and $K_m$. How does this value compare with your previous estimate from the nonlinear plot?

Another method to obtain a linear relationship between substrate concentration and initial velocity is to use an Eadie-Hofstee plot. The Michaelis-Menten equation can be rearranged to a linear form,

$$v_o = -K_m \frac{v_o}{[S]} + V_{max}$$

Calculate your data and complete Table 4.6 and then plot your data using an Eadie-Hofstee Plot.

Compare the $K_m$ and $V_{max}$ values calculated by the three different methods discussed above and record your data in the Table 4.7.

## Week 3: Effect of Inhibitors on an Enzymatic Reaction

Most enzymes can be inhibited by certain chemical reagents. Much of the current drug therapy is based on inhibition of specific enzymes with a substrate analog. Therefore, it is important to discuss briefly enzyme inhibition prior to examining MAO inhibitors. Basically, there are three major classes of inhibitors: competitive, noncompetitive, and uncompetitive. In addition to inhibition by the immediate product, products of other enzymes can also inhibit or even activate a particular enzyme. Our brief discussion in this module will be limited to the basic three major classes.

### Competitive Inhibitors

Competitive inhibitors are defined as inhibitors whose action can be reversed by increasing amounts of substrate. Competitive inhibitors are usually structurally enough like the substrate that they bind at the substrate binding site and compete with the substrate for the enzyme. Once bound, the enzyme cannot convert the inhibitor to products. Increasing substrate

**TABLE 4.5**  Lineweaver-Burk Plot Data
$1/[S]$ versus $1/v_o$

| Tube Number | $[S]$ | $1/[S]$ | $v_o$(μmoles/min) | $1/v_o$ |
|---|---|---|---|---|
| 1 | $4.4 \times 10^{-1}$ M | | | |
| 3 | $9.1 \times 10^{-1}$ M | | | |
| 5 | $1.7 \times 10^{-3}$ M | | | |
| 7 | $4.5 \times 10^{-3}$ M | | | |
| 9 | $6.5 \times 10^{-3}$ M | | | |

**TABLE 4.7**  Comparison of $K_m$ and $V_{max}$ Determined by Michaelis-Menten's Plot, Lineweaver-Burk Plot and Eadie-Hofstee Plot

| Graphical Method | $K_m (M \times 10^{-3})$ | $V_{max}$(μmoles/min) |
|---|---|---|
| $V_o$ versus $[S]$ | | |
| $1/V_o$ versus $1/[S]$ | | |
| $V_o$ versus $V/[S]$ | | |

**TABLE 4.6**  Eadie-Hofstee Plot Data ($V_o$ versus $V_o/[S]$)

| Tube Number | $[S]$ (mM) | $V$(μmoles/min) | $v_o/[S]\dfrac{(\mu moles\ liters)}{mole\ min}$ |
|---|---|---|---|
| 1 | $4.4 \times 10^{-1}$ M | | |
| 3 | $9.1 \times 10^{-1}$ M | | |
| 5 | $1.7 \times 10^{-3}$ M | | |
| 7 | $4.5 \times 10^{-3}$ M | | |
| 9 | $6.5 \times 10^{-3}$ M | | |

concentration will displace the reversibly bound inhibitor. Since the substrate and inhibitor are competing for the same site on the enzyme, the $K_m$ for the substrate shows an apparent increase in the presence of inhibitor. This can be seen in a double-reciprocal plot as a shift in the $x$-intercept $(-1/K_m)$ and in the slope of the line $(K_m/V_{max})$. If we first establish the velocity at several levels of substrate and then repeat the experiment with a given but constant amount of inhibitor at various substrate levels, two different straight lines will be obtained as shown in Figure 4.1A. As can be seen, the $V_{max}$ does not change; hence the intercept is no longer the negative reciprocal of the true $K_m$, but is $-1/K_{mapp}$. Where

$$K_{mapp} = K_m \times \left[ 1 + \frac{[I]}{K_I} \right]$$

Thus the inhibitor constant, $K_I$, can be determined from the concentration of inhibitor $[I]$ used and the $K_m$, which was obtained from the $x$-intercept of the line obtained in the absence of inhibitor.

### Noncompetitive Inhibitors

An noncompetitive inhibitor binds at a site other than the substrate binding site. Its inhibition is not reversed by increasing concentration of substrate. It behaves as though it were removing active enzyme from the solution, resulting in a decrease in $V_{max}$. This effect is seen graphically in the double reciprocal plot (Figure 4.1B), where $K_m$ does not change but $V_{max}$ changes.

### Uncompetitive Inhibitors

An uncompetitive inhibitor binds the enzyme with an apparent equivalent change in $K_m$ and $V_{max}$, which is reflected in the double reciprocal plot as a line parallel to that of the uninhibited enzyme (Figure 4.2A).

## Effect of a Competitive Inhibitor on MAO Activity

Para-hydroxyamphetamine acts as a reversible competitive inhibitor.[11] In this experiment, a low concentration of this competitive inhibitor will be used in the

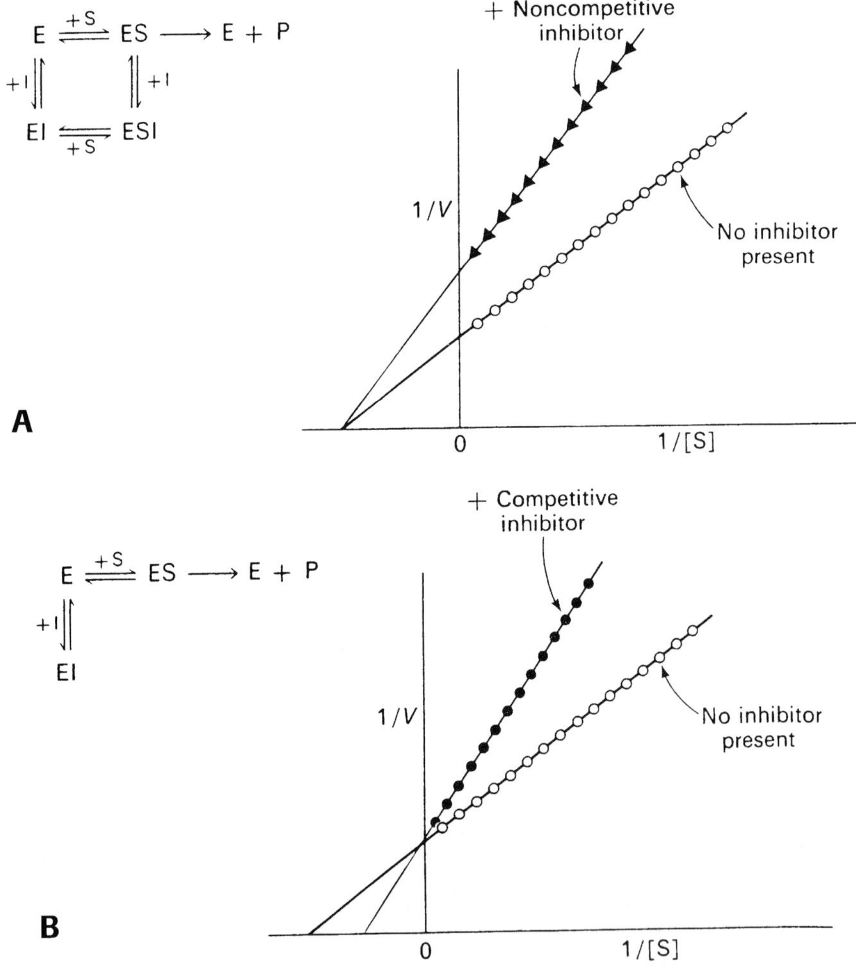


**FIGURE 4.1** **(A)** A double reciprocal plot of enzyme kinetics in the presence and absence of a noncompetitive inhibitor; $K_m$ is unrelated by the noncompetitive inhibitor, whereas $V_{max}$ is decreased. **(B)** A double-reciprocal plot of enzyme kinetics in the presence and absence of a competitive inhibitor; $V_{max}$ is unrelated, whereas $K_m$ is increased.

From Stryer, Lubert. 3rd edition. W. H. Freeman Co.

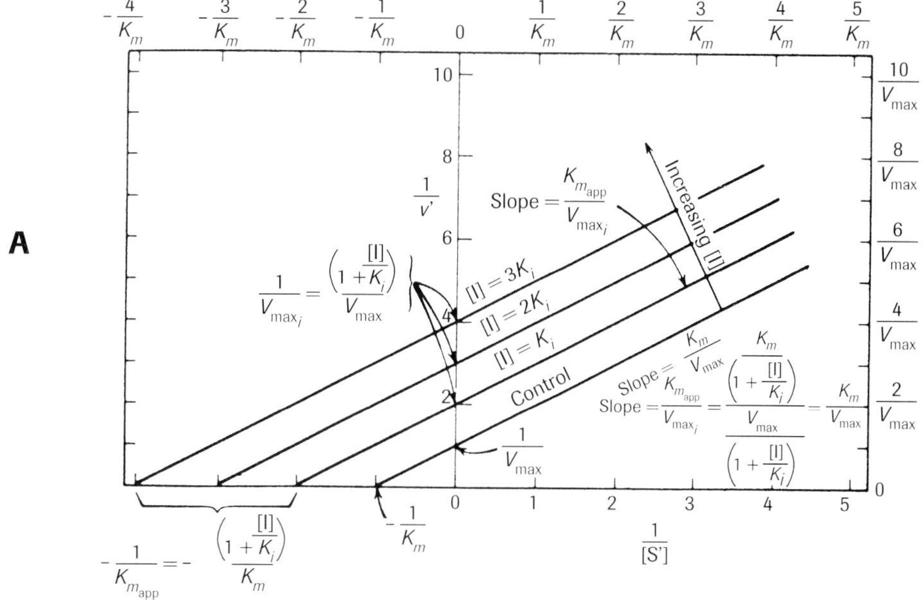

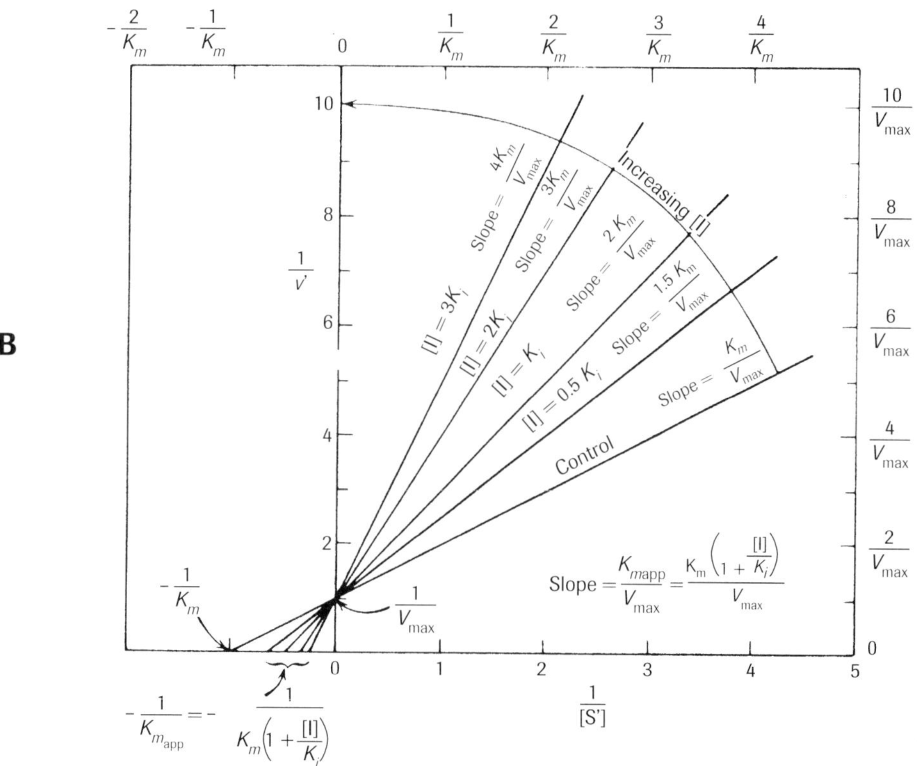

**FIGURE 4.2**   **(A)** The $1/V$ versus $1/[S]$ plot in the presence of different fixed concentrations of an uncompetitive inhibitor. **(B)** The $1/V$ versus $1/[S]$ plot in the presence of different fixed concentrations of a competitive inhibitor.

From Segel, I. H. 1975. *Enzyme Kinetics.* Reprinted by permission of John Wiley and Sons, Inc.

presence of increasing concentrations of [S] to study the effects of the inhibitor on the reaction rate. Prepare a set of ten test tubes as described in Table 4.4. Add 0.1 ml of 0.005 M para-hydroxyamphetamine, pH 7.6 to each tube. Otherwise, the experimental procedure is exactly the same in Table 4.4. Fill in Table 4.8 and Table 4.9 with your results.

Using the data of Table 4.8, plot $v_o$ versus [S] and estimate $V_{max}$ in the presence of the inhibitor. Use Figures 4.1A, 4.1B, and 4.2A,B as a guide to plot your data. Then, make a double reciprocal plot of the same data ($1/v_o$) versus $1/[S]$ and find the $V_{max}$ and apparent $K_m$ under these conditions. How does this line compare with the one obtained in the absence of added para-hydroxyamphetamine? Plot the line obtained previously in the absence of the inhibitor on this same graph for purposes of comparison. How does para-hydroxyamphetamine affect the affinity of the enzyme for its substrate? What are the natural substrates for MAO.

For enzymatic reactions in the presence of a competitive inhibitor, the Michaelis-Menten equation can be written:

$$v_o = \frac{V_{max}}{1 + \dfrac{K_m}{[S]}\left(1 + \dfrac{(I)}{K_i}\right)}$$

The Lineweaver-Burke plot of $1/v_o$ versus $1/[S]$,

$$1/v_o = \frac{K_m}{V_{max}}\frac{\left(1 + (I)\right)}{K_i} \times 1/[S] + 1/V_{max}$$

**TABLE 4.8**   Effect of a Competitive Inhibitor (Para-hydroxyamphetamine) on Rate of Catalysis

| Tube Number | Concentration [S] Added | Rate at 250 nm ($v_o$) |
|---|---|---|
| 1 | $4.4 \times 10^{-4}$ M | |
| 2 | $4.4 \times 10^{-4}$ M | |
| 3 | $9.1 \times 10^{-4}$ M | |
| 4 | $9.1 \times 10^{-4}$ M | |
| 5 | $1.7 \times 10^{-3}$ M | |
| 6 | $1.7 \times 10^{-3}$ M | |
| 7 | $4.5 \times 10^{-3}$ M | |
| 8 | $4.5 \times 10^{-3}$ M | |
| 9 | $6.5 \times 10^{-3}$ M | |
| 10 | $6.5 \times 10^{-3}$ M | |

in which $v_o$ = the rate of the reaction at any given [S] in the presence of the inhibitor. The maximum velocity ($V_{max}$) in the presence of the competitive inhibitor should be as in the absence of the inhibitor (determined in Table 4.8).

$K_m$ = The Michaelis-Menten constant which you determined from data in Table 4.4 and 4.5

$(I)$ = the concentration of the inhibitor

$K_i$ = the dissociation constant of the enzyme-inhibitor complex

$[S]$ = the substrate concentration

Note that $K_m$ in the original Lineweaver-Burk equation has been replaced by

$$K_m + \frac{K_m(I)}{(I)}$$

or

$$1 + (I)/K_I$$

which means that the slope of your line, and hence the apparent $K_m$, has been increased by the value $1 + I/K_I$ in the presence of the inhibitor. As $1/[S]$ becomes 0 however, this term will also become 0, so that the intercept on the $1/v_o$ axis (i.e., $1/V_{max}$) will be the same with or without inhibitor, and $V_{max}$ should, therefore, be the same in both cases.

Since you know the values for all the quantities in the above form of the Michaelis-Menten equation except that for $K_p$, calculate $K_I$. $K_I$ is a useful constant since it is a quantitative measure of the inhibitor. Determination of $K_I$'s permits comparisons to be made between different inhibitors.

The relationship between $K_I$ and $K_m$ is as follows:

$$K_p = K_m(1 + 1/K_I)$$

The apparent $K_m$ can also be calculated by an Eadie-Hofstee plot. Substitute the other known values into the following equation and solve for $K_I$:

$$\frac{v_o}{[S]} = \frac{V_{max}}{K_m\left(1 + \dfrac{(I)}{K_I}\right)} - \frac{v_o}{K_m\left(1 + \dfrac{(I)}{K_I}\right)}$$

and a plot of $V/[S]$ versus $v$ has a slope of $-1/K_m$ $(1 + 1/K_I)$. Using the data in Table 4.8 plot $V/[S]$ versus $V$. Calculate the apparent $K_m$ and the slope of your plot.

**TABLE 4.9**   Effect of Para-hydroxyamphetamine on Monoamine Oxidase Activity

| Tube Number | [S] | 1/[S] | Rate at 250 nm ($v_o$) | $1/v_o$ |
|---|---|---|---|---|
| 1 | $4.4 \times 10^{-4}$ M | | | |
| 3 | $9.1 \times 10^{-4}$ M | | | |
| 5 | $1.7 \times 10^{-3}$ M | | | |
| 7 | $4.5 \times 10^{-3}$ M | | | |
| 9 | $6.5 \times 10^{-3}$ M | | | |

# Review Questions

1. Discuss the biological significance of monoamine oxidase.
2. Describe briefly the physiochemical properties of monoamine oxidase.
3. Explain in detail both the kinetic and clinical significance for the use of competitive inhibitors such as para-hydroxyamphetamine (use References listed on page 71).
4. From the plot of $v_0$ versus $[S]$, is monoamine oxidase an allosteric enzyme? Explain your answer.
5. What are the common antidepressant agents?

## Reagents Preparations

1. 0.01 M potassium phosphate.
    a. First prepare two stock buffer solutions at 0.1 M concentration, the first is monobasic potassium phosphate ($KH_2PO_4$) buffer and the second is dibasic potassium phosphate ($K_2HPO_4$).
    b. Prepare stock solution at the desired pH by titrating the $KH_2PO_4$ buffer with $K_2HPO_4$ buffer until you achieve the desired pH.
    c. Dilute the buffer you prepared above 1:10 (dilute 100 ml to one liter with deionized $H_2O$).
2. Prepare 1% stock Triton X-100 by disolving 100 ml of Triton X-100 to one liter.
3. Prepare 2% potassium cholate solution at pH 7.6.
4. To prepare one liter of 0.01 M potassium phosphate buffer containing 0.01% Triton X-100 and 0.2% potassium cholate, 0.002 M β-mercaptoethanol, at pH 7.6 add the following:
    a. 100 μl of 0.1 M potassium phosphate buffer at pH 7.6.
    b. 10 ml of 1% Triton X-100.
    c. 100 ml of potassium cholate at 7.6.
    Mix a, b, and c and dilute to one liter. To prepare 2, 3, or 4 liter, scale up the reagents accordingly.

# POLYPHENOL OXIDASE ASSAY

### *Supplies and Equipment*
Electrophoretic cell
Spectrophotometer
Glass tubes
Water bath
1.5 ml semi-micro plastic cuvettes or 1.0 ml glass cuvette

### *Chemicals*
100 mM catechol
50 mM catechol
100 mM acetate buffer
20 mM diethylthiocarbamate

# Introduction

In addition to the enzymatic assay of monoamine oxidase (MAO), this module will also quantify the specific activity (spc. act.) of another oxidase, i.e., polyphenol oxidase (PPO). This quantification can be accomplished within a single three hour laboratory period and is necessary to perform the purification of PPO.

Polyphenol oxidase occurs within certain mammalian tissues as tyrosinase[12] as well as both lower[13, 14] and higher[15–22] plants. In mammalian systems, the enzyme is important in melanin synthesis. In contrast, fungal PPO is involved in the metabolism of certain tree-synthesized phenolic compounds that have been implicated in disease resistance.[14]

The PPO oxidase complex of higher plants consists of a cresolase, a catecholase and a laccase. These copper metalloproteins catalyze the one and two electron oxidations of phenols to quinones at the expense of $O_2$. Whereas cresolase mediates the o-hydroxylation of a monophenol followed by its oxidation to an o-diquinone, catecholase carries out the oxidation of an o-dihydroxyphenol to an o-diquinone. In contrast, laccase converts p-hydroquinone to either p-phenyldiamines or p-quinonediamines.[13] Polyphenol oxidase has been proposed to be involved in chloroplast $O_2$ levels, biosynthesis of phenolics, wound healing, and anti-nutritive modification of plant proteins to discourage herbivory.[20,22]

In this portion of the enzyme assay module, purified, mushroom (*Agaricus bisporus*) PPO (EC 1:10:3:1) and PPO from *Coriolus versicolor*, a basidiomycete important in the disease wood decay will be assayed. The PPO's spc. act, will be expressed as Worthington units,[23]

$$\frac{\Delta A\ min^{-1} \times 1,000}{mg\ protein}$$

rather than μmoles/min/mg protein.[24] The reason for the Worthington expression is that while PPO catalyzes the conversion of o-diphenols to o-diquinones (Figure 4.3), the precise quinone has not been established, although the product of fungal PPO's action appears to be a benzoquinone.[25] Thus, the employment of a quinone standard curve to quantify the reaction product is not possible.

Because the concepts of temperature and pH optima, increasing assay time, and varying enzyme amount in relation to enzyme activity have been explained in the section regarding MAO assay, these will not be re-examined here. Instead, the type of PPO inhibitor that diethylthiocarbamate[16] is will be established after the assay of PPO is mastered. This exercise also illustrates the effect of increasing substrate concentration on enzyme activity.

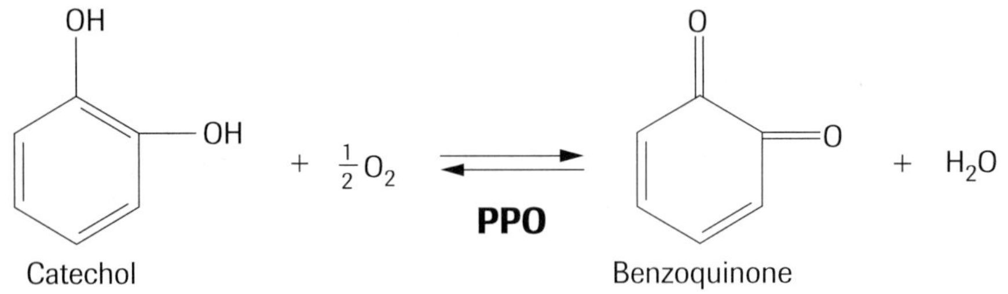

**FIGURE 4.3**   Conversion of o-diphenols to o-diquinones by polyphenol oxidase.


**TABLE 4.10**   Worksheet for the Calculation of PPO Units

| Time (minutes) | $^A$440 nm | $^{\Delta A}$440 nm | $^{\Delta A}$440 nm min$^{-1}$ × 1,000 | mg Protein | Worthington Units |
|---|---|---|---|---|---|
| 0 | | | | | |
| 2 | | | | | |
| 4 | | | | | |
| 6 | | | | | |
| 8 | | | | | |
| 10 | | | | | |


# Procedures

## Sources of PPO

*Coriolus versicolor*'s PPO occurs both intracellularly and extracellularly.[13] The enzyme can be obtained by dialysis of the liquid growth medium that supports *Coriolus versicolor*'s *in vitro* growth. The enzyme begins to appear in the medium seven days post-inoculation of the medium with mycelium. In addition, purified *Agaricus bisporus*' PPO can be obtained from Sigma Chemical Co. (St. Louis, MO).

## Assay of PPO

The assay of either *Agaricus* or *Coriolus* PPOs consists of pipetting 1 ml of 100 mM catechol in 100 mM acetate buffer, pH 5.0 (pH optimum) into five glass tubes with a subsequent 2 minute equilibration in a shaking water bath at 30° C (PPO's temperature optimum). Following the equilibration, a 10 µl aliquot (1.1 mg protein) of PPO is added to each tube. Note: protein amount can be quantified by UV-spectroscopy utilizing bovine serum albumin for standard curve construction (see Module 2). Next, the reaction mixture is incubated for 2, 4, 6, 8 and 10 minutes, after which the mixture is transferred to either a 1.5 ml semi-micro plastic cuvette or 1.0 ml glass cuvette for determination of its absorbance at 440 nm using a spectrophotometer.[26] A 0 time absorbance is performed by pipetting 1 ml of substrate-buffer mixture into a cuvette with the subsequent recording of the absorbance at 440 nm immediately upon addition of 10 µl PPO. Alternatively, a heat-denatured enzyme sample can be employed. To derive units of PPO, the worksheet in Table 4.10 can be employed.

### *Is Diethylthiocarbamate a Competitive or Noncompetitive Inhibitor of PPO?*

To ascertain what type of PPO inhibitor diethylthiocarbamate[15] is, perform the following experiment. Pipette the reagents into glass, disposable assay tubes, as in Table 4.11.

After performing the assay as above and calculating PPO units, plot 1/Worthington units on the *y*-axis and 1/[S] on the *x*-axis to derive a Lineweaver-Burk plot, and compare the plots for the control and diethylthiocarbamate to those for uncompetitive and noncompetitive inhibitors. In addition, determine both the slope and *y*-intercept to assess the type of inhibition (Table 4.12).

## Review Questions

1. Discuss the minimal parameters required to demonstrate that a reaction is enzymatic.
2. Describe how you would perform an enzyme assay resulting in expression of enzyme activity as either Lehninger specific activity or Worthington units.

**TABLE 4.11**  Summary of Procedure for Determining
Type of PPO Inhibitor Diethylthiocarbamate Is[a]

| Tube Number | Constituents[b] |
| --- | --- |
| 1–6 | 1 ml of 100 mM acetate buffer, pH 5.0 containing 50 mM catechol |
| 7–12 | 1 ml of 100 mM acetate buffer, pH 5.0 containing 50 mM catechol and 20 mM diethylthiocarbamate |
| 13–18 | 1 ml of 100 mM acetate buffer, pH 5.0 containing 100 mM catechol |
| 19–24 | 1 ml of 100 mM acetate buffer, pH 5.0 containing 100 mM catechol and 20 mM diethylthiocarbamate |
| 25–30 | 1 ml of 100 mM acetate buffer, pH 5.0 containing 150 mM catechol |
| 31–36 | 1 ml of 100 mM acetate buffer, pH 5.0 containing 150 mM catechol and 20 mM diethylthiocarbamate |
| 37–42 | 1 ml of 100 mM acetate buffer, pH 5.0 containing 200 mM catechol |
| 43–48 | 1 ml of 100 mM acetate buffer, pH 5.0 containing 200 mM catechol and 20 mM diethylthiocarbamate |

[a] In each series, the six tubes correspond to 0, 2, 4, 6, 8, and 10 minutes of incubation. If a water jacketed cuvette is available to maintain the temperature at 30° C, the number of tubes could be reduced by performing a continuous 2–10 minute assay with a single tube.
[b] Each tube should contain 10 µl (1.1 mg protein) of PPO.

**TABLE 4.12**  Summary of the Effects of Inhibitors
on Lineweaver-Burk Plots $1/V$ versus $1/[S]$

| | Slope | Intercept on Ordinate |
| --- | --- | --- |
| No inhibitor | $\dfrac{K_m}{V_{max}}$ | $\dfrac{1}{V_{max}}$ |
| Competitive | $\dfrac{K_m}{V_{max}}\left(1+\dfrac{[I]}{K_1}\right)$ | $\dfrac{1}{V_{max}}$ |
| Uncompetitive | $\dfrac{K_m}{V_{max}}$ | $\dfrac{1}{V_{max}}\left(1+\dfrac{[I]}{K_i}\right)$ |
| Noncompetitive | $\dfrac{K_m}{V_{max}}\left(1+\dfrac{[I]}{K_1}\right)$ | $\dfrac{1}{V_{max}}\left(1+\dfrac{[I]}{K_i}\right)$ |

3. Explain how the Michaelis-Menten constant $(K_m)$ can be obtained and then discuss its value to enzyme catalyzed reactions. Include in your answer Michaelis-Menten plots as well as their double reciprocals, i.e., Lineweaver-Burk plots.

4. Design an enzyme kinetic experiment to differentiate between competitive, uncompetitive and noncompetitive inhibitors. Explain what changes occur in the slope and $y$-intercept for each type of inhibitor.

5. Detail how you would differentiate between Michaelis-Menten and allosteric enzymes.

# REFERENCES

## MAO


1. Singer, T. P., R. W. Von Korff, and D. L. Murphy (eds.). 1979. *Monoamine Oxidase: Structure, Function and Altered Functions*. New York: Academic Press. 1–57.
2. Usdin, E. (ed.). 1976. *Neuropsychopharmacology of Monoamines and Their Regulatory Enzymes*. New York: Raven Press. 1–6.
3. Tipton, K. F., M. D. Housely, and T. J. Mantle. 1976. "Monoamine oxidase and its inhibition." *Ciba Foundation Symp.* 39:5–31.
4. Jain, M. 1977. *Monoamine Oxidase*. Life Science Co. 1925–33.
5. White, H. L. and R. L. Tanski. 1977. *Monoamine Oxidase, Structure, Function, and Altered Functions*. New York: Academic Press. 129–44.
6. White, H. L. and D. K. Stine. 1982. *J. Neurochem.* 3:1429–36.
7. Haung, H. K. 1981. *Functions and Regulations of Monoamine Enzymes*. New York: MacMillan. 489–501.
8. Haung, K. H. 1980, *Mol. Pharmcol.* 17:192–98.
9. Fowler, C. J. and S. B. Ross. 1984. *Med. Res. Rev.* 4:323–58.
10. Zeidan, H., K. Watanabke, L. H. Piette, and K. Yasunobu. 1980. *Frontiers in Protein Chemistry*. Tech-Yung Liv, Gunji Mamiya and Kerry T. Yasunsbu (eds.). Amsterdam: Elsevier. 133–48.
11. Yasunobu, K. T. and S. Oi. 1972. *Adv. Biochem. Psychopharmacol.* 5:91.

## PPO

12. Menter, J. M., C. L. Moore, and I. Willis. 1986. *Photochem. Photobiol.* 43:255.

13. Moore, N. L., D. H. Mariam, A. L. Williams, and W. V. Dashek. 1989. *J. Indust. Microbiol.* 4:349–63.

14. Dashek, W. V., N. L. Moore, A. C. Williams, A. L. Williams, C. E. O'Rear, and G. C. Llewellyn. 1990. *Biodeterioration Research III.* New York: Plenum Press. 391–404.

15. Henry, E. H., M. N. O'Conor, S. M. Shrak, and J. M. Denouvow. 1981. *Photobiochem. Photobiophys.* 3:69–76.

16. Vaughn, K. C. and S. O. Duke. 1981. *Protoplasma.* 108:319–27.

17. Ryan, I. D., R. Gregory, and W. M. Tingey. 1983. *Phytochem.* 21:1885–87.

18. Flurkey. W. H. 1985. *Plant Physiol.* 79:564–67.

19. Flurkey, W. H. 1986. *Plant Physiol.* 86:614–18.

20. Mayer, A. M., and E. Harel. 1990. *Enzymes in Foods.* Amsterdam: Elsevier.

21. Moore, B. M., and W. H. Flurkey. 1990. *J. Biol. Chem.* 265:4982–88.

22. Kowalaski, S. P., J. Bamberg, W. M. Tingey, and J. C. Steffens. 1990. *J. Heredity.* 81:475–78.

23. *Worthington Enzymes and Related Biochemicals.* 1982. Freehold, NJ: Worthington Diagnostic Systems, Inc.

24. Lehninger, A. L. 1982. *Principles of Biochemistry.* New York: Worth Publishers.

25. Taylor, R., J. E. Mayfield, W. C. Shortle, G. C. Llewellyn, and W. V. Dashek. 1987. *Biodeterioration Research I.* New York: Plenum Press. 43–62.

26. Evans, C. S. and J. M. Palmer. 1983. *J. Gen. Microbiol.* 129:2103–08.

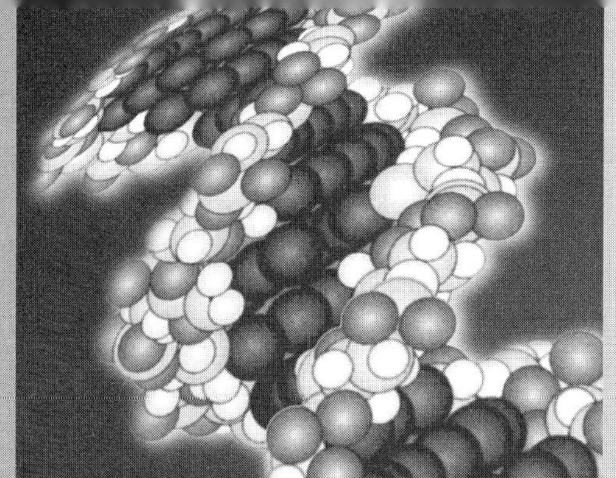



# Enzyme Characterization

## Outline of Module

CHARACTERIZATION OF MONO AND PLASMA AMINE OXIDASES (MAO, PAO) BY FLUORESCENCE SPECTROPHOTOMETRY

Monoamine oxidase (MAO)
Spectral measurements
Fluorescence spectrum on MAO
*Duration:* 2 hours

Results and spectral analysis
*Duration:* 2 hours

Modification of monomine oxidase
*Duration:* 4 hours

Fluorescence labeling of the essential sulfhydryl groups
*Duration:* 4 hours

Fluorescence labeling of purified (native) monoamine oxidase
*Duration:* 4 hours

Effect of urea on the pyrenemaleimide spectrum
*Duration:* 2–2.5 hours

Effect of urea on MAO activity
*Duration:* 2–2.5 hours

Effect of ionic strength on pyrenemaleimide-MAO
*Duration:* 2–2.5 hours

Effect of ionic strength on MAO activity
*Duration:* 2–2.5 hours

Effect of pH on NPM-MAO spectrum
  *Duration:* 2–2.5 hours

Effect of pH on MAO activity
  *Duration:* 2–2.5 hours

## PLASMA AMINE OXIDASES (PAO)

Fluorescence labeling of either purified or commercially available plasma amine oxidase
  *Duration:* 2.5 hours

Fluorescence labeling of the essential sulfhydryl groups
  *Duration:* 2.5 hours

Spectral measurements
Results and spectral analysis
  *Duration:* 2 hours

Effect of urea on NPM-PAO spectrum
  *Duration:* 2.5 hours

Effect of urea on PAO activity
  *Duration:* 2.5 hours

Effect of pH on the pyrenemaleimide-PAO spectrum
  *Duration:* 2.5 hours

Effect of pH on enzyme activity
  *Duration:* 2.5 hours

Effect of ionic strength on PAO-NPM fluorescence spectra
  *Duration:* 2.5 hours

Effect of ionic strength on PAO activity
  *Duration:* 2.5 hours

UV-visible PAO spectrum
  *Duration:* 1–1.5 hours

## CHARACTERIZATION OF MONO AND PLASMA AMINE OXIDASES (MAO, PAO) BY ESR—SPIN LABELING METHOD

Laboratory exercises based upon 2 hour durations.

Methodology
Modification of enzyme
Spin labeling the essential sulfhydryl groups
Effect of the side chain length on the environment of the essential sulfhydryl groups
Effect of pH on spin labeled MAO spectrum
Effect of ionic strength on ESR spectrum of spin labeled MAO

## INTERACTION OF SPIN-LABELED TRYPTAMINE WITH MONOAMINE OXIDASE: PROBING THE MICROENVIRONMENT OF THE ACTIVE SITE BY SPIN PROBE-SPIN LABEL TECHNIQUES

Laboratory exercises based upon 2 hour durations.

Spin labeling the essential sulfhydryl groups with [$^2$H–$^{15}$N] MSL
ESR spectrum of tryptamine spin label (TrySL)-MAO interaction
Determination of the distance between the tryptamine binding site and the essential sulfhydryl group
Plasma amine oxidase (PAO)
ESR spin labeling of plasma amine oxidase

# LABORATORY EXERCISES

## CHARACTERISTICS OF MONO AND PLASMA AMINE OXIDASES BY FLUORESCENCE SPECTROPHOTOMETRY

### Supplies and Equipment
Whatman filter paper #2
Perkin-Elmer MPF–4 fluorescence
 spectrophotometer
Test tubes
Water bath incubator

### Chemicals
Molecular Probe
 Maleimide N-(1-pyrine)
Sigma Chemical Company
 Benzylamine sulfate
 Bovine serum albumin
 Potassium phosphate (mono and dibasic)
 Phenylhydrazine
 Sodium chloride
 Urea
 β-mercaptoethanol
 Triton X-100
Aldrich Chemical Company
 Dimethyl sulfoxide
 Ethanol 95%
 Methanol
 Plasma amine oxidase

## Introduction

Fluorescence probes are useful in studying structure and dynamics of macromolecules. The flourescence of a probe (extrinsic fluorescence), as opposed to the intrinsic fluorescence of proteins due to their complements of Flavin and aromatic amino acids is a useful tool for studying protein structure and interaction.[1-3] The covalent introduction of fluorescence and spin probes into biological macromolecules has led to an understanding of many of the structural parameters underlying their biological functions. In the following series of experiments, the student will first report the fluorescence spectrum of the native enzyme alone, then she or he will use an extrinsic covalently attached label, N-(1-pyrene) maleimide, a specific sulfhydryl reagent that becomes fluorescent only after reacting with sulfhydryl groups.[4-8] The fluorophore offers the advantages of high quantum yield for easily detectable signal, clear excitation and emission spectra, and a sufficiently long fluorescent lifetime.[2,5,9] This probe will be used in this study to monitor any conformational change that takes place during pH change, urea denaturation, and increasing ionic strength.

## Monoamine Oxidase (MAO)

Monoamine oxidase is involved in the control of the biogenic amines degradation in the brain and central nervous system.[10-14] This involvement has stimulated considerable interest in the physiochemical and enzymatic properties of the enzyme. The biological role of the amine oxidases is clearly not restricted to the inactivation of toxic amines that enter the blood stream from the gastro-intestinal tract. In the course of the enzymatic deamination of amines, aldehydes are formed which possess very different biological activities from those of the amines.[15,16] There are even claims in the literature of the participation of mitochondrial monoamine oxidases in the transport across membranes (particularly across the blood-brain barrier) of certain monoamines and related compounds.[17,18]

In the early 1980s, new information began to merge on the possible role of the amine oxidases in psychiatric illness. The report of Murphy and Buchsbaum, (1979) of decreased platelet monoamine oxidase activity in schizophrenia has attracted much attention. The decreased activity may be genetically determined in index individual's predisposition to develop mental illness of this type.[16,18]

Physiochemical studies have indicated that the enzyme has a molecular weight of 104,000 per protomer, and is made up of two identical subunits and one mole of FAD per two subunits.[11,14] The enzyme contains four sulfhydryl groups per subunit of enzyme, only one of which is thought to be essential.[14] From the effect of pH on kinetic properties of the enzyme, investigations concluded that a cysteine residue in the enzyme is catalytically important in the breakdown of amines.[11,19,20] Previous investigations by many researchers including the author have shown that only one sulfhydryl residue was essential per subunit. It was suggested that this sulfhydryl group may be a component of the active site. The student can use either the purified enzyme that she or he isolated by following the detailed procedure in Module 3 or use a commercial enzyme. In either case, the student is required to determine the activity of the enzyme as discussed in Module 4 and also assess the purity of the protein by either SDS-polyacrylamide gel electrophoresis and/or UV-visible spectroscopy, as explained in Modules 3 and this module, respectively.

# Procedures

## I. Spectral Measurements

Fluorescence excitation and emission spectra can be recorded with a Perkin-Elmer MPF-4 fluorescence spectrophotometer or other fluorescence spectrophotometer using 1.0 cm path cuvette. Excitation can be recorded between 300–500 nm for native enzyme and at 330 nm for NPM fluorescence. Emission-spectra between 400–700 nm both native and NPM fluorescence and should be recorded at room temperature.

## II. Fluorescence Spectrum on MAO

We can take advantage of the inherent properties of MAO. The enzyme has one FAD moiety per two subunits. Students can perform directly both the emission and excitation spectra on the native enzyme.

a. Add 1 ml of 1 mg MAO to 1.0 m fluorescence curette.
b. Run fluorescent excitation spectrum between 300–500 nm (emission at 530 nm).
c. Run emission spectrum for MAO between 400–700 nm (excitation at 340 nm).
d. Compare the excitation spectrum with the emission spectrum.
e. Dissolve 10 mg of Flavin adenine dinucleotide (FAD) in 10 ml of 0.05 M potassium phosphate buffer, pH 7.4.
f. Add 1.0 ml of FAD solution you prepared in e in a 1.0 ml fluorescence curette.
g. Run excitation spectrum between 300–500 nm (emission at 530 nm).
h. Run emission spectrum between 400–700 nm (excitation at 340 nm).
i. Compare spectra obtained in steps g and h with spectra obtained in c and d.

# Review Questions

1. What is the wavelength of the maximum emission?
2. What is the wavelength of the maximum excitation?
3. How can you use the above spectra to determine the purity of the enzyme?
4. What can you conclude about monoamine oxidase (MAO)?
5. Select an enzyme which has an interinsic probe and propose a fluorescence experiment similar to MAO.

## III. Modification of Monoamine Oxidase

A modified enzyme derivative (in which the nonessential sulfhydryl groups were blocked using iodoacetate) can be synthesized as follows.[8,11]

a. Dilute the enzyme first to 1 mg/ml using 0.05 M potassium phosphate buffer, pH 7.4.
b. Add 0.1 ml of 0.5 M iodoacetic acid pH 7.4 to 0.9 ml of the diluted enzyme and shake the enzyme at room temperature for 5 hours.
c. Check the activity of the enzyme after 5 hours; then dialyze the modified enzyme against 0.05 M potassium phosphate buffer, pH 7.4, containing 0.001 M β-mercaptoethanol.
d. Change the first buffer after 4 hours, and dialyze overnight with 0.05 M potassium phosphate buffer pH 7.4.

## Results and Spectral Analysis

By following enzyme modification with iodoacetate as described in experiment I, the nonessential sulfhydryl groups can be blocked by reacting the enzyme with 0.05 M iodoacetate for 5 hours. The student is encouraged to run this experiment as a separate experiment following the procedure described previously.[11,12]

## IV. Fluorescence Labeling of the Essential Sulfhydryl Groups

a. Prepare a stock solution of $5 \times 10^{-3}$ M N-(1-pyrene) maleimide (NPM) in absolute 100% dimethyl sulfoxide.
b. Add 0.1 ml of the stock solution to a Whatman filter paper #2 in a test tube.
c. Dry the test tube by a stream of nitrogen.
d. Add 1 ml of the modified enzyme dissolved in 0.05 M potassium phosphate buffer, pH 7.4, to the test tube.
e. Allow the reaction to proceed with gentle stirring at room temperature and under anaerobic condition for 4 hours at room temperature.
f. Dialyze the label enzyme overnight with a change of buffer against 0.05 M potassium phosphate buffer, pH 7.5.
g. Only departments that have access to an amino acid analyzer and wish to run amino acid analysis can perform this analysis on 24 hour hydrolyzates (110° C), using an automatic amino acid analyzer as described by Tarr;[21] otherwise the student can proceed safely to perform fluorescence spectra.

Since our main objective is to concentrate on fluorescence studies, we assume that students will closely follow fluorescence labeling of the essential sulf-hydryl group as described in experiment III. The authors selected NPM labeling because NPM does not fluoresce or form a fluorescent adduct in solution alone. Under the conditions described in the experimental methods, the enzyme is first synthesized, where the six nonessential sulfhydryl groups per dimer were blocked. When this derivative reacts with NPM, the label will then be attached on the essential sulfhydryl groups only. After dialysis to remove excess reagent, the fluorescence spectrum can be recorded. A typical spectrum for this study will be similar to Figure 5.2 on page 83. The student should run the spectrum and make sure it is identical to the spectrum in Figure 5.2.

# Review Questions

1. From your spectrum, what are the wavelengths where the two major vibronic bands appear?
2. How do you interpret the existence of the two spectroscopic components?
3. What are the major advantages for selecting fluorescence spectroscopy over UV-visible?
4. Compare the spectra obtained from experiment III with those obtained from experiment II. Discuss your results in detail.

## V. Fluorescence Labeling of Purified (Native) Monoamine Oxidase

In the previous experiment, you attached the fluorescence label only to the essential sulfhydryl groups after blocking all the nonessential sulfhydryl groups by reacting the enzyme with 0.05 M iodoacetate for 5 hours. In this experiment, you will add the purified or native enzyme directly to the label and compare the spectra. Repeat all the steps in experiment IV. with only one exception: use the directly purified (native, nonmodified) enzyme and carry out the labeling procedure (a–g) in experiment II. In this experiment, you expect to label all the essential and nonessential sulfhydryl groups of the enzyme.

## VI. Effect of Urea on the Pyrenemaleimide Spectrum

a. In this experiment, compare the emission spectra of two molecules. The first is a monoamine oxidase derivative in which all the nonessential sulfhydryl groups are blocked first with iodoacetate and then the essential groups are labeled with NPM fluorescence probe. The second molecule is β-mercaptoethanol directly labeled with NPM fluorescence label using the same procedure described above for labeling MAO.
b. Set up five test tubes and add solid urea to each tube with increasing concentration from 0 M, 2 M, 4 M, 6 M, and 8 M urea in a total volume of 1 ml.
c. Add 1 ml of modified MAO that has been labeled on the essential sulfhydryl group to each tube.

d. Shake the tubes in a water bath incubator for 2 minutes, then dialyze the enzyme against 0.01 M potassium phosphate buffer, pH 7.5, for 24 hours with a change of buffer, and then record the spectrum.
e. Repeat the above procedure on β-mercaptoethanol by setting up two test tubes. The first should have β-mercaptoethanol at concentration of 0.02 M reacted with NPM as described in experiment I. To the second add solid urea so that the final concentration is 6 M–8 M urea in 1 ml.
f. In this experiment, add to each of two test tubes 0.1 ml of 0.05 M iodoacetate, pH 7.5, to 0.9 ml of 0.02 M β-mercaptoethanol. Repeat the above procedure on β-mercaptoethanol as mentioned in step e.
g. Analyze any change in the spectra obtained in experiment VI.
h. Compare the spectra obtained in steps e and f. Are they the same? If not why? Explain your results.
i. Dialyze all the high and low molecular weight derivatives overnight against 0.05 M potassium phosphate buffer, pH 7.5, and then record the spectra.

# Review Questions

1. Do you observe any spectral changes when you add 8 M urea to β-mercaptoethanol?
2. Do you see any spectral changes when you add 2 M, 4 M, 6 M, and 8 M urea to the MAO-NPM derivative? If you see a change, indicate at what concentration.
3. If your answer on question 2 is positive, do the different emission bands quench equally?
4. How would you explain the changes (if any) of the emission spectra of MAO-NPM?
5. What is the effect of iodoacetate on β-mercaptoethanol? Explain your results.
6. Why is fluorescence spectroscopy the preferred biophysical technique over UV-visible spectroscopy in detecting structural changes?
7. Was this denaturation reversible? How could you conclude whether the denaturation is reversible or not? If it is irreversible, at what urea concentration did you conclude that the effect was irreversible?
8. Do you detect any spectral parameters for β-mercaptoethanol? If not, explain.

## VII. Effect of Urea on MAO Activity

a. Check the activity of native (purified) MAO.
b. To four test tubes add 2 M, 4 M, 6 M, 8 M urea in a total volume of 1 ml.

c. Add 1 ml of native (purified) MAO to each of the four test tubes in step b.

d. Mix well and check the activity of MAO.

e. Dialyze all five tubes prepared in steps a and c against 0.05 M potassium phosphate buffer, pH 7.4, containing 0.002 M β-mercaptoethanol.

f. Recheck the activity on dialyzed enzyme.

## Review Questions

1. Is the denaturation reversible?
2. Draw your conclusions by comparing fluorescence with activity studies.

### VIII.  Effect of Ionic Strength on Pyrenemaleimide-MAO

a. Set up five test tubes. Add to each test tube sodium chloride, increasing the total concentration from 0 M to 0.2 M, 0.4 M, 0.6 M, and 0.8 M sodium chloride in 1 ml volume.

b. Add 1 ml of MAO-NPM enzyme derivative prepared by the procedure described in experiments I and II.

c. Shake the tubes in a water bath at 25° C for 2 minutes, and then record the emission spectra.

## Review Questions

1. Do you see any spectral changes when you added sodium chloride to MAO-NPM derivative?
2. Explain your results.

### IX.  Effect of Ionic Strength on MAO Activity

a. Set up five test tubes. Add O M, 0.2 M, 0.4 M, 0.6 M, and 0.8 M sodium chloride to each tube respectively in a total volume of 1 ml.

b. Mix well and check the activity in all five test tubes prepared in step a.

c. Dialyze all five MAO preparations against 0.05 M potassium phosphate buffer, pH 7.4, containing 0.0002 M β-mercaptoethanol.

d. Check the activity on dialyzed enzyme. The student should draw conclusions by comparing fluorescence and activity studies.

### X.  Effect of pH on the NPM-MAO Spectrum

a. To 1 ml of NPM-MAO derivative prepared as described previously, decrease the pH by adding an increment of concentrated phosphoric acid to achieve pHs of 6.0, 4.0 and 3.0.

b. Record the emission spectrum at each of pH 7.5, 6.0, 4.0, and 3.0. Make sure to check the pH by a pH meter or litmus paper.

c. To 1 ml of NPM-MAO derivative, increase the pH by adding an increment of NaOH to achieve a pH of 8.0, 9.5, and 12.0.

d. Record the emission spectra at pH 8.0, 9.5, and 12.0.

e. Dialyze all the MAO-NPM derivatives prepared in steps a and c against 0.05 M potassium phosphate buffer, pH 7.5, overnight with a change of buffer.

f. Record the spectra for all the derivatives after dialysis.

## Review Questions

1. Do you see any spectral changes when you added phosphoric acid or NaOH?
2. With your background in biochemistry, how would you explain the changes (if any)?
3. Is the change reversible? Explain your answer.

### XI.  Effect of pH on MAO Activity

a. Check the activity of the native (purified) MAO.

b. To the native enzyme, add an increment of phosphoric acid to change the pH to 6.0, 5.0, and 4.0.

c. Check the activity on MAO treated with phosphoric acid in step b.

d. To the native enzyme, add an increment of NaOH to change the Ph to 9.0, 10.0, and 12.0.

e. Check the activity on MAO treated with NaOH in step d.

f. Dialyze all the preparations listed in steps b and d against 0.05 M potassium phosphate, pH 7.4, containing 0.002 M β-mercaptoethanol.

g. Check the activity on dialyzed MAO preparations.

## Review Questions

1. Do you observe any change in MAO activity by changing the pH?
2. Is the change reversible?
3. Draw your conclusions by comparing fluorescence and activity studies.

# PLASMA AMINE OXIDASE

*Supplies and Equipment*

SDS-polyacrylamide gel electrophoresis cell
UV-visible spectrometer
Test tubes
Whatman filter paper #2
Perkin-Elmer MPF-4 fluorescence spectrophotometer

### *Chemicals*

Molecular Probe
  Maleimide N-(1-pyrene)
Sigma Chemical Company
  Benzylamine sulfate
  Bovine serum albumin
  Iodoacetic acid
  Potassium phosphate (mono and dibasic)
  Plasma amine oxidase
  Sodium chloride
  Urea
  β-mercaptoethanol
  Triton X-100
Aldrich Chemical Company
  Dimethyl sulfoxide
  Ethanol 95%
  Methanol

# Introduction

The role of plasma amine oxidase (diamino: 02 oxi-doreductase; EC 1.4.3.6) in the catabolism of various amines found in association with DNA and its function in one of the catabolic pathways of histamine has led to considerable interest in this enzyme.[20] Physiochemical studies of bovine plasma amine oxidase indicated that the enzyme is made up of two identical subunits and has a molecular weight of 110,000 KD.[20,22] From the effect of the pH on the kinetic properties of the enzyme, it was concluded that the enzyme contains four sulfhydryl groups per mole that are not essential for activity. Several investigators, including the author, have disclosed that addition of substrate liberates two sulfhydryl groups per enzyme dimer.[20,22–25] These investigators postulated that the liberated sulfhydryl groups are covalently linked to the unknown cofactor in the oxidized enzyme.

In this experiment, students can use either the purified enzyme that was isolated by following the detailed procedures described in Module 4 or use a commercial enzyme. In either case, students are required to determine the activity of the enzyme as discussed in Module 4 and also assess the purity of the protein by either SDS-polyacrylamide gel electrophoresis and or UV-visible spectroscopy as explained in Module 3.

# Procedures

## I. Fluorescence Labeling of Either Purified or Commercially Available Plasma Amine Oxidase

a. Assay the native enzyme either purified or obtained from a commercial source utilizing the procedure described in Module 4.

b. Prepare a stock solution of $5 \times 10^{-3}$ M N-(1-pyrene) maleimide (NPM) in absolute dimethyl sulfoxide.

c. Add 0.1 ml of the stock solution to Whatman filter paper #2 in a test tube.

d. Dry the test tube by a stream of nitrogen.

e. Add 1 ml of 1 mg/ml PAO dissolved in 0.01 M potassium phosphate buffer, pH 7.4, to the test tube.

f. Allow the reaction to proceed with gentle stirring at room temperature and under anaerobic conditions for 4 hours.

g. Dialyze the labeled enzyme overnight with a change of buffer against 0.05 M potassium phosphate buffer, pH 7.5.

h. Now determine the activity of the labeled enzyme using the procedure described in Module 4.

# Review Question

1. Do you see a change in PAO activity after labeling? Explain your results.

## II. Fluorescence Labeling of the Essential Sulfhydryl Groups

a. Determine PAO activity as described in experiment I, step a.

b. Add 12.0 μmoles of the substrate benzylamine at pH 7.6 to 1.0 ml of PAO at 1.0 mg/ml concentration.

c. Repeat experiment I, steps b, c, and d.

d. Add 1 ml of the enzyme you prepared in step b.

e. Allow the reaction to proceed with gentle stirring at room temperature and under anaerobic conditions for 4 hours.

f. Dialyze the labeled enzyme overnight with a change of buffer against 0.05 M potassium phosphate buffer, pH 7.5.

g. Now determine the activity of the modified labeled enzyme.

# Review Question

1. Did the activity of the enzyme change after you added first the substrate (benzylamine) and then labeled the enzyme? Explain your results in detail.

## III. Spectral Measurements

Fluorescence excitation and emission spectra can be recorded with a Perkin-MPF-4 fluorescence spectrophotometer using 1.0 cm path cuvette. Excitation can be recorded at 330 nm for NPM. Emission spectra should be recorded at room temperature.

## IV. Results and Spectral Analysis

Under the conditions described in experiment I and II and by Zeidan et al.,[22,25] PAO can first be reduced by addition of 12.5 µmoles of benzylamine at pH 7.6 to 1.0 ml of the enzyme. The reduced enzyme can then be reacted with fluorescent label NPM, which will then attach the label on the essential liberated sulfhydryl group per monomer. The reaction of NPM with the liberated sulfhydryl group is presented in Figure 5.1. NPM does not fluoresce or form a fluorescent product in solution alone. Record fluorescence spectra as follows:

a. Record fluorescence emission spectrum of NPM in the absence (alone) and the presence of PAO (after addition of substrate).
b. Fluorescence spectrum can be recorded in 1.0 ml of 0.01 M potassium phosphate (KBP), pH 7.5.
c. Excitation and emission slits can be set at 7.0 mm.
d. The sample sensitivity is normally 30 if the protein concentration is 1.0 mg/ml.
e. Determine the protein concentration.
f. Be certain that the labeled enzyme is completely dialyzed against 0.01 M KBP, pH 7.5, for 24 hours with a change of buffer before recording the spectra.

## Review Questions

1. From your spectrum, what are the wavelengths where the two major vibronic bands appear?
2. How do you interpret the existence of the two spectroscopic components?
3. Compare the spectra obtained from experiment IV. Discuss your results in detail.

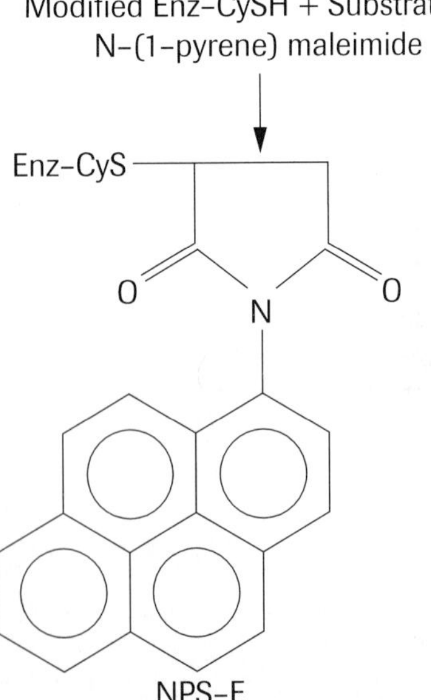


**FIGURE 5.1**   Fluorescent labeling of the essential sufhydryl group on PAO.

Zeidan, H., and S. Oyouga. 1994. "The active site of plasma amine oxidase. The key enzyme of biogenic amines degradation in plasma as seen by fluorescence investigation." *Biodeterioration Research 4.* New York: Plenum Press.

## V. Effect of Urea on NPM-PAO Spectrum

a. In this experiment, students are asked to compare the emission spectra of two derivatives. The first is plasma amine oxidase after addition of the substrate, then NPM fluorescence probe and complete dialysis; and the second molecule is β-mercaptoethanol directly labeled with NPM using the same procedure described for labeling PAO.
b. Set up five test tubes and add solid urea to each tube, increasing the concentration from 0 to 2 M, 4 M, 6 M, and 8 M urea in a total volume of 1 ml.
c. Add 1 ml of PAO that has been labeled on the essential liberated sulfhydryl group after addition of the substrate benzylamine and complete dialysis overnight.
d. Shake the tubes in a water bath incubator for 2 minutes then dialyze the enzyme against 0.01 M potassium phosphate buffer, pH 7.5, for 24 hours with a change of buffer, and then record the spectrum.
e. Repeat the above procedure on β-mercaptoethanol by setting up two test tubes. The first has β-mercaptoethanol at 0.002 M reacted with NPM as described in step a; and to the second add solid urea so that the final concentration is 6 M– 8 M urea in 1 ml.
f. Compare and analyze any change in the spectra obtained in steps b and/or e. Are they the same? If not, why? Explain your results.
g. Dialyze all the high and low molecular weight derivatives overnight against 0.05 M potassium phosphate buffer, pH 7.5, and then record the spectra.

## Review Questions

1. Do you observe any spectral changes when you add 8 M urea to β-mercaptoethanol? Explain your results.
2. Do you see any spectral changes when you add 2 M, 4 M, 6 M, and 8 M urea to PAO-NPM derivative? If you see a change, indicate at what concentration.
3. If your answer in question 2 is positive, did the different emission bands quench equally?
4. How would you explain the changes (if any) of the emission spectra of PAO-NPM?
5. Why is fluorescence spectroscopy the preferred biophysical technique over UV-visible spectroscopy in detecting structural changes?
6. Is the denaturation reversible? How could you conclude whether or not the denaturation was reversible? If it was reversible, at what concentration did you conclude that the effect was irreversible?
7. Compare the effect of urea on MAO spectra with PAO spectra.

## VI. Effect of Urea on PAO Activity

a. Check the activity of native purified PAO.
b. To five test tubes, add 2 M, 4 M, 6 M, and 8 M urea in a total volume of 1 ml.
c. Add 1 ml of native purified PAO.
d. Mix well and check the activity as described previously.
e. Dialyze all the denatured PAO against 0.01 M potassium phosphate buffer, pH 7.5, and recheck the activity.

# Review Questions

1. Is the denaturation reversible?
2. Draw your conclusions by correlating fluorescence with activity studies.

## VII. Effect of pH on the Pyrenemaleimide-PAO Spectrum

a. To 1 ml of NPM-PAO derivative prepared as described previously, lower the pH by adding an increment of concentrated phosphoric acid to achieve pHs of 6, 4, and 3.
b. Record the emission spectrum at each of pH 7.5, 6.0, 4.0, and 3.0. Make sure to check the pH by a pH meter or litmus paper.
c. To 1 ml of NPM-PAO derivative, increase the pH by adding an increment of NaOH to achieve a pH of 8.0, 9.5, and 12.0.
d. Record the emission spectra at pH 8.0, 9.5, and 12.0.
e. Dialyze all the PAO-NPM derivatives prepared in steps a and c against 0.05 M potassium phosphate buffer, pH 7.5, overnight with a change of buffer.
f. Record the spectra for all the derivatives after dialysis.

# Review Questions

1. Do you see any spectral changes when you add phosphoric acid or NaOH?
2. With your background in biochemistry, how would you explain the changes (if any)?
3. Is the change reversible? Explain your answer.

## VIII. Effect of pH on Enzyme Activity

a. Using a native purified enzyme check the activity as described earlier at pH 7.5.
b. Change the pH of the native enzyme by adding an increment of phosphoric acid to achieve a pH of 6.0, 4.0, and 3.0. Check the activity of PAO at the corresponding pHs.
c. Change the pH of the native enzyme by adding an increment of NaOH to achieve a pH of 8.0, 9.0, 11.0, and 12.0. Check the activity of PAO at their corresponding pHs.
d. Correlate the activity study with the changes in fluorescence spectra (if any) by changing the pH.
e. What is your conclusion?

## IX. Effect of Ionic Strength on PAO-NPM Fluorescence Spectra

a. Set up five test tubes. Add to each test tube sodium chloride, increasing the total concentration from 0 M to 0.2 M, 0.4 M, 0.6 M, and 0.8 M sodium chloride in 1 ml volume.
b. Add 1 ml of PAO-NPM enzyme derivative prepared by the procedure described previously.
c. Shake the tubes in a water bath at 25° C for 2 minutes, then record the emission spectra.

# Review Question

1. Do you see any spectral changes when you add sodium chloride to PAO-NPM derivative?

## X. Effect of Ionic Strength on PAO Activity

a. Set up five test tubes. Add to each test tube sodium chloride with increasing total concentration from 0 M to 0.2 M, 0.4 M, 0.6 M, and 0.8 M sodium chloride in 1 ml volume.
b. Add 1 ml of native purified enzyme to each of the five test tubes prepared in step a.
c. Mix well and then check the activity of the enzyme as described previously.
d. Correlate the activity study with fluorescence spectra study.

# Review Questions

1. Do you observe any changes in PAO activity?
2. Explain and correlate your results.

## XI. UV-Visible PAO Spectrum

a. Run the absorption spectrum for PAO over the wavelengths of 300 nm–600 nm.
b. Add phenylhydrazine with a molar concentration of 1:1 to PAO.
c. Run the absorption spectrum for PAO-phenylhydrazine derivative. Draw your conclusion about the enzyme and UV-visible spectroscopy.

# CHARACTERIZATION OF MONO AND PLASMA AMINE OXIDASES BY ESR—SPIN LABELING METHOD

## *Supplies and Equipment*

Electron spin resonance spectrometer, Brucker or Varian E-4
pH meter
UV-visible spectrophotometer
Cold room, 4° C
Aluminum foil
Beakers (50 ml, 100 ml, 500 ml, 1,000 ml, 2,000 ml)
Dialysis tubing, MW cut of 14,000
Eppendorf pipettes and microtips
Erlenmeyer flasks (250 ml, 500 ml, 1,000 ml)
Flat EPR tube
Graduate cylinders
Graph paper
Kimwipes
Magnetic stirring bar
Magnetic stirrer
Micropipettes
Nitrogen tank
Pasteur pipettes
Pipettes
Quartz cuvettes
Spin labels for N-ethylmaleimide (Molecular Probes, Inc.)
Spin label tryptamine (Medimpex, Budapest, Hungary)
Spin label [$^2$H–$^{15}$N] MSL molecular isotopes

## *Chemicals*

Acetone
Ammonium hydroxide
Benzylamine sulfate
Ethanol
β-mercaptoethanol
Sodium chloride
Potassium phosphate
Phosphoric acid
pH buffer standards
Urea
Monamine oxidase
Plasma amine oxidase

# Introduction

This module is written with two goals in mind: to introduce the student to the fundamentals of electron spin resonance (ESR) and its application to problems in biochemistry and to acquaint a current user with new developments in the use of this technique. A module of this length cannot treat the subject in depth and focus on theory as well as practice; only a broad view can be presented to show how ESR is currently being employed in the study of biological systems. What has been sacrificed to a large extent is the theory at the level of mathematical description of the various physical phenomena discussed.[26–29]

To help the student and the user compensate for the brevity of this module, the end-of-module reference section is organized in standard citation format.[26–41]

## Methodology

### *Spin Probes*

A spin probe is a molecule labeled with a stable free radical. The interaction of the magnetic moment of the unpaired electron of the free radical with an applied magnetic field gives rise to the electron spin resonance (ESR) spectrum. In biological studies, this unpaired electron usually belongs to the nitroxide group of the pyrolidine ring (Figure 5.2). The alkyl side chains serve to stabilize the radical, and R is a reactive group that can be used to attach the label to the enzyme, protein, or membrane.[30–34] A library of protein-reactive spin labels is available, consisting of a nitroxide coupled to an maleimide, or iodoacetamide group, substrates. The first two agents react primarily with sulfhydryl groups and the last binds at the active site of an enzyme.[9,10]

### *ESR Spectroscopy*

ESR and spin-labeling techniques have contributed significantly to investigations of structural, conformational, functional, and kinetic information of macromolecules, membrane, and tissues of biological systems. The spin label method offers several advantages. In general terms, the ESR spectra of nitroxide spin labels are sensitive to molecular motion, orientation, and electric and magnetic environments. It potentially offers the ability to probe all of these features in a biomolecular environment. In practice, the method has been found to be most sensitive to the first two properties: motion and orientation. That is, a small paramagnetic reporter group (spin label) incorporated at a specific site into a protein can relay information about the structural nature and the conformational changes in its physical environment from an analysis of its motional freedom.[35–39]

The utility of the spin label as a probe results from two factors. First, spin label and/or probe introduced into a biological system generally exhibits a unique ESR spectrum—i.e., there is no background. Second, although the nitroxide itself is essentially invariant (Figure 5.3), the organic framework of which the paramagnetic center is a part can be tailored to conform to nearly any structural design.[39] As such, labels have been synthesized as enzyme substrates, inhibitors, and

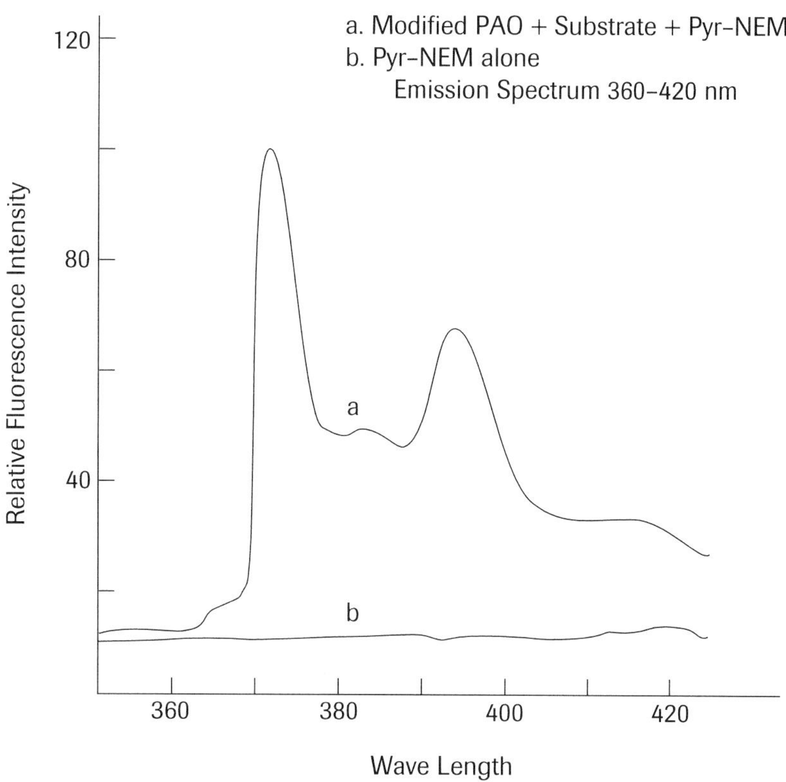



**FIGURE 5.2**   Fluorescence spectrum of plasma amine oxidase; **(a)** Modified plasma amine oxidase after addition of substrate and Pyr–NEM; **(b)** Pyr–NEM alone.

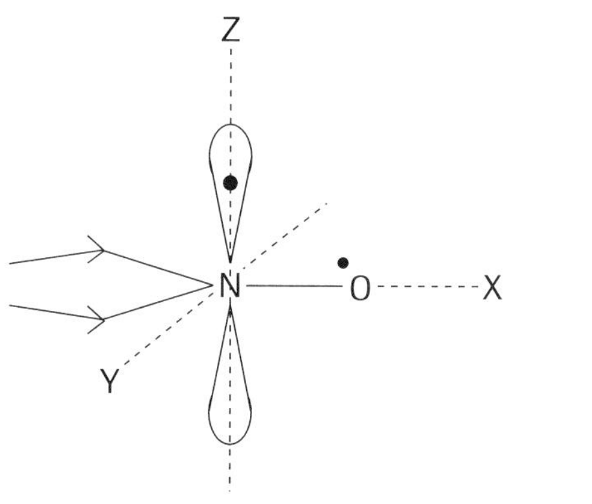



**FIGURE 5.3**   ESR spectrum of the nitroxide radical.

From *Structural and Resonance Techniques in Biological Research.* 1984. Academic Press. Reprinted by permission from Dr. Daniel J. Kosman.

affinity labels; probes of conformational changes in proteins, nucleic acids, and lipids; probes of protein-protein, protein-nucleic, acid, protein-lipid, and lipid-lipid interactions; and probes of the structure fluidity of biological and model membranes. The remainder of this module will present selected examples of spin labels that illustrate both the applicability and the use of this type of probe, with emphasis on those examples that illustrate the fundamental aspects of spin-label behavior.

These labels and probes can be used to study changes in protein conformation caused by perturbans—i.e., changes in pH and ionic strength.

One very important feature of the nitroxide spin label is the very great biochemical specificity that can be achieved through the use of a suitable chemical group R. In other words, even though the N $\rightarrow$ $\cdot$0 group and the surrounding carbon atoms and methyl groups can hardly be considered to have a negligible size, it has now been clearly demonstrated[31,39–42] that this group is nonetheless an entirely tolerable perturbation in a great variety of selected biochemical situations, including, for example, the active site of enzymes. A crucial property of the sterically protected nitroxide group is that it is neither strongly hydrophobic nor hydrophilic and this important characteristic of the molecule is dominated by the group R. A second important feature of the nitroxide spin labels is the comparative simplicity and high accuracy of the spin quantum mechanics that relate the physical state of the label to its paramagnetic resonance spectrum. Here the "physical state" of the label signifies its orientation in space; it has fixed orientation and the nature of its motion if it is moving. A third important feature of spin labels in biological systems is that their paramagnetic resonance spectra are usually completely free from any interference. This is because the vast majority of the molecules are diamagnetic and not paramagnetic.[25–30]

ESR spectra are obtained by inserting the sample into a constant microwave field, usually at 9.5 gigahertz (GHz), and scanning it with a magnetic field at right angles to the microwave field. The spectrum consists of the microwave energy absorption peaks occurring over the length of this scan. The $N \rightarrow \cdot O$ spectrum lies between 3000–3500 gauss (G) (Figure 5.4).[33,41–45]

When a spin label is freely rotating in solution (Figure 5.4A), the hyperfine interaction between the unpaired electron and $^{14}N$ nucleus of the $N \rightarrow \cdot O$ group splits the spectrum into three sharp lines. The dielectric constant in the vicinity of the label affects the hyperfine interaction (the distance between the lines). ESR spectra are normally recorded as the first derivative of the absorption spectrum (see Figure 5.4A).

When a spin label is partly or completely immobilized, a spectrum with broadened peaks is produced (Figure 5.4B and 5.4C). When the probe is partly or strongly immobilized, a specific spectral parameter change occurs and can be determined to probe in detail the environment where the nitroxide is attached. For example, for partially immobilized nitroxide, the width of the midfield line ($W_o$) increases and the height of the high field line ($H_{-1}$) decreases. For strongly immobilized nitroxide, the hyperfine splitting constant—the distance between the outer two peaks increases, and the height of the high field line ($H_{-1}$) significantly decreases. (Figure 5.4C).[25,34]

The simplest use of the spin label is as a detector and an indicator of macromolecular conformational change. In the following laboratory exercises, experiments are carried out without regard to quantitation of the change in the structural constraints placed on the nitroxide, reflected in the changes in the ESR spectra.[30,43–54]

# Procedures

## I. Modification of Enzyme

a. Add a total concentration of 0.05 M iodoacetic acid to 1–2 mg of the enzyme; and let the reaction run for 6 hours to block the nonessential sulfhydryl groups as reported in Modules 3 and 5.

b. Dialyze the enzyme against 0.05 M potassium phosphate buffer, pH 7.4, containing 0.002 M β-mercaptoethanol and 0.1% Triton X-100 overnight with a change of buffer.

## II. Spin Labeling the Essential Sulfhydryl Groups

a. Prepare a stock of spin labeled N-ethylmaleimide with a total amount of 1 mg to 3 ml ethanol, then

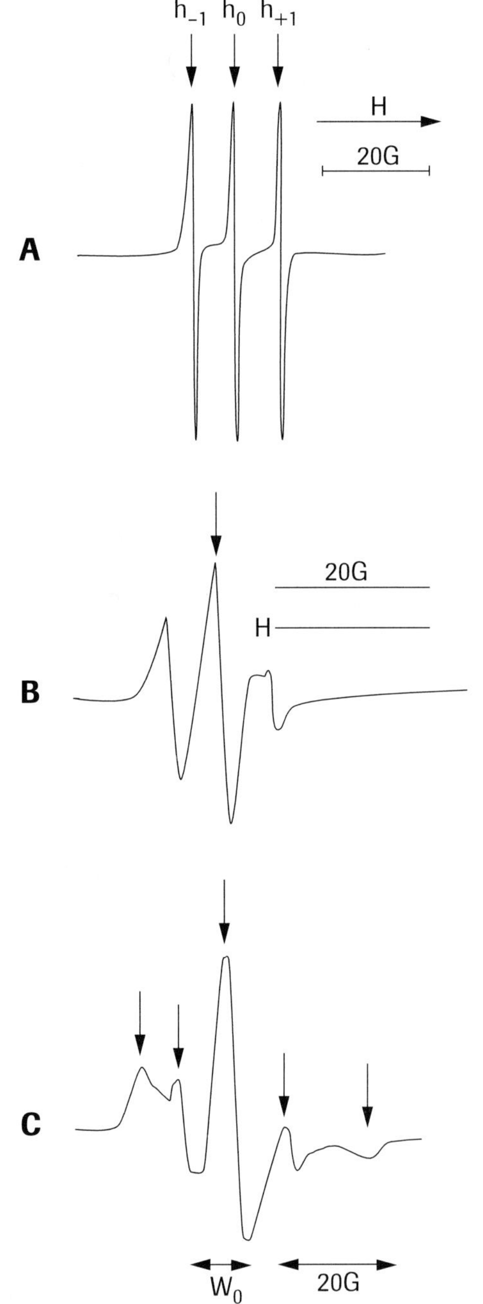


**FIGURE 5.4** ESR spectra recorded at room temperature of a soluble spin label (TEMPO) or N-ethylmaleimide. **(A)** tumbling freely in aqueous solution; **(B)** partially immobilized N-ethyl maleimide spin label attached to an enzyme at SH group. The change in spectral parameters $W_o$ and $h_o$ = width and height of the midline, respectively, $h_{-1}$ = height of high field line can be used to monitor conformational changes; **(C)** strongly immobilized N-ethyl maleimide spin label attached to an enzyme at SH group. The primary changes in spectral parameters are: **(A)** the hyperfine splitting constant (distance between outer two peaks) increases; **(B)** $W_o$ increases; **(C)** $h_{-1}$ decreases significantly.

Zeidan et al.[11,12,41,51] and Zeidan, H.[42,61,62]

flush it with nitrogen, cover it with aluminum foil, and keep it at 5° C (see the structure of spin labeled N-ethylmaleimide).

b. Add 50 µl of the spin labeled enzyme prepared in step a to Whatman # 2 paper in a test tube, then dry with a stream of nitrogen.

c. Add 1.0 ml–1.5 ml of the modified enzyme directly to the test tube, then flush with nitrogen for 1 minute, and shake the test tube overnight at room temperature.

d. Dialyze the spin labeled enzyme overnight against 0.05 M potassium phosphate buffer, pH 7.4, containing 0.002 M β-mercaptoethanol and 0.1% Triton X-100 with a change in buffer.

e. Record the spectra for both spin label N-ethylmaleimide in ethanol and spin labeled MAO using an electron spin resonance spectrometer.

Program the spectrometer parameters at the following settings:

1. the microwave frequency at 9.5 gigahertz (GHz)
2. field setting 3415 gauss
3. modulation amplitude at 4.0 gauss
4. sweep range 200 gauss
5. sweep rate or time 4 minutes
6. response 0.4 seconds

f. The student must use a flat EPR tube. Make sure you tune the ESR spectrometer prior to taking the spectrum. You can carry on the tuning utilizing a pitch as described in any of the listed references.

g. Students should be able to obtain a spectrum identical or very similar to the spectrum in Figure 5.5. At this time, students are encouraged to read the first chapter of one of the references listed at the end of this module and understand the terminology listed in step e.[34–36] It is highly recommended that this reading take place prior to starting the experiment.

h. You are expected to see only highly bound N-ethylmaleimide covalently attached to the essential sulfhydryl group. If you suspect any free unbound label (increase in the amplitude of the low and high field peaks), dialyze the modified enzyme for about 10 hours. This experiment can be highly quantitative if students interface the computer with the ESR, but they can easily understand the spectrum and analyze the data without resorting to the computer.

## Review Questions

1. Why is the EPR spectrum for spin labeled MAO asymmetrical?
2. Identify each peak and indicate which peaks represent a strongly, partly immobilized, and free label.
3. Measure the ratio of strongly immobilized to partially immobilized.
4. Draw your conclusions about the environment of the essential sulfhydryl group from the spectral parameters.
5. Compare your results with the fluorescence spectrum results you obtained in previous experiments.

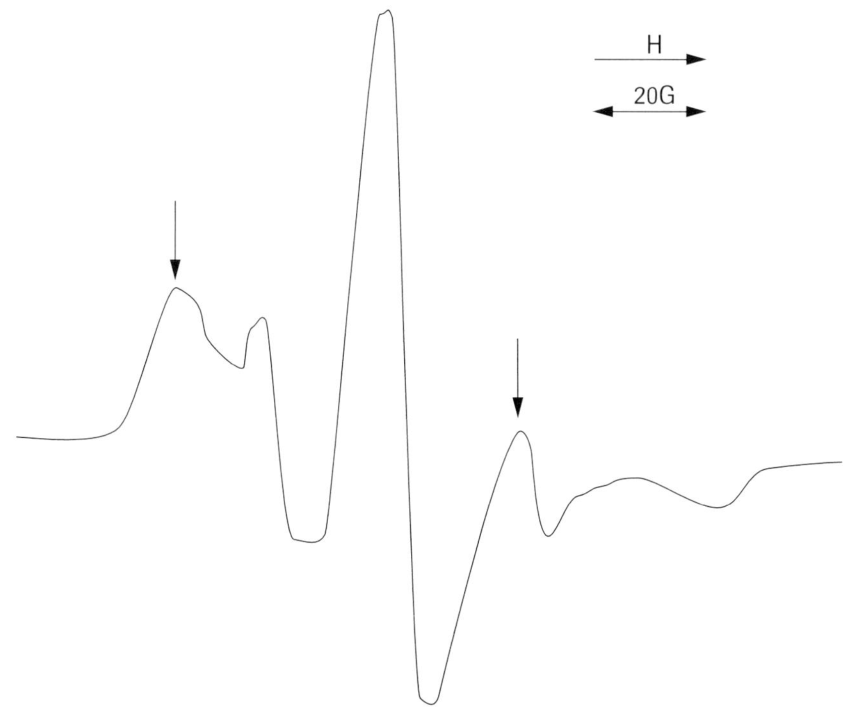


**FIGURE 5.5**   EPR spectrum for MAO labeled at the essential SH group.
Zeidan, H et al. (1990).

## III. Effect of the Side Chain Length on the Environment of the Essential Sulfhydryl Group*

a. Repeat the previous experiment, but this time use a side chain of 5.8° A instead of 4.5° A.

b. Record the spectrum as described in experiment II, step e.

c. Analyze the spectrum in terms of the hyperfine splitting constant (the distance between the outer two peaks) and compare spectrum 2 (5.8° A) with spectrum 1 (4.5° A).

## Review Questions

1. Determine the hyperfine splitting constant in gauss (the difference between the outer two peaks).
2. Compare the hyperfine splitting constant obtained in spectrum 1 with the one you determined in spectrum 1.
3. Draw your conclusions from the previous comparison.
4. What kind of information can the hyperfine splitting constant provide?
5. Discuss the advantages of using spin label techniques.
6. Repeat the labeling experiment using N-ethylmaleimide spin labels, which have side chain lengths of 9.3° A, 10.5° A, and 12.9° A.

## IV. Effect of pH on Spin Labeled MAO Spectrum

a. Prepare five test tubes of spin labeled MAO as described in experiment I and II, and then increase the pH of the enzyme by adding ammonium hydroxide in test tube no. 2 to 8.0, no. 3 to 9.0, no. 4 to 10.0, and no. 5 to 12.0.

b. Record the spectra for all the derivatives using the same parameters listed in experiment I.

c. Compare all the spectral parameters of the above derivatives.

d. Dialyze all the derivatives against 0.05 M potassium phosphate buffer, pH 7.4, containing 0.002 M β-mercaptoethanol and 0.1% Triton X-100.

e. Record the spectra for all derivatives.

f. With native enzyme quantify enzyme activity at pH 7.5, 8.0, 9.0, 10.0, and 12.0. You can use the results of the effect of pH on MAO activity that you obtained in your fluorescence experiment.

## Review Questions

1. Determine the ratio of the strongly immobilized to partially immobilized label by measuring the amplitude of both strongly and partly immobilized species.
2. Is the change of pH reversible? If not, at what pH is the effect irreversible?
3. Determine the ratio of the strongly immobilized to partially immobilized label by measuring the amplitude of both as a function of increasing urea concentration.
4. Did you see any spectral alteration when the urea concentration increased?
5. If there is a change is this change reversible? If so, at what urea concentration did you observe that the change was irreversible?
6. Did the change in native MAO activity with increasing urea concentration correlate with the spectral changes?
7. Discuss in detail the effect of urea on protein structure.
8. Discuss the advantage of using fluorescence or spin label methodology in determining conformational changes in protein structure.

## V. Effect of Ionic Strength on ESR Spectrum of Spin Labeled MAO

a. Repeat the labeling experiment at the essential sulfhydryl group. Make sure to prepare five derivatives in five separate test tubes.

b. Add sodium chloride to give a final concentration of 0.1 M, 0.2 M, 0.4 M, 0.6 M, and 0.8 M in the five test tubes prepared in step a, respectively.

c. Record the spectra of the five derivatives using the parameters listed on the labeling procedure.

d. Dialyze all the derivatives against 0.05 M potassium phosphate buffer, pH 7.4, containing 0.002 M β-mercaptoethanol and 0.1% Triton X-100.

e. Determine the activity of native MAO while increasing the ionic strength from 0.0 M to 0.8 M sodium chloride, or use the data obtained in your fluorescence studies.

## Review Questions

1. Do you observe any change in spectral parameters by increasing the ionic strength from 0.0–0.8 M sodium chloride?
2. If there is a change, is it reversible?
3. Is the change correlated with the activity studies? Explain your results.

---

*This experiment requires more spin labels. For departments that wish to eliminate the high cost, they may eliminate the effect of side chain.

# INTERACTION OF SPIN LABELED TRYPTAMINE WITH MONOAMINE OXIDASE: PROBING THE MICROENVIRONMENT OF THE ACTIVE SITE BY SPIN PROBE-SPIN LABEL TECHNIQUES

### Supplies and Equipment

Electron spin resonance spectrometer, Brucker or Varian E-4
pH meter
UV-visible spectrophotometer
Cold room, 4° C
Aluminum foil
Beakers (50 ml, 100 ml, 500 ml, 1,000 ml, 2,000 ml)
Dialysis tubing, MW cut of 14,000
Eppendorf pipettes and microtips
Erlenmeyer flasks (250 ml, 500 ml, 1,000 ml)
Flat EPR tube
Graduate cylinders
Graph paper
Kimwipes
Magnetic stirring bar
Magnetic stirrer
Micropipettes
Nitrogen tank
Pasteur pipettes
Pipettes
Quartz cuvettes
Spin labels for N-ethylmaleimide (Molecular Probes, Inc.)
Spin label tryptamine (Medimpex, Budapest, Hungary)
Spin label [$^2$H–$^{15}$N] MSL molecular isotopes

### Chemicals

Acetone
Ammonium hydroxide
Benzylamine sulfate
Ethanol
β-mercaptoethanol
Monamine oxidase
Sodium chloride
Plasma amine oxidase
Potassium phosphate
Phosphoric acid
pH buffer standards
Urea

# Introduction

This laboratory exercise is designed only for those departments that have graduate courses in biochemistry or biophysics. Some of the labels are of high cost and thus it is recommended that either only students with advanced training who have already completed their undergraduate degrees or advanced senior students perform this exercise.

The sensitivity and resolution of EPR spectroscopy were substantially improved for biological studies by deuteration of the frequently used spin label [$^2$H–$^{15}$N] maleimide spin label (MSL), which binds covalently to proteins.[47] Significant gains in detectability and resolution were observed with [$^2$H] MSL in both the fast motion limit for freely tumbling spin label (t$^{-10}$) and in the slow to very slow correlation time range exhibited by labeled proteins. The improvement resulted from a reduction in the nonhomogenous line broadening because of the weaker superhyperfine interactions of the unpaired electron with deuterium than with hydrogen. In this experiment, the use of doubly substituted [$^2$H–$^{15}$N] MSL to probe the environment of the essential sulfhydryl groups of the enzyme and to determine the distance between the essential sulfhydryl groups and the substrate tryptamine binding site will enhance the capability for interpretation of labeled MAO spectra. It was already known that $^{15}$N, or deuterium substitution, increased sensitivity for freely tumbling spin label and that $^{15}$N increased sensitivity for freely tumbling spin label and that $^{15}$N simplified the spectrum by reduction of the number of nuclear mainfolds from three to two lines.[28–35] Also significant was that in many applications the signal-to-noise ratio was increased by a factor of 3 with $^{15}$N deuterium labels over a wide correlation time range facilitating studies on smaller amounts of biological materials.[42,48]

# Procedures

## I. Spin Labeling the Essential Sulfhydryl Groups with [$^2$H–$^{15}$N] MSL

a. Prepare a stock of [$^2$H–$^{15}$N] MSL with a total concentration of 1 mg to 3 ml ethanol, then flush it with nitrogen and cover it with aluminum foil and keep at 5° C (see structure of [$^2$H–$^{15}$N] MSL in Figure 5.6).

b. Add 20 µl of [$^2$H–$^{15}$N] MSL prepared in step a to Whatman #2 paper in a test tube, then dry with a stream of nitrogen.

c. Add 1.0 ml–1.5 ml of the modified enzyme (prepared as described under exercises I and II) to the test tube and then flush with nitrogen for 1 minute, and shake the test tube overnight at room temperature.

d. Dialyze the spin labeled enzyme overnight against 0.05 M potassium phosphate buffer, pH 7.4, containing 0.002 M β-mercaptoethanol and 0.1% Triton X-100 with a change in buffer.

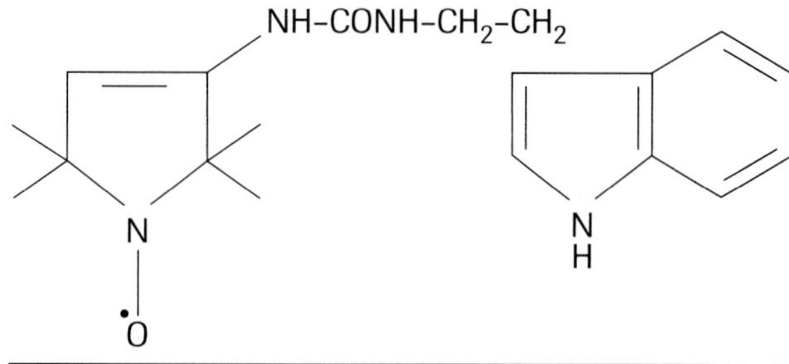



**FIGURE 5.7**   Structure of spin labeled tryptamine.
Zeidan, H., and S. Oyouga. 1994. "The active site of plasma amine oxidase. The key enzyme of biogenic amines degradation in plasma as seen by fluorescence investigation." *Biodeterioration Research 4.* New York: Plenum Press.

**FIGURE 5.6**   Structure of $^2$H–$^{15}$N maleimide spin label.
Zeidan, H., and S. Oyouga. 1994. "The active site of plasma amine oxidase. The key enzyme of biogenic amines degradation in plasma as seen by fluorescence investigation." *Biodeterioration Research 4.* New York: Plenum Press.

e. Record the spectra for both $^2$H–$^{15}$N MSL in ethanol and [$^2$H–$^{15}$N] MSL labeled enzyme using an electron spin resonance spectrometer. Program the spectrometer parameters at the following settings:
   1. the microwave frequency at 9.5 gigahertz (GHz)
   2. field setting 3415 gauss (G)
   3. modulation amplitude at 4.0 gauss
   4. sweep range 200 gauss (G)
   5. response 0.4 seconds
f. Repeat this labeling procedure using native MAO (without modification by iodoacetate).
g. Record the spectrum of native MAO labeled with $^2$H–$^{15}$N MSL (Figure 5.9).

## II. ESR Spectrum of Tryptamine Spin Label (TrpSL)-MAO Interaction

The objective of this experiment is to probe directly the active site of MAO with the spin labeled substrate tryptamine.

a. Prepare a stock solution of spin labeled tryptamine with a total concentration of 1 mg to 3 ml ethanol, then flush it with nitrogen, cover it with aluminum foil, and keep at 5° C (see the structure of spin labeled tryptamine, TrpSL, in Figure 5.7).
b. Add 20 μm of TRpSL prepared in step a to Whatman #2 paper in a test tube, then dry with a stream of nitrogen.
c. Add 1.0 ml–1.5 ml of native MAO to the test tube and then flush with nitrogen for 1 minute.
d. Record the TrpSL-MAO spectrum. Program the spectrometer parameters as described earlier.
e. Students should be able to obtain a spectrum similar to Figure 5.8.
f. Recover the TrpSL-MAO from the EPR tube and dialyze the derivative against 0.01 M potassium

phosphate, pH 7.4, containing 0.1% Triton X-100 and 0.002 M β-mercaptoethanol with a change of buffer overnight.
g. Record the spectrum after dialysis.

## Review Questions

1. Explain the spectral parameters for TrpSL-MAO interaction.
2. What can you tell about the active site of MAO from TrpSL-MAO spectrum?
3. What is the effect of dialysis on TrpSL-MAO derivative? Explain your results.

## III. Determination of the Distance between the Tryptamine Binding Site and the Essential Sulfhydryl Group

In the following experiment, the student will apply the use of $^{14}$C:$^{15}$N spin probe-spin label pairs methodology to examine intermolecular interactions between probes located in diverse environments. Thus information on molecular distance between the two paramagnetic sites can be obtained from the spectra. The student is encouraged to read the paper by Zeidan, H. (1988)[42] and references 55–57 for further detail. Interaction of spin label tryptamine with monamine oxidase probing the active site of spin probe-spin label technique.

Clear resolution of $^{14}$N and $^{15}$N spectral features allows assessment of interaction between the different isotropic spin systems without interference from interaction among probes within each population.[29–33] Dual-label ELDOR studies appear possible in any samples in which both spin labels are rotationally mobile and in systems where the $^{14}$N probe is mobile and the $^{15}$N population is rotationally immobile.[33]

The monoamine oxidase spin probe, tryptamine, at the substrate-binding site where $^{14}$N probe is mobile or partially immobile and [$^2$H–$^{15}$N] MSL covalently bound on the essential sulfhydryl groups is immobile meet these requirements.

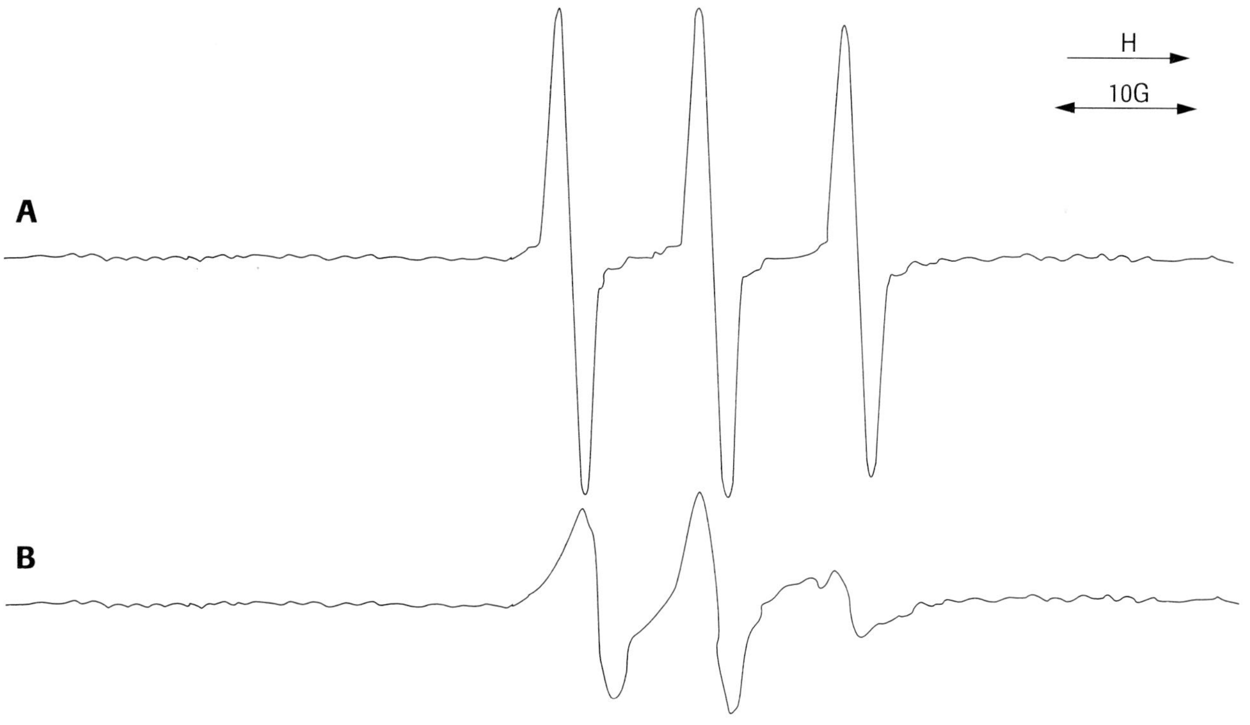



**FIGURE 5.8**  TrpSL attached to MAO at the active site. **(A)** TrpSL in buffer; **(B)** MAO + TrpSL.

Zeidan, H. (1988)[42] (1994)[62]

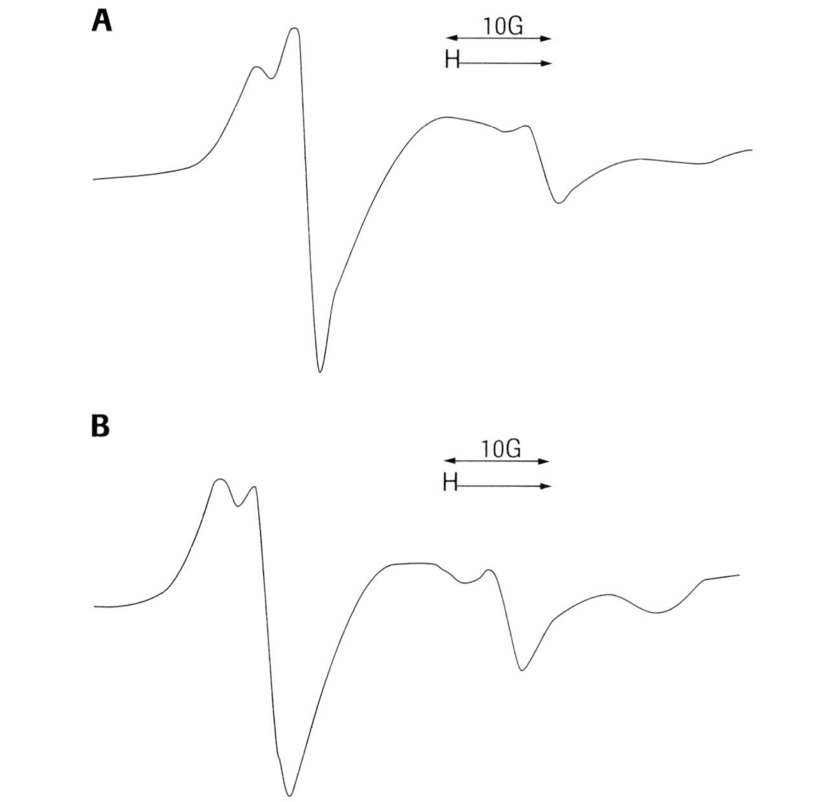



**FIGURE 5.9**  $^2$N–$^{15}$H MSL attached to MAO at the essential SH group. **(A)** MAO labeled with $^2$N–$^{15}$H MSL; **(B)** MAO modified at only the essential SH group with $^2$N–$^{15}$H MSL.

Zeidan, H. 1988. *Biochem. Biophys. Acta.* 955:111–18.

a. Prepare [$^2$H–$^{15}$N] MSL-MAO derivative as described in experiments I and II.

b. Record the spectrum.

c. Recover [$^2$H–$^{15}$N] MSL-MAO derivative from the EPR tube and add TrpSL at low concentration of $5 \times 10^{-6}$ M.

d. Record the spectrum for [$^2$H–$^{15}$N] MSL-TrpSL MAO. DO NOT CHANGE SENSITIVITY OR ANY OTHER PARAMETERS.

e. Recover and dialyze the [$^2$H–$^{15}$N] MSL-TrpSL MAO against 0.01 M potassium phosphate buffer containing 0.1% Triton X-100 and 0.002 M β-mercaptoethanol with a change of buffer overnight.

f. Record the spectrum after dialysis.

## Review Questions

1. What is the effect of addition of TrpSL on [$^2$H–$^{15}$N] MSL-MAO spectrum? Explain your results in detail.

2. From your results, would you consider that the essential sulfhydryl group is an active site group? Explain your results.

3. From your results, what are your conclusions about the minimum distance between the essential sulfhydryl group and the tryptamine binding site? Explain your results.

4. What is the effect of dialysis on [$^2$H–$^{15}$N] MSL-TrpSL MAO? Explain your results.

5. It is highly recommended that students write a complete report on this experiment using their own data along with information listed in many references.[25,42,57–69] The report should include the following topics:
   a. background on monoamine oxidase
   b. background on ESR-spin labeling techniques and dual spin label methodology
   c. results of all experiments listed above

### IV. ESR Spin Labeling of Plasma Amine Oxidase

Plasma amine oxidase (PAO) is a glycoprotein containing 7% to 8% carbohydrate and has a molecular weight of 110,000. It is composed of two identical subunits covalently linked by disulfide bonds.[39] It contains 2 g atoms of copper/mole of enzyme that are essential for activity.[40] The contributions of many investigators in this field have resulted in a significantly greater understanding of the mode of action of the enzyme. However, many of the details of the catalytic mechanisms have not been highly explained; of these, the role of the metal ion and the sulfhydryl groups are of particular interest.

Wang and Yasunobu[41] have reported that plasma amine oxidase contains four sulfhydryl groups per mole of enzyme that are nonessential for activity. Many investigators have suggested that there is another unknown cofactor, *X*, involved in catalysis, which has been identified tentatively as pyridoxal phosphate.[42,43] Suva and Abeles,[23] Castelov et al.,[24] and Zeidan et al.[25] disclosed that the addition of substrate liberates two essential sulfhydryl groups per monomer. It was postulated that these essential sulfhydryl groups were covalently linked to the unknown cofactor.

In this laboratory exercise, students will use ESR spin labeling to probe the environment of the essential sulfhydryl groups.

a. Prepare a stock of spin labeled N-ethylmaleimide with a total concentration of 1 mg to 3 ml ethanol.

b. Add 50 μl of spin labeled N-ethylmaleimide prepared in step a to Whatman #2 paper in a test tube, then dry with a stream of nitrogen.

c. Add 1 mg/ml of native PAO dissolved in 0.01 M potassium phosphate buffer, pH 7.4, to the dried label, then shake for 6 hours at room temperature.

d. Dialyze the spin labeled enzyme derivative against 0.01 M potassium phosphate buffer, pH 7.4, overnight with a change of buffer.

e. Record spectrum 1 using the same parameters listed in the MAO section.

f. Assay the enzyme for activity as described in the assay procedure in the PAO section.

g. Now, repeat steps a and b.

h. Add 25 μl of 0.1 M benzylamine, pH 7.4, to 1 mg/ml PAO dissolved in 0.01 M potassium phosphate buffer, pH 7.4. Benzylamine acts as a substrate for the enzyme.

i. Add the enzyme prepared in step h to spin label N-ethylmaleimide prepared in step g. Shake the enzyme gently at room temperature for 6 hours.

j. Repeat steps d and e and record spectrum #2.

## Review Questions

1. What can you conclude from spectrum #1?
2. Compare spectrum #1 with spectrum #2. Explain your results in detail.

## REFERENCES


1. Beddorat, G. S., and M. A. West. 1981. *Fluorescent Probes*. New York: Academic Press.

2. Dewey, T. G. 1991. *Biophysical and Biochemical Aspects of Fluorescence Spectroscopy*. New York: Plenum Press.

3. Gillespie, A. M. 1985. *A Manual of Fluorometric and Spectrophotometric Experiments*. Gordon and Sieve Publishers.

4. Jamesor, D. M., and G. D. Reinhart. 1989. *Fluorescent Biomolecules: Methodologies and Applications.* New York: Plenum Press.

5. Lakowiz, J. P. 1991. *Topics in Fluorescence Spectroscopy.* New York: Plenum Press.

6. Tsien, A. 1992. "Fluorescent and potential probes of dynamic behavioral signals inside living cells." *Abstracts of the American Chemical Society.* 204:293.

7. Weltman, J. K., R. P. Szaro, R. E. Cathou, R. M. Dowben, and J. R. Bunting. 1973. "N-(3-pyrene) maleimide: A long life time fluorescent sulfhydryl reagent." *J. Biol. Chem.* 248:3173–77.

8. Wu, C-W, L. Yarbrough, and F. Wu. 1976. "N-(1-Pyrene) maleimide: A fluorescent cross linking reagent." *Biochem.* 15:2863–67.

9. Wolfbeis, Q. S. 1993. *Fluorescence Spectroscopy: New Methods and Applications.* New York: Springer-Verlag.

10. Orland, L., and B. A. Callinglan (eds.). 1987. *Monoamine Oxidase Enzymes: A Review and Overview.* New York: Springer-Verlag.

11. Zeidan, H., K. Watanabe, L. Piette, and K. Yasunobu. 1980. "ESR studies of spin labeled bovine liver monoamine oxidase B." *Frontiers in Protein Chemistry.* Amsterdam: Elseviere. 133–48.

12. Zeidan, H., and S. Buchanan. 1990. *Spectral Studies on Monoamine Oxidase Modified with N-(1-Pyrene) Maleimide Fluorescent Probe, Biodeterioration Research 3.* New York: Plenum Press.

13. Mov, J. C. S., and R. B. Silverman. 1993. "Fluorescence studies of monoamine oxidase B—Observation of two different chromophores in the resting state." *Abstracts of the American Chemical Society.* 206:316.

14. Singer, T. P. 1979. *Monoamine Oxidase: Structure, Function and Altered Functions.* New York: Academic Press.

15. Murphy, D. L., N. A. Garrlick, and R. M. Cohen, 1982. *Monoamine Oxidase, Biochemical* and *Physiological Aspects Relevant to Human Psychopharmacology in Antidepressants.* Burrows, N., and D. Davies (eds.). Amesterdam: Elsevier.

16. Murphy, D. L. and M. S. Buchsbaum, 1978. *Critical Issues in Psychiatric Diagnosis.* Spitzer, R. L., and D. F. Klein (eds.). New York: Raven Press. 305–321.

17. Usidin, E. (ed.). 1976. *Neurophyschopharmacology of Monoamines and their Regulatory Enzymes.* New York: Raven Press.

18. Youdim, M. M. H. and J. P. M. Finberg, 1982. *Monoamine Oxidase Inhibitor Antidepressants in Psychopharmacology.* Graham-Smith, D. G., H. Hippins, and G. Winoker (eds.). Amesterdam: Elsevier.

19. Oi, S., K. T. Yasunobu, and J. Westley. 1971. "The effect of pH on the kinetic parameters and mechanisms of monoamine oxidase." *Arch. Biochem. Biophys.* 145:557.

20. Yasunobu, K. T., H. Ishizaki, and N. Minamiura. 1976. "The amine oxidase." *Molecular Cell. Biochem.* 13:3.

21. Tarr, G. E. 1985. *Microcharacterization of Polypeptides: A Practical Manual.* Clifton, GA: Humana Press.

22. Zeidan, H., and S. Oyouga. 1994. "The active site of plasma amine oxidase. The key enzyme of biogenic amines degradation in plasma as seen by fluorescence investigation." *Biodeterioration Research* 4. New York: Plenum Press.

23. Suva, R. H., and R. H. Abeles. 1978. *Biochemistry.* 17:3538.

24. Castellov, F. M., Z. W. He, and F. T. Grehavoy. 1993. "Hydroxyl radical production in the reaction of copper-containing amine oxidases with substrate." *Biochem. Biophys. Acta.* 1157: 162–64.

25. Zeidan, H., K. Watanabe, L. H. Piette, and K. Yasunobu. 1980. "ESR spin labelling of plasma amine oxidase." *J. Biol. Chem.* 255:7621–26.

26. Bersohn, M., and J. C. Baird. 1966. *An Introduction to Electron Paramagnetic Resonance.* New York: Benjamin.

27. Poole, C., Jr. 1967. *Electron Spin Resonance—A Comprehensive Treatise on Experimental Techniques.* New York: Wiley Interscience.

28. Ingram, D. J. E. 1967. *Biological and Biochemical Applications of Electron Spin Resonance.* New York: Plenum Press.

29. Feher, G. 1970. *Electron Paramagnetic Resonance with Applications to Selected Problems in Biology.* New York: Gordon and Breach.

30. Wertz, J., and J. Bolton. 1972. *Electron Spin Resonance—Elementary Theory and Applications.* New York: McGraw-Hill.

31. Swartz, H. M., J. R. Bolton, and D. C. Borg (eds.). 1972. *Biological Applications of Electron Spin Resonance.* New York: Wiley Interscience.

32. Atherton, N. M. 1973. *Electron Spin Resonance.* New York: John Wiley and Sons.

33. Berliner, L. J. 1976. *Spin Labeling: Theory and Applications*, Vols. I and II. New York: Academic Press.

34. Lichtenstein, G. I. 1976. *Spin Labeling Methods in Molecular Biology.* New York: Wiley Interscience.

35. Jost, P. C., and D. H. Griffith. 1978. "The spin-labeling technique." *Metho. Enzymol.* 69:369–418.

36. Berlinger, L. J. 1978. "Spin labeling in enzymology: spin-labeled enzymes and proteins." *Meth. Enzymol.* 69:381–480.

37. March, D. 1981. "Electron spin resonance: spin labels." *Membrane Spectroscopy.* New York: Springer-Verlag.

38. Schreier-Muccillo, S., C. F. Polnaszek, and I. C. P. Smith. 1978. *Biochem. Biophys. Acta.* 515: 375–436.

39. Kosman, D. J. 1984. "Electron spin resonance." *Structural and Resonance Techniques in Biological Research.* New York: Academic Press. 89–244.

40. Curtain, C. O., F. D. Looney, and L. M. Gordon. 1987. *Methods in Enzymology,* vol. 150. 418–46.

41. Zeidan, H., K. Watanabe, L. W. Piette, and K. T. Yasunobu. 1980. *Frontiers in Protein Chemistry.* Amsterdam: Elsevier. 131–48.

42. Zeidan, H. M. 1988. *Biochem. Biophys. Acta.* 955: 111–18.

43. Gaffiney, B. J., and H. M. McConnell. 1974. "The paramagnetic resonance spectra of spin labels in phospholipid membranes." *J. Mag. Reson.* 16:1–28.

44. Hisa, J. C., and L. Piette. 1969. "Spin-labeling as a general method in studying antibody active site." *Arch. Biochem. Biophys.* 129:296–307.

45. McConnell, H. M., and B. C. McFarland. 1970. *Quart. Rev. Biophys.* 3:91.

46. Freed, J. H., G. V. Burno, and C. Polanasazek. 1971. *J. Phys. Chem.* 75:3385.

47. Taylor, M. G., and I. C. P. Smith. 1981. "Reliability of nitroxide spin probes in reporting membrane properties: A comparison of nitroxide and deuterium-labeled steroids." *Biochemistry.* 20:5252–55.

48. Schimschick, E. J., and H. M. McConnell. 1972. *Biochem. Biophys. Res. Commun.* 46:341.

49. Stone, T., T. Buckman, P. Norido, and H. M. McConnell. 1965. *Proc. Nat. Acad. Sci.* 54:1010.

50. Smith, I. C. P. 1972. *Biological Applications of ESR Spectroscopy.* New York: Wiley Interscience. 489.

51. Zeidan, H. M., K. H. Pearson, S. G. Brown, and P. F. Han. 1986. *Biochem. Biophys. Acta.* 870:141–47.

52. Meier, P., A. Blume, E. Ohmes, F. A. Neugebauer, and G. Kothe. 1982. "Structure and dynamics of phospholipid membranes: An electron spin resonance study employing biradical probes." *Biochemistry.* 21:526–34.

53. Hendrick, W. R., A. Matheew, J. D. Zimbrick and T. W. Whaley. 1979. *J. Magn. Resonance.* 36: 207–14.

54. Bienvenue, A., P. Herve, and P. Devanx. 1978. *Hebd. Seances Acad. Sci. Serv. D.* 287:1247–50.

55. Devanx, P., J. P. Davonust, and A. Rousselet. 1981. *Biochem. Soc. Symp.* 46:207–22.

56. Seigneuret, M., J. Davonust, P. Herve, and P. Devaux. 1981. *Biochemie.* 63:867–70.

57. Stetter, E., H. M. Vieth, and K. H. Hausser. 1976. *J. Mag. Reson.* 23:493–504.

58. Krugh, T. R. 1976. *Spin Labeling: Theory and Applications.* New York: Academic Press. 339–72.

59. Hyde, J. S., L. R. and Dalton. 1979. *Spin Labeling: Theory and Applications,* vol. 2. New York: Academic Press. 1–70.

60. Morrisett, J. D. 1976. *Spin Labeling: Theory and Applications.* New York: Academic Press. 273–338.

61. Zeidan, H. M. 1990. "The contact sites of sickle hemoglobin as seen by spin probe-spin label techniques." *Clinical Physiology and Biochemistry.* 8:81–90.

62. Zeidan, H. 1994. "Monoamine oxidase: Some new findings." *Biodeterioration Research 4.*

63. Zeidan, H., K. Watanabe, L. Piette, and K. Yasunobu, 1980. "ESR studies on plasma amine oxidase: probing the substitute liberated essential sulfhydryl group by spin label techniques." *J. Biol. Chem.* 255:7621–26.

64. Watanabe, K., and K. T. Yasunobu. 1970. *J. Biol. Chem.* 245:4612–17.

65. Yasunobu, K. T., H. Ishizaki, and N. Minamiura. "The amine oxidase." 1976. *Mol. Cell. Biochem.* 13:3–29.

66. Wang, T. M., F. M. Achee, and K. T. Yasunobu. 1968. *Arch. Biochem. Biophys.* 128:106–12.

67. Suya, R. H., and R. H. Abeles. 1978. *Biochemistry.* 17:3538–46.

68. Klutz, M. D., and P. G. Schmidt. 1977. *Biochemistry.* 9:3378–84.

69. Castellov, F. M., Z. W. He, and F. T. Grehavoy. 1993. *Biochem. Biophys. Acta.* 1157:162–64.

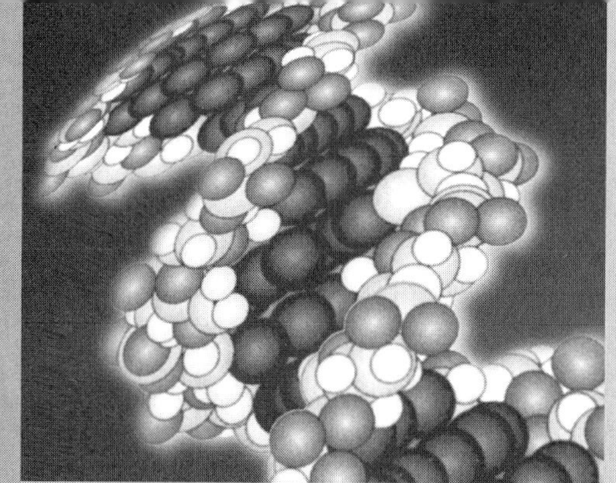



# Antibody Production

## Outline of Module

Theory of antigen—antibody interaction
   Antigenic sites—epitopes
   Classes of antibodies
Immunization

### PREPARATION OF POLYCLONAL ANTIBODY TO MONOAMINE OXIDASE (MAO) FROM HUMAN, BEEF, OR RAT LIVERS

Laboratory exercises based upon two 2–3 hour laboratories per week for which prior laboratory preparation was accomplished.

Preparation of mitochondria from human, beef, or rat livers
Extraction of MAO with Triton X-100
Ammonium sulfate fractionation
Antibody production
Purification of MAO B-polyclonal antibody complex
Immunotitrations

### ANTI-POLYPHENOL OXIDASE—POLYCLONAL ANTIBODY PRODUCTION AND USE

Laboratory exercises based upon two 2–3 hour laboratories per week for which prior laboratory preparation was accomplished.

Immunization
Detection of antibody in antiserum and quantification of antibody titer
Purification of antibody from antiserum
Antibody tagging and uses of tagged antibodies

# MODULE INTRODUCTION

An antigen is a molecule foreign to an individual. While they can be proteins, antigens can be low molecular weight compounds. When either injected or ingested, antigens elicit the production of antibodies, part of the immune response in vertebrates. Antibodies belong to a class of proteins known as immunoglobulins.

An appropriate antigen can combine with its antibody yielding an antigen/antibody complex. Because of this specificity, various analytical methodologies employed in biochemistry and/or molecular biology capitalize on ability to complex.

The two most often used antibody preparations in research are polyclonal and monoclonal. While polyclonal antibodies are those synthesized by various antibody-producing cells, monoclonal antibodies are produced by a clone (a population of identical cells). Thus, a polyclonal antibody preparation can contain a mixture of proteins, but a monoclonal preparation is homogeneous.

Because antibodies can be "tagged" with various reagents—e.g., enzymes, fluorescent labels, spin lables, and colloidal gold—they are beneficial in localizing antigens within or upon cells through the application of light and electron microscopic immunocytochemical procedures. In addition, antibodies are extensively employed in biochemistry and molecular biology—for example, in the detection of *in vitro* translation products and the fusion products of transcribing cDNA (see Module 7). Here the student will become acquainted with both monoclonal and polyclonal production procedures. Several modules in this manual have noted that monoamine oxidase exists in two forms, A and B, which differ in their substrate and inhibitor specificity, tissue distribution, and immunological characteristics. In this experiment, students will isolate a highly pure and active MAO B: antibody complex that is ideal for raising antisera and also can be used to perform all the kinetics and physicochemical studies presented in previous modules. Students will gain a highly rewarding experience regarding the use of immunoaffinity chromatography as a powerful technique for MAO purification. This immunoaffinity resin can bind to the enzyme from crude extracts. The methodology presented in this module offers a unique advantage in the use of a highly specific monoclonal antibody that binds to Protein A-Sepharose chromatography to achieve a highly pure and catalytically active MAO antibody complex.

There are two essential requirements for this methodology: first, the monoclonal antibody must bind to an epitope on the enzyme that is not involved in the catalytic site; and second, the antibody class can be eluted easily from Protein A at a mild pH. Figure 6.1 presents a summary of the procedures for monoclonal antibody[1,2,3,4] production, and Figure 6.2 shows those for polyclonal antibody generation.

# LABORATORY EXERCISES

## PREPARATION OF POLYCLONAL ANTIBODY TO MONOAMINE OXIDASE (MAO) FROM HUMAN, BEEF, OR RAT LIVERS*,**

### Supplies and Equipment

Conical centrifuge tubes
IEC-M52 centrifuge or equivalent (radius = 35 cm)
Microplates (Linbro FB 96TC)
96 wells and plates with 16 mm or 50 mm wells
Micropipettes and μl tips
Pasteur pipettes
Pipettes, 0.2 ml, 5.0 ml
60 mm Petri dish
Syringes plus 26 gauge needles

### Chemicals

0.17 M $NH_4Cl$
Complete Freund's adjuvant, Difco Laboratories, Detroit, MI
30% polyethylene glycol solution [heat sterile PEG 1000 and DMEM without serum (DMEM-SO) pH 7.4–7.6 to 41° C and mix 3 ml of PEG with 7 ml of DMEM by repeatedly drawing the mixture into a warmed pipette to assume complete mixing]
Maintain the solution at 37° C until use

---

*Antisera can be prepared commercially using sheep.
**It is highly recommended that preparation of the antisera takes place one semester or quarter prior to performing this experiment.

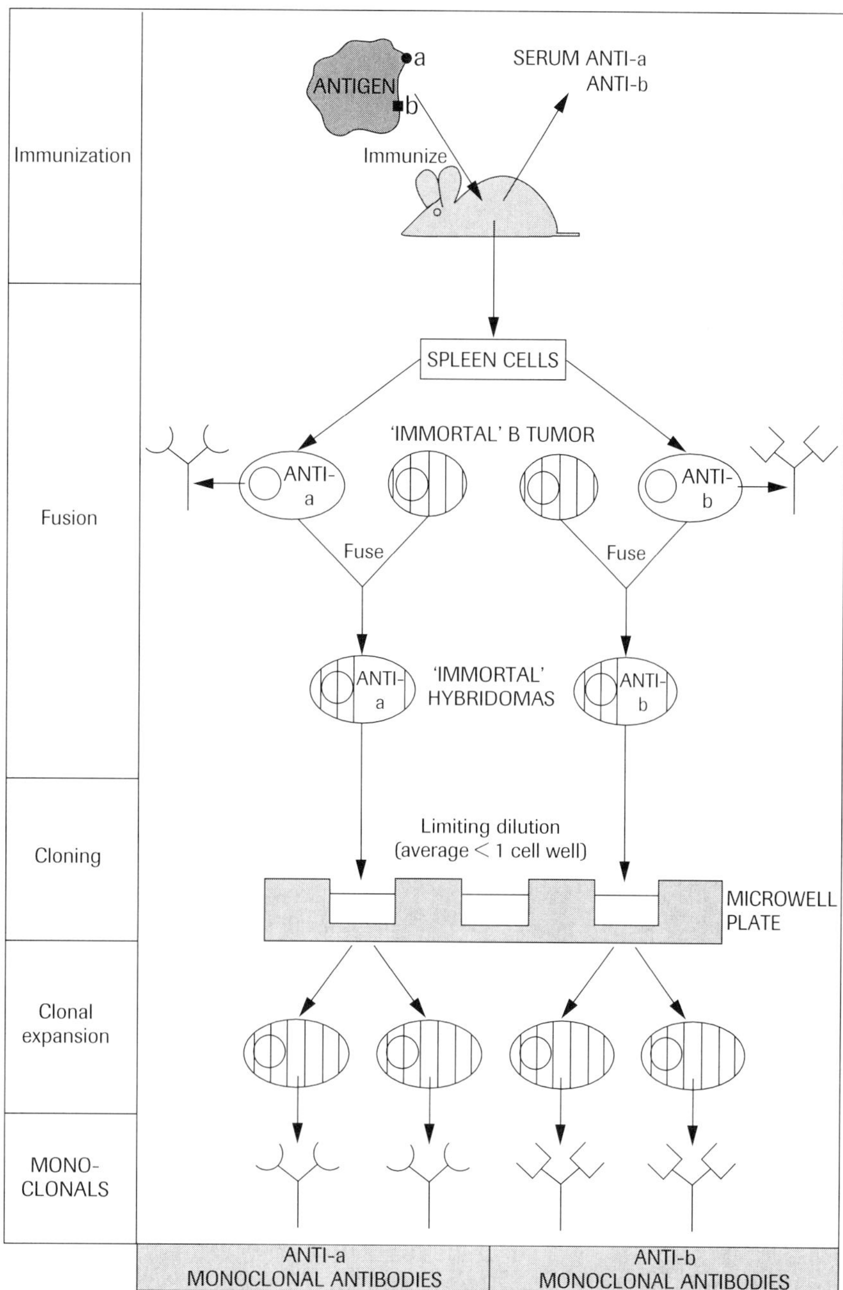


**FIGURE 6.1**  Production of monoclonal antibodies. Mice immunized with an antigen bearing (shall we say) two epitopes, a and b, develop spleen cells making anti-a and anti-b, which appear as antibodies in the serum. The spleen is removed and the individual cells fused in polyethylene glycol with constantly dividing (i.e. 'immortal') B-tumour cells selected for a purine enzyme deficiency and often for their inability to secrete Ig. The resulting cells are distributed into micro-well plates in HAT (hypoxanthine, aminopterin, thymidine) medium which kills off the perfusion partners, at such a high cell dilution that on average each well will contain less than one hybridoma cell. Each hybridoma being the fusion product of a single antibody-forming cell and a tumour cell will have the ability of the former to secrete a single species of antibody and the immortality of the latter enabling it to proliferate continuously, clonal progeny providing an unending supply of antibody with a single specificity— the monoclonal antibody. In this example, we considered the production of hybridomas with specificity for just two epitopes, but the same technique enables monoclonal antibodies to be raised against complex mixtures of multiepitopic antigens. Fusions using rat cells instead of mouse may have certain advantages in giving a higher proportion of stable hybridomas, and monoclonals which are better at fixing human complement, a useful attribute in the context of therapeutic applications to humans involving cell depletion.

Roitt, I. 1991. *Essential Immunology.* London: Blackwell Scientific Publications.

---

Immunization of New Zealand rabbits

↓

Bleed

↓

Quantify antibody by ELISA

↓

Purify antibody by immunoaffinity chromatography

---

**FIGURE 6.2**   Procedure for the production of poly-clonal antibody.

# Procedures

## Preparation of Mitochondria from Human, Beef, or Rat Livers

Human, beef, or rat mitochondria can be prepared by the method described in Module 4 or by modification of the method of Johnson and Lardy[5] as follows:

a. Wash the liver with tap water and place it on ice and then remove connective tissue and fatty materials.
b. Carry out the following steps at 4° C.
c. Mince a portion of the liver (200 g) with scissors and then suspend in homogenization medium (500 ml of 0.25 M-sucrose, 1 mM EDTA, 5 mM HEPES, pH adjusted to 7.2 with KOH) and homogenize in a Waring blender at maximum speed for 1 minute.
d. Dilute the homogenate with homogenization medium (1500 ml) and filter through two layers of Miracloth (Calbiochem-Boehringer).
e. Centrifuge the filtrate at 700 $xg$ for 15 minutes.
f. Centrifuge the supernatant obtained from step e at 8,000 $xg$ for 15 minutes.
g. Discard the supernatant obtained from step f and suspend the sediment in 1200 ml of homogenization medium using a loosely fitted glass-Teflon homogenizer, and centrifuge at 8,000 $xg$ for 15 minutes.
h. Wash the sediment again with homogenization medium, then resuspend in buffer (0.15 M KCl and 10 mM HEPES, pH 7.2), and centrifuge at 8,000 $xg$ for 15 minutes.
i. Store the sedimented mitochondrial preparation at –20° C for subsequent extraction of MAO.

## Extraction of MAO with Triton X-100

a. Thaw out the washed mitochondrial preparation at room temperature and then suspend in 10 volumes of ice-cold distilled water using a glass-Teflon homogenizer.
b. Centrifuge the mitochondrial suspension (550 ml) at 105,000 $xg$ for 15 minutes.
c. Homogenize the pellet in 100 mM potassium phosphate buffer, pH 8.0, containing 1.5% (v/v)

Triton X-100 and 0.002 M β-mercaptoethanol stir the mixture at 4° C for 1.5 hours.
d. Centrifuge the homogenate at 150,000 $xg$ for 2 hours and store the extract at –20° C

## Ammonium Sulfate Fractionation

a. Carry out ammonium sulfate fraction as detailed in purification of MAO, Module 4.
b. DEAE Column Chromatography. Carry out this step directly after step a or after calcium phosphate titration as detailed in Module 4.

## Antibody Production: Preparation of Antisera to MAO from Human, Beef, or Rat Livers

An IgG class immunoglobulin can be developed and the antibody can be purified by ammonium sulfate precipitation, DEAE column chromatography,[6] and Protein A-Sepharose column chromatography. The procedure is detailed as follows:

a. Dialyze MAO overnight against 0.01 M potassium phosphate buffer, pH 7.4, containing 0.1% Triton X-100 and 0.002 M β-mercaptoethanol, and then freeze-dry the enzyme.
b. Dissolve the freeze-dried MAO in water (12 ml).
c. Raise the antisera in rabbit by injecting 240 µg of purified enzyme with Freund's complete adjuvant on four occasions.
d. Inject MAO at six sites (four subcutaneously and two intramuscularly) on each occasion at 14-day intervals.
e. At each site, inject a total volume of 1.0 ml (0.5 ml of antigen solution and 0.5 ml adjuvant).
f. At one week after the final injection, bleed the animal by jugular cannulation.
g. Leave the blood to clot and centrifuge at 1,500 $xg$ for 30 minutes.
h. Add an equal volume of neutralized saturated $(NH_4)_2SO_4$ to the serum at 0° C and then stir for 30 minutes.
i. Collect the precipitate by centrifugation at 25,000 $xg$ for 20 minutes.
j. Wash the precipitate twice with $(NH_4)_2SO_4$ solution (291 g/l) at 0° C, dissolved in a volume of 10 mM-potassium phosphate buffer, pH 8.0, equal to one-half the initial serum volume and dialyze overnight against 5 liters of the same buffer.
k. Centrifuge the suspension at 2,500 $xg$ for 20 minutes.
l. Dialyze the supernatant overnight against 5 liters of 20 mM-sodium phosphate, pH 7.0, containing 0.15 M NaCl.
m. Heat the preparation at 70° C for 30 minutes.
n. Centrifuge at 25,000 $xg$ for 20 minutes, and store the supernatants at –15° C.

### Purification of MAO B-Polyclonal Antibody Complex*

a. Determine the activity and total protein content of MAO-B in Triton X-100 extract following the assay.

1. MAO activity can be measured spectrophotometrically by monitoring the increase in absorbance at 250 nm on the oxidation of benzylamine with formation of benzylaldhyde. The measurement should be carried out at room temperature in 0.01 M potassium phosphate buffer, pH 7.6, containing 0.1% Triton X-100, plus 0.2% potassium cholate.

2. Substrate: 0.1 M benzylamine, pH 8.2.

3. Add 50 µl of benzylamine to 1 ml of buffer solution. Mix well and then add 0.1 ml of MAO. Follow the increase in absorbance at 250 nm.

4. Determine protein concentration by Lowry et al.[8] and then determine specific activity. Read relevant material in Modules 2 and 4.

b. Add a five-fold molar excess of antibody (IgG) to enzyme to the Triton X-100 extract.

c. Shake the mixture gently for 15 hours at 4° C.

d. Equilibrate a protein A-Sepharose column (5 ml) with 100 mM potassium phosphate buffer, pH 8.0, containing 0.5% Triton X-100 (100 ml).

e. Pass the mixture from step c through the column and adjust the flow rate at 20 ml/h.

f. Wash the column with 100 ml of equilibration buffer.

g. Elute the MAO B-polyclonal antibody complex with 100 mM citrate-potassium phosphate buffer, pH 5.8, containing 0.5% Triton X-100 (100 ml).

h. Collect fractions (4 ml) in tubes containing 1 M potassium phosphate ($K_2PO_4$) (0.6 ml) and mix immediately to raise the pH to 7.0–7.5.

i. Assay each fraction for MAO B activity as described above.

j. Pool the fractions with high enzyme activity.

k. Remove Triton X-100 by passage through Bio-Beads SM-2 (Bio-Rad).

l. Dialyze the active fraction against 100 mM potassium phosphate buffer, pH 7.4, for 36–40 hours.

m. Centrifuge the MAO B-polyclonal antibody complex at 12,000 $xg$ for 10 minutes.

n. Suspend the precipitate in 10 mM potassium phosphate buffer, pH 7.4, containing 50% (v/v) glycerol and 0.1 mM β-mercaptoethanol (1.2 ml).

o. Store the MAO B-polyclonal antibody complex at –20° C.

### Immunotitrations

Immunotitrations can be carried between partially purified enzyme (ammonium sulfate fractions) with antisera that has been processed as described previously as follows:

---

*The fold purified of the enzyme should be close to 154.

a. Mix samples (0.2 ml–0.3 ml) of Triton X-100 depleted extract (after Bio-rad column) with processed antisera and adjust the volume with 0.1 M potassium phosphate, pH 7.0.

b. Keep the solution overnight and then centrifuge at 1,400 $xg$ for 6 minutes.

c. Assay for MAO activity as described in the previous experiment, step a.1.

The student is encouraged to read references 7–21 for additional details, with emphasis on 21.

## Review Question

1. What are the differences in preparation of monoclonal versus polyclonal antibodies?

## ANTI-POLYPHENOL OXIDASE— POLYCLONAL ANTIBODY PRODUCTION AND USE

### *Supplies and Equipment*

Amicon ultrafiltration—stirred cell with YM10 filters
Ammonium chloride
Assay tubes
Dialysis tubing MW cut-off 14,000
ELISA microtiter plates—96 wells
ELISA reader
Eppendorf pipettes and microtips
Fraction collector
Glass or diamond knife
Gloves
Graph paper
Kimwipes
Lyophilization flasks
Microtome—ultramicrotome
Nitrogen tank
Nickel grids
Pasteur pipettes
Pipettes
Propipettes
Ringstands and clamps
Quartz cuvettes
Syringes and needles
Test tubes

### *Chemicals*

Immunoelectron microscopy
Ammonium chloride
Dimethylformamide
Lowicryl $K_4M$
Paraformaldehyde

Potassium phosphate mono and dibasic—0.1 M,
pH 7.2, phosphate buffer
Sucrose
Immunization
Freund's complete adjuvant
Freund's incomplete adjuvant
New Zealand rabbits
Sigma Chemical Company's tyrosinase
ELISA Reagents
Pierce BupH carbonate buffer packet
Pierce BSA blocking buffer
Pierce BuPH Dulbecco's PBS
Pierce surfactant amps Tween 20
Pierce alkaline phosphatase substrate kit
Bio-Rad EIA grade affinity
Purified goat anti-rabbit IgG
Antibody purification
Pharmacia Mab trap G kit
Colloidal gold tagging
Immunopre gold
Bovine serum albumin
Pierce's Dulbecco phosphate buffered saline
pH standardization buffer
0.2 N $K_2CO_3$
PEG carbowax 20
Polycarbonate tubes
10% NaCl

### Other Needs

Dialysis
Cold room
Dialysis tubing
Magnetic stirring bar
Magnetic stirrer
Pasteur pipettes
String
Towels for tubing blotting
Two l beaker
Tagging
Clear tubes
pH
pH meter
Magnetic stirring bar
Magnetic stirrer
Protein quantification
Quartz cuvettes
UV-visible spectrophotometer

# Introduction

The preparation of polyclonal antibody to monoamine
oxidase (MAO) polyphenol oxidase (PPO) involves
immunization and bleed of rabbits, detection and
quantification of antibody within antiserum, and subsequent purification of the antibody by immunoaffinity chromatography of the antiserum.

# Procedures

## Immunization

To obtain an antibody to PPO, purchase two male
New Zealand rabbits of 1816 g each, and then accli-
mate in your institution's approved animal quarters
(comply with your vet's and animal care facilities
manager's rules and regulations regarding optimal
animal care). Subsequent to acclimatizing, employ the
immunization protocol in Figure 6.3.[8] Because the
time to obtain an antibody is lengthy (80 days), the
immunization should be performed well in advance of
the ELISA quantification of antibody titer, immunoaffinity chromatography of the antiserum, and subsequent tagging of the purified antibody with colloidal
gold, peroxidase, or fluorescein.[22]

---

Bleed two New Zealand rabbits (1 and 2)
for normal serum

**Inoculate rabbit #1**

1.5 mg *Agaricus bisporous* PPO (Sigma)
into two sites, IM

| | | |
|---|---|---|
| Dilution | 150 µl | PPO (10 mg ml$^{-1}$) |
| | 450 µl | Phosphate—buffered saline (PBS) |
| Emulsion | 600 µl | Diluted PPO |
| | 600 µl | Freund's *complete* adjuvant |
| Use 500 µl hind muscle$^{-1}$ | | |

**Inoculate rabbit #2**

2.0 mg in two sites, IM

| | | |
|---|---|---|
| Dilution | 200 µl | PPO (10 mg ml$^{-1}$) |
| | 500 µl | PBS |
| Emulsion | 700 µl | PPO |
| | 700 µl | Freund's *complete* adjuvant |
| Use 500 µl hind muscle$^{-1}$ | | |

Boost rabbits 1 and 2
Each received 500 µl total in twos

| | | |
|---|---|---|
| Dilution | 50 µl | PPO (10 mg ml$^{-1}$) |
| | 650 µl | Phosphate-buffered saline |
| Emulsion | 700 µl | Diluted PPO |
| | 700 µl | Freund's *incomplete* adjuvant |

Re-inoculate
Bleed (14 days)
Re-inoculate
Bleed (11 days)

---

**FIGURE 6.3**  Summary of immunization protocol.
Moore, N. L., L. A. Brako, C. Claussen, B. R. Jones, and W. V. Dashek.
1994. "Distribution of polyphenol oxidase in organelles of hyphae of
the wood-deteriorating fungus, *Coriolus versiocolor*." *Biotoxins,
Biodeterioration, and Biodegradation Research.* New York: Plenum Press.

## Detection of Antibody in Antiserum and Quantification of Antibody Titer

Both the detection and quantification of putative antibody within crude serum from New Zealand rabbits immunized with authentic PPO (e.g., Sigma's tyrosinase) can be accomplished by an indirect ELISA assay (enzyme-linked immunoabsorbent assay[23]), utilizing an ELISA Starter Kit (Pierce, Rockford, IL). In this technique, antigen, antiserum, and a secondary antibody are added to microtiter plate test wells.[23] The procedure for performing an indirect ELISA is presented in Figure 6.4. In this technique, antigen diluted with coating buffer (Figure 6.5) is bound to the wells of microtiter plates and incubated overnight.[24–26] After the addition of a BSA-blocker solution, antiserum diluted with PBS-tween is applied to the wells followed by the addition of goat anti-rabbit phosphatase "tagged" secondary antibody. Finally, the antigen-antibody interaction is monitored through the assay of phosphatase activity by supplying p-nitrophenyl phosphate in DEA buffer (Pierce Phosphatase Substrate Kit, Rockford, IL). A "typical" indirect ELISA microtiter plate "set-up" to detect and quantify antibody titer within antiserum derived from immunized rabbits is depicted in Table 6.1. Students should construct the antigen and antiserum dilutions depicted in Table 6.1. Note that adequate controls—i.e., wells reflecting deletion of antigen (NA), antiserum (NAS),

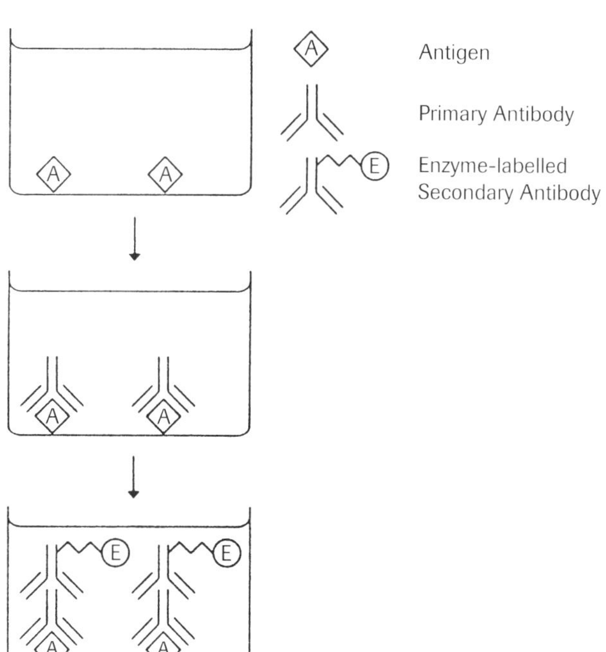


**FIGURE 6.4** Procedure for an indirect antibody ELISA.

From Pierce Chemical Co. 1990. *Pierce ELISA Starter Kit Instructions.* Rockford, IL: Pierce Chemical Co.

## Method

### Coating of antigen onto microtiter plate wells

1. Place 100 µl antigen solution into each well of a microtiter plate.
2. Incubate the plate for 1 hour at room temperature.
3. Empty plate and tap residual liquid on a paper towel.
4. Rinse each well with 3 × 100 µl wash buffer.
5. Empty plate and tap residual liquid on a paper towel.
6. Incubate plate for 1 hour at room temperature with 100 µl of blocking buffer.
7. Empty plate and tap residual liquid on a paper towel.

### Addition of primary antibody

1. Add 100 µl of primary antibody to each well.
2. Incubate the plate for 1 hour at room temperature.
3. Rinse each well with 3 × 100 µl wash buffer.
4. Empty plate and tap residual liquid on a paper towel.

### Addition of peroxidase/phosphatase-labeled secondary antibody

1. Add 100 µl of labelled secondary antibody to each well.
2. Incubate the plate for 2 hours at room temperature.
3. Rinse each well 3 × 100 µl with wash buffer.
4. Add 100 µl of wash buffer to each well and incubate for 5 minutes.
5. Empty plate and tap residual liquid on a paper towel.

### Addition of substrate solution

1. Add 100 µl of the appropriate enzyme substrate solution.
2. Incubate at room temperature for 30 minutes or until sufficient color develops.
3. Add 50 µl of the appropriate stop solution (if desired).
4. Measure absorbance of reagents at appropriate wavelength.

**FIGURE 6.5** Procedure for performance of an indirect ELISA assay.

From Pierce Chemical Co. 1990. *Pierce ELISA Starter Kit Instructions,* and *Pierce Immunopre Rabbit Anti-goat IgG (H + L) Alakaline Phosphatase Conjugated* and *Alkaline Phosphatase Labeled Antibodies.* Rockford, IL: Pierce Chemical Co.

**TABLE 6.1**  Construction of ELISA Assay Microtiter Plate

antigen (top) antiserum (bottom)

Pipet antigen          100 µl/well of each dilution

Pipet antiserum →  (right side) ; 100 µl/well

Each cell is shown as antigen (top) / antiserum (bottom).

| | 1 | 2 | 3 | 4 | 5 | 6 | 7 | 8 | 9 | 10 | 11 | 12 |
|---|---|---|---|---|---|---|---|---|---|---|---|---|
| A | 1/10 / 1/10 | 1/10 / 1/100 | 1/10 / 1/200 | 1/10 / 1/400 | 1/10 / 1/800 | 1/10 / 1/1600 | 1/10 / 1/3200 | 1/10 / 1/NAS | 1/10 / 1/NRS | 1/10 / 1/NSA | 1/10 | 1/10 |
| B | 1/100 / 1/100 | 1/100 / 1/100 | 1/100 / 1/200 | 1/100 / 1/400 | 1/100 / 1/800 | 1/100 / 1/1600 | 1/100 / 1/3200 | 1/100 / 1/NAS | 1/100 / 1/NRS | 1/100 / 1/NSA | NA / 1/10 | |
| C | 1/200 / 1/10 | 1/200 / 1/100 | 1/200 / 1/200 | 1/200 / 1/400 | 1/200 / 1/800 | 1/200 / 1/1600 | 1/200 / 1/3200 | 1/200 / 1/NAS | 1/200 / 1/NRS | 1/200 / 1/NSA | NA / 1/100 | |
| D | 1/400 / 1/10 | 1/400 / 1/100 | 1/400 / 1/200 | 1/400 / 1/400 | 1/400 / 1/800 | 1/400 / 1/1600 | 1/400 / 1/3200 | 1/400 / 1/NAS | 1/400 / 1/NRS | 1/400 / 1/NSA | NA / 1/200 | |
| E | 1/800 / 1/10 | 1/800 / 1/100 | 1/800 / 1/200 | 1/800 / 1/400 | 1/800 / 1/800 | 1/800 / 1/1600 | 1/800 / 1/3200 | 1/800 / 1/NAS | 1/800 / 1/NRS | 1/800 / 1/NSA | NA / 1/400 | |
| F | 1/1600 / 1/10 | 1/1600 / 1/100 | 1/1600 / 1/200 | 1/1600 / 1/400 | 1/1600 / 1/800 | 1/1600 / 1/1600 | 1/1600 / 1/3200 | 1/1600 / 1/NAS | 1/1600 / 1/NRS | 1/1600 / 1/NSA | NA / 1/800 | |
| G | 1/3200 / 1/10 | 1/3200 / 1/100 | 1/3200 / 1/200 | 1/3200 / 1/400 | 1/3200 / 1/800 | 1/3200 / 1/1600 | 1/3200 / 1/3200 | 1/3200 / 1/NAS | 1/3200 / 1/NRS | 1/3200 / 1/NSA | NA / 1/1600 | |
| H | | | | | | | | | | | | |

NAS = *No* antiserum  
NRS = Normal rabbit serum  
NSA = *No* secondary antibody  
NA = *No* antigen

From Moore, N. L., L. A. Brako, C. Claussen, B. R. Jones, and W. V. Dashek. 1994. "Distribution of polyphenol oxidase in organelles of hyphae of the wood-deteriorating fungus, *Coriolus versicolor.*" *Biotoxins, Biodeterioration, and Biodegradation Research.* New York: Plenum Press.

**TABLE 6.2**  Quantification of PPO Antibody in Crude New Zealand Rabbit Serum by ELISA Assay[a]

| Well Row Down Microtiter Plate | \multicolumn{12}{c}{Well Row Across Microtiter Plate} |

| Well Row Down Microtiter Plate | 1 | 2 | 3 | 4 | 5 | 6 | 7 | 8 | 9 | 10 | 11 | 12 |
|---|---|---|---|---|---|---|---|---|---|---|---|---|
| A | 0.613[b] | 0.861 | 0.864 | 0.805 | 0.691 | 0.682 | 0.438 | 0.073 | 0.109 | 0.078 | 0.084 | 0.093 |
| B | 0.530 | 0.725 | 0.764 | 0.669 | 0.553 | 0.489 | 0.378 | 0.079 | 0.099 | 0.076 | 0.118 | 0.082 |
| C | 0.515 | 0.668 | 0.656 | 0.587 | 0.549 | 0.438 | 0.357 | 0.075 | 0.099 | 0.073 | 0.088 | 0.075 |
| D | 0.484 | 0.689 | 0.606 | 0.643 | 0.491 | 0.365 | 0.296 | 0.072 | 0.084 | 0.072 | 0.090 | 0.081 |
| E | 0.459 | 0.530 | 0.516 | 0.443 | 0.376 | 0.321 | 0.222 | 0.081 | 0.099 | 0.079 | 0.089 | 0.099 |
| F | 0.396 | 0.439 | 0.404 | 0.276 | 0.266 | 0.181 | 0.151 | 0.074 | 0.091 | 0.074 | 0.087 | 0.086 |
| G | 0.333 | 0.329 | 0.260 | 0.205 | 0.151 | 0.142 | 0.123 | 0.072 | 0.090 | 0.075 | 0.081 | 0.077 |
| H | 0.073 | 0.068 | 0.072 | 0.074 | 0.069 | 0.075 | 0.068 | 0.071 | 0.087 | 0.074 | 0.075 | 0.069 |

[a]See Table 6.1 for identification of rows.  
[b]A490nm

From Moore, N. L., L. A. Brako, C. Claussen, B. R. Jones, and W. V. Dashek. 1994. "Distribution of polyphenol oxidase in organelles of hyphae of the wood-deteriorating fungus, *Coriolus versicolor.*" *Biotoxins, Biodeterioration, and Biodegradation Research.* New York: Plenum Press.

and secondary antibody—should be constructed. In addition, wells containing normal rabbit serum (NRS) rather than antiserum should be constructed. After the addition of the antigen to the wells and its incubation overnight, the procedure depicted in Figure 6.5 should be followed. To visualize the antigen/antibody interactions, p-nitrophenyl phosphate should be added to the wells. Subsequent to terminating the reaction with 3N NaOH, quantify the developed yellow color with an ELISA reader or utilize a spectrophotometer at 405 nm. Record your results as in Table 6.2.

## Purification of Antibody from Antiserum

After detection of anti-PPO and quantification of its titer by ELISA, the antibody can be purified by subject-

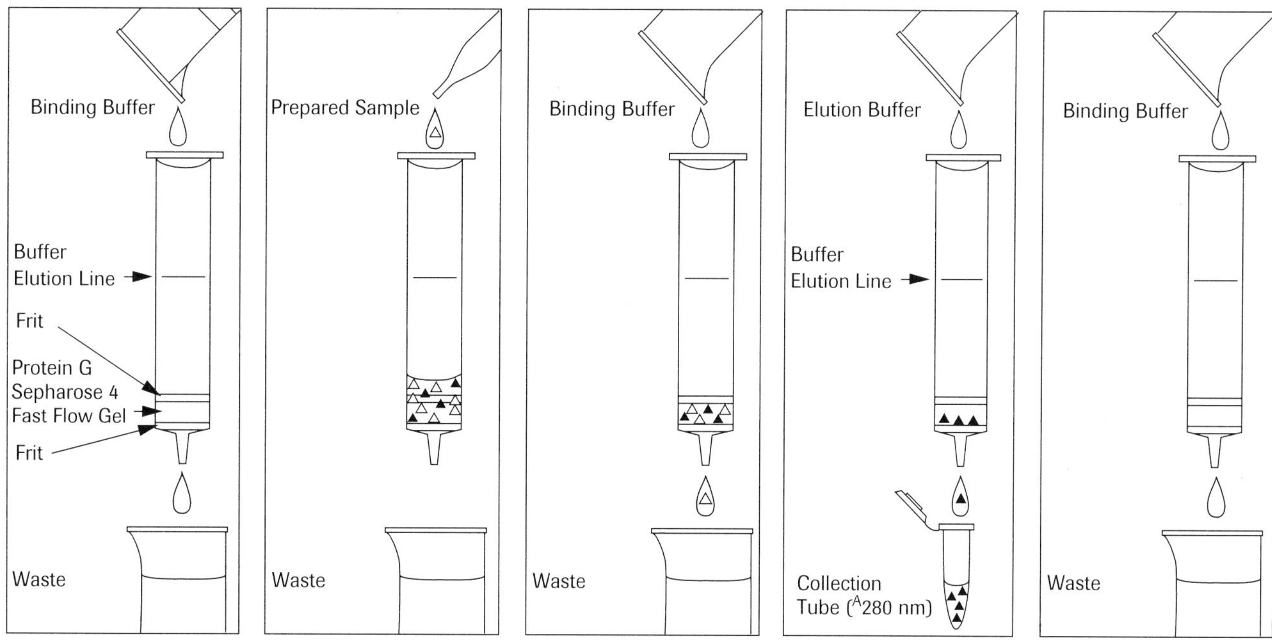


**FIGURE 6.6**   Procedure for immunoaffinity chromatographic purification of anti-polyphenol oxidase.

From Pharmacia Biotech. Sepharose is a registered trademark of Pharmacia, Piscataway, NJ, USA.

ing the antiserum from immunized rabbits to chromatography on Mab trap G (Figure 6.6). To achieve the purification of anti-PPO, dilute 2 ml of anti-serum (280 nm absorbing units) with 2 ml of binding buffer (Pharmacia Mab trap G kit), and layer the diluted antiserum onto a Mab trap G column pre-equilibrated with binding buffer (Figure 6.7). Then, elute the unbound serum proteins with binding buffer. Next add 20 ml of 1X elution buffer and desorb the bound antibody by collecting 1 ml fractions with an automated fraction collector. Quantify the proteins in the fractions by UV-spectroscopy at 280 nm. Plot $^A$280 nm on the $y$-axis and ml eluted on the $x$-axis. Compare your results to that in Figure 6.8.

## Antibody Tagging and Uses of Tagged Antibodies

To colloidal gold tag the immunoaffinity chromatographically purified anti-PPO, follow the protocol in Figure 6.9. Once the immunoaffinity purified anti-PPO has been "tagged" with colloidal gold, the tagged antibody can be employed to localize PPO within *Coriolus versicolor* fungal hyphae by immunoelectron microscopy.[27–33] To accomplish this, the student should utilize the procedure in Figure 6.10.

## Review Question

1. Design an experiment that would result in the production of a polyclonal antibody to polyphenol oxidase.

### Buffer preparation

1. Dilute the 10 × buffer concentrate **for one separation** as follows:
   a. Add 10 ml Binding Buffer concentrate to 90 ml high quality water for a total volume of 100 ml.
   b. Add 2 ml Elution Buffer concentrate to 18 ml for a total volume of 20 ml.
2. Prepare collection tubes by adding 60 to 100 μl of Neutralizing Buffer per ml of fraction to be collected. This allows for immediate renaturing of the purified sample. Neutralizing buffer should not be added once the purified fraction is collected.

### Sample preparation

The Protein G Sepharose 4 FF column allows purification of up to 5 ml ascites fluid or 25 ml cell culture supernatant. The minimum sample volume should be 250 μl.

1. For serum, ascites fluid or cell culture supernatant samples, centrifuge (10,000 $xg$ for 10 minutes) and filter (0.22 μm).
2. Dilute sample 1: with Binding Buffer to ensure proper ionic strength and pH for optimal binding. Cell culture supernatant should not be diluted with Binding Buffer.

*Cont'd*

**FIGURE 6.7**   Procedure for Mab-trap G immunoaffinity chromatography of PPO antibody.

From Pharmacia Biotech. Sepharose is a registered trademark of Pharmacia, Piscataway, NJ, USA.

## Purification protocol

1. Open the Protein G Sepharose 4 FF column by removing the top cap first. This will avoid air bubbles being drawn into the gel. Pour off the 20% ethanol storage solution.
2. Equilibrate the Protein G Sepharose 4 FF column by filling it to the top with Binding Buffer (~30 ml). Allow the column to drain. The column will stop flowing automatically as the meniscus reaches the top frit, preventing the column from drying out.
3. Apply the prepared sample to the top frit, allowing it to absorb into the gel.
4. Wash away unbound proteins by filling the column to the top with Binding Buffer (~30 ml). Allow the buffer to pass through the column, eluting unbound materials.
5. Elute the bound IgC by filling the column with Elution Buffer to the black line (~15 ml) on the column. Collect the antibody fraction into the prepared tubes. To obtain concentrated samples, collection is best done in 1 ml fractions. If collecting 1 ml fractions, collection may be easier by reducing the flow rate (dependent upon the height of the eluent above the gel bed). This can be accomplished by dispensing three separate 5 ml aliquots of elution buffer, or by placing a syringe needle onto the column tip. Fractions can be monitored by absorbance at 280 nm.

**FIGURE 6.7** Continued

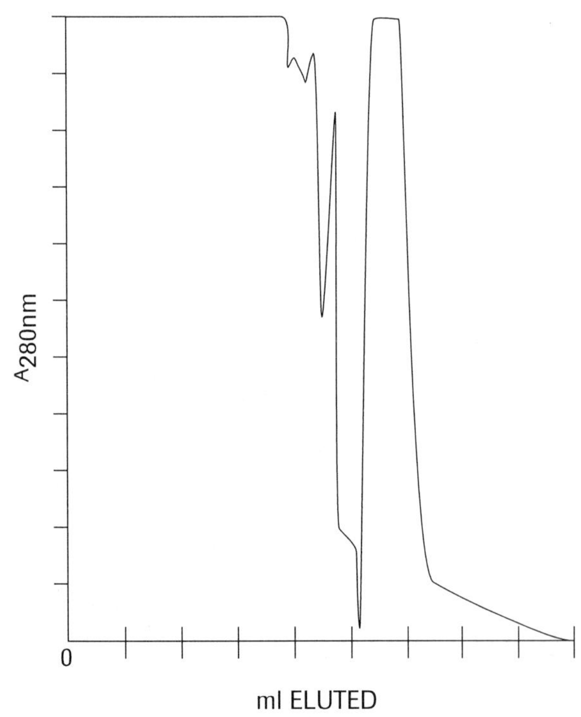

**FIGURE 6.8** Elution of anti-PPO from Mab trap G.

From Moore et al. 1994. *Biotoxins, Biodegradation and Biodeterioration Res 4.* Llewellyn, G.C., C.E. O'Rear and W.V. Dashek (eds.). New York: Plenum Press.

## Adsorption Isotherms and Colloidal Gold Procedures

a. General procedures:
   Dialyze antibody against 2 liters of phosphate buffer prepared as follows:
   2.16 g $Na_2HPO_4$–$7H_2O$ or 1.12 g $Na_2HPO_4$ anhydrous
   0.266 g $NaH_2PO_4$
   2 liters $H_2O$, pH 7.4
   Dialyze at 4°C overnight
   Adjust pH of gold to 7.4 using 0.2N $K_2CO_3$

b. Isotherm:
   *Add gold to the antibody
   0.1 ml antibody (Start with a low concentrate of Ab)
   +0.5 ml gold sol
   Wait 2 minutes
   + 0.1 ml 10% NaCl
   Wait 5 minutes
   Blue = Unstable
   Pink = Stable
   Use the concentration of antibody to stabilize a known amount of gold in the last pink tube in series of concentrations.

c. Conjugation:
   1. Add protein to tube.
   2. Add $H_2O$ (10% of volume), spin 2800 rpm, 25 minutes, 4° C, to get rid of clumps.
   3. Add gold to tube (enough to stabilize the antibody).
   4. Invert 6 times and let sit 5 minutes.
   5. Add 1% PEG (Carbowax 20) and invert 6 times.
   6. Spin at 10K for 30 minutes at 4° C.
   7. Aspirate as much supernatant as possible.
   8. Resuspend pellet in 1 ml 0.22 µm filtered buffer, 20 mM Tris buffered saline, pH 8.2 + 0.1% BSA, and 0.05% Na Azide.
   9. Standardize $O.D._{.525}$ for $Au_{18}$
      $O.D._{.540}$ for $Au_5$
   Commercial preps-O.D. at 520 nm is 3.5 for 15 nm and 7.0 for 30 nm.

   NOTE: Use only acid-washed glassware, glass-distilled water, and 0.22 µm filtered reagents.

   Store at 4–8° C. DO NOT FREEZE!

**FIGURE 6.9** Protocol for colloidal gold tagging of PPO antibody.

From Clausen, C. Forest Products Laboratory, Madison, WI.

a. Embedding in Lowicryl
  1. Fix tissue in 1%–4% formaldehyde (freshly prepared from paraformaldehyde) or 1% paraformaldehyde in 0.01% glutaraldehyde in 0.1 M phosphate buffer, 2 hours at 0–4° C.
  2. Wash in several changes of 0.5 M ammonium chloride or 0.1 M phosphate plus 3.5% sucrose.
  3. Dehydrate in a graded series of dimethylformamide (DMF) in $dH_2O$:
     50% DMF: 10 minutes
     75% DMF: 10 minutes
     90% DMF: 10 minutes
  4. Infiltrate tissue in:
     1 part Lowicryl, 2 parts DMF: 10 minutes
     1 part Lowicryl, 1 part DMF: 10 minutes
     100% Lowicryl: 20 minutes
     100% Lowicryl: 25 minutes

Dehydration/infiltration should take place on a rotating platform at room temperature.

To prepare Lowicryl (Polysciences) (done at room temperature):
     2 g Crosslinker "A"
     13 g Monomer "B"
     Add 75 mg Initiator "C," and gently mix with a stir paddle
     Makes approximately 20 mls.

  5. Embedding. Place tissue in GELATIN capsules that are approximately 1/2 full with fresh Lowicryl. Fill until brimming with additional Lowicryl and cover with capsule top. Blocks can be labeled after hardening—do not include labels now, as the paper may affect polymerization.
  6. Polymerization. Select a cardboard box large enough to hold the UV lamp and apparatus described below, and line it on all six sides (including the top and bottom) with aluminum foil, shiny side facing the inside of the box. (Our box measures 80 cm long, 45 cm tall, and 27 cm wide). Cut a hole at the bottom of one of the sides of the box for the lamp cord and place the UV lamp (25 Watts, General Electric #F25T8/BL single 18-inch tube) inside. Place filled Lowicryl capsules in a flexible plastic ELISA plate where the wells have been cut out so that the base of the capsule protrudes beneath the bottom of the plate. The ELISA plate can then be attached to a clamp on a ring stand.

Position the bottom of the capsules directly above the lamp with a lamp-to-tissue distance of exactly 10 cm. Place the box in a cold room. Close the top of the box and turn on the UV lamp. Polymerize for 2 hours at 4° C.

b. Sectioning. Lowicryl blocks can be trimmed and sectioned using conventional procedures except that the meniscus in the knife boat should be as low as possible to minimize wetting of the block face. The first few thin sections cut should not be used for labeling. For two-sided labeling, pick up sections on uncoated 400 mesh nickel grids.

c. Immunolabeling. All solutions, except gold, should be filtered through a 0.22 μm cellulose acetate (Corning) filter.
  1. Treat grids with attached sections with 1 M ammonium chloride in PBS, 1 hour, room temperature.
  2. Rinse in a gentle stream of PBS.
  3. Treat with 0.1% BSA in PBS, 1 1/2–2 hours, room temperature.
  4. Rinse by passage through 2–3 drops of BSA in PBS, 30 sec/drop.
  5. Wash with a gentle stream of PBS.
  6. Incubate sections with primary antibody diluted with 0.1% BSA in PBS (we use affinity purified IgC at a final concentration of 15 μg/ml), OVERNIGHT–24 hours at 4° C.
  7. Wash thoroughly with PBS.
  8. Incubate with appropriate colloidal gold-antibody conjugate (10 nm, Janssen Pharmaceutica, Beerse, Belgium) diluted 1:3 with PBS, 2–3 hours, room temperature.
  9. Wash thoroughly with PBS.
  10. Wash thoroughly with $dH_2O$.
  11. Stain with 1% uranyl acetate in $dH_2O$, 10–30 seconds, wash and stain with lead citrate for 30 seconds. Wash with $dH_2O$.

**FIGURE 6.10**   Procedure for post-embedding immunolabeling in Lowicryl K4M.

From Abrahamsom, D. R. 1986. "Post embedding gold immunolocalization of laminin to the lamina rara interna, lamina densa and lamina rara externa of renal glomerular basement membranes." *J. Histochem. Cytochem.* 34:847–53; and Altman, L. G., B. G. Schneider, and D. S. Papermaster. 1984. "Rapid embedding of tissue in lowicryl K4M for immunoelectron microscopy." *J. Histochem. Cytochem.* 34:847–53.

# REFERENCES

1. IR International Symposium on Biotechnology: Monoclonal Antibodies in the Treatment of Human Disease. 1987. "Human monoclonal antibodies: Current technologies and future perspectives." Oxford: IRL Press.
2. Denney, R. M., N. T. Patel, R. R. Fritz, and C. W. Abeli. 1982. *Mol. Pharmacol.* 22:500–08.
3. Eye, P. L., S. J. Prowsw, and C. R. Jenkin. 1978. *Immunochem.* 15:429–36.
4. Kennett, R. H. 1980. *Monoclonal Antibodies: Hybridomas, a New Dimension in Biological Analyses.* New York: Plenum Press.
5. Johnson, D., and H. Lardy. 1967. *Methods in Enzymol.* 10:94–96.
6. Liddell, J. E., and A. Cryer. 1991. *A Practical Guide to Monoclonal Antibodies.* New York: John Wiley and Sons.
7. Libert, M. A. 1990. *Monoclonal Antibodies.* New York.
8. Lowry, O. H., N. J. Rosenbrough, A. L. Farr, and R. I. Randall. 1951. *J. Biol. Chem.* 193:265–75.
9. McCullough, K. C., and R. E. Spier. 1990. *Monoclonal Antibodies in Biotechnology Theoretical and Practical Aspects.* New York: Cambridge University Press.
10. Parham, P., M. J. Androlewicz, F. M. Brodosky, N. J. Holmes, and J. P. Ways. 1982. *J. Immnol. Methods.* 53:133–73.
11. Roitt, I. 1991. *Essential Immunology.* London: Blackwell Scientific Publications.
12. Salvaggio, J. E. 1985. *Immunology and Monoclonal Antibodies.* Mount Kisco, NY: Futura.
13. Seaver, S. S. 1987. *Commercial Production of Monoclonal Antibodies: A Guide for Scale-Up.* New York: Marcel Dekker, Inc.
14. Soldano-Ferrone, M. P. 1985. *Handbook of Monoclonal Antibodies: Applications in Biology and Medicine.* Park Ridge, NJ: Noyes Publications.
15. Zola, H. 1987. *Monoclonal Antibodies: A Manual of Techniques.* Boca Raton, FL: CRC Press.
16. Dennick, R. G., and R. J. Mayer. 1977. "Purification and immunochemical characterization of monamine oxidase from rat and human liver." *Biochemical J.* 161:167–74.
17. Denny, R. M., R. R. Fitz, N. T. Patel, and C. W. Abell. 1982. "Human liver MAO-A and MAO-B separated by immunoaffinity chromatography with MAO-B specific antibody." *Science.* 215:1400–03.
18. Kennett, R. H., K. A. Dennis, A. S. Tung, and M. R. Klinmar. 1978. "Hybrid plasmocytoma production fusions with adult spleen cells." *Curr. Topics Microbiol. Immunology.* 81:77.
19. Kohler, G., and C. Milstein. 1975. "Continuous cultures of fused cells secreting antibody of predefined specificity." *Nature.* 256:496.
20. Kohler, G., and C. Milstein. 1976. "Derivation of specific antibody-producing tissue culture and tumor lines by cell fusion." *Eur. J. Immunol.* 6:511–19.
21. Patel, N. T., R. R. Fritz, and C. W. Abell. 1984. "Isolation of pure, catalytically active human liver monoamine oxidase B: Antibody complex." *Biochem. Biophy. Res. Commun.* 125:748–54.
22. Knox, R. B., and A. Clarke. 1978. "Localization of proteins and glycoproteins by binding to labeled antibodies and lectins." *Electron Microscopy and Cytochemistry of Plant Cells.* Amsterdam: Elsevier. 150–85.
23. Voller, A., D. Bidvell, and A. Ballett. 1976. "Microplate enzyme immunoassay for the immunodiagnosis of virus infections." *Manual of Clinical Immunology.* Alexandria, VA: Dynatech Laboratories, Inc.
24. Pierce Chemical Co. 1990. *Pierce ELISA Starter Kit Instructions.* Rockford, IL: Pierce Chemical Co.
25. Pierce Chemical Co. 1990. *Pierce Immunopre Rabbit Anti-goat IgG (H + L) Alkaline Phosphatase Conjugated.* Rockford, IL: Pierce Chemical Co.
26. Pierce Chemical Co. 1990. *Pierce Immunopre Alkaline Phosphatase Labeled Antibodies.* Rockford, IL: Pierce Chemical Co.
27. Abrahamsom, D. R. 1986. "Post embedding gold immunolocalization of laminin to the lamina rara interna, lamina densa and lamina rara externa of renal glomerular basement membranes." *J. Histochem. Cytochem.* 34:847–53.
28. Altman, L. G., B. G. Schneider, and D. S. Papermaster. 1984. "Rapid embedding of tissue in lowicryl K4M for immunoelectron microscopy." *J. Histochem. Cytochem.* 34:847–53.
29. Bozolla, J. J. and L. J. Russell. 1991. *Electron microscopy: Principles and Techniques for Biologists.* Boston: Jones and Bartlett Publishers.
30. Day, A. W., R. B. Gardner, R. Smith, A. M. Svircen, and W. E. McKeen. 1986. "Detection of fungal fumbriae by Protein A-gold immunocytochemical labeling in host plants infected with *Ustilago herflori* or *Permspora hyosayani* f. sp. talacina." *Can. J. Bot.* 32:577–84.
31. Flurkey, W. H. 1986. "Polyphenoloxidase in higher plants: immunological detection and analysis of *in vitro* translation products." *Plant Physiol.* 86:614–18.
32. Green, F., C. A. Clausen, M. J. Larsen, and T. L. Highley. 1991. "Ultrastructural characterization of the hyphal sheath of *Postia placenta* by selective removal and immunogold labeling." *Biodeterioration and Biodegradation.* 8:530–32.
33. Moore, N. L., L. A. Brako, C. Claussen, B. R. Jones, and W. V. Dashek. 1994. "Distribution of polyphenol oxidase in organelles of hyphae of the wood-deteriorating fungus, *Coriolus versicolor.*" *Biotoxins, Biodeterioration, and Biodegradation Research.* New York: Plenum Press.

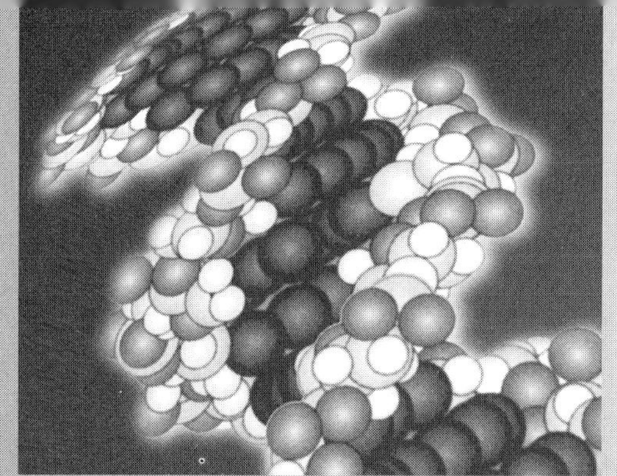



# Cloning Protocols

## Outline of Module

### DNA Chemistry

Monosaccharide
Phosphate
Nitrogenous bases
   Purines
   Pyrimidines
Nucleosides
Nucleotides
   Phosphodiester bonds
Polynucleotides
Duplex structure
   Supercoiling

### DNA Synthesis

Semi-conservative replication
Directionality
Role of polymerases
Role of other enzymes and protein factors
   DNA replisome system
   Replisome
   Helicases
   Topisomerases and Supercoiling
   DNA-binding proteins
   Primers
   DNA ligases

# Recombinant DNA

Genetic recombination
Reciprocal exchange of genetic homology
  Sequence homology
Non-reciprocal exchange of genetic material
  Little or no sequence homology
Techniques
  Genetics of DNA fragments appropriate for cloning
  Linkage of fragments to be cloned to a vector
  Plasmid or viral genomic via treatment of the DNA fragment/vector mixture with DNA ligase—
    review of plasmids, phage, cosmids and expression vectors
  Introduction of the DNA ligation products into a suitable host cell system by transformation,
    transfection, or infection
  Detection of transformed clones containing hybrid DNA species


# cDNA Cloning of Mono Amine Oxidase

Isolation of total RNA from liver
  *Duration:* One laboratory period.

Isolation of liver poly (A)+ RNA
  *Duration:* One laboratory period.

  Construction of cDNA libraries
Screening of cDNA clones
  *Duration:* One laboratory period.

Isolation of phage DNA
  *Duration:* One laboratory period.

Subcloning
  *Duration:* One laboratory period.

Mini-preparation of plasmid DNA
  *Duration:* One laboratory period.

DNA sequencing
  *Duration:* One laboratory period.

  Sequencing reaction
  Electrophoresis on 6% sequencing gel


# cDNA cloning of Polyphenol Oxidase

Chloroplast preparation
  *Duration:* One laboratory period, then maintain on ice.

Isolation of RNA
  *Duration:* One laboratory period for bulk RNA

Construction of cDNA libraries
  *Duration:* Week long exercises with incubations as long as overnight (see UNI-ZAP and Pico Blue
    kits' instructions for careful planning.

DNA sequencing
  *Duration:* One laboratory period and one day for sequencing gel

  DNA sequencing gels
  Migration of marker dyes
  Autoradiography of sequencing gels

Oligonucleotide synthesis and protein sequencing
   *Duration:*   One laboratory period

DNA language

| | |
|---|---|
| Chromosomes | Semiconservative replication |
| Genes | Replication forks |
| Plasmids | Denaturation mapping |
| Recombinant DNAs | Okazaki fragments |
| Phenotype | DNA polymerase I, II, and III |
| Mutations | Klenow fragment |
| Structural gene | Template |
| Regulatory sequences | Primers |
| Satellite DNA | Proofreading |
| Centromere | DNA replicase system |
| | Replisome |
| Introns | Helicases |
| Exons | |
| DNA supercoiling | DNA-binding proteins |
| Linking number | Primases |
| Topoisomers | DNA ligases |
| Topoisomerases | Primosome |
| | Mutations |
| Nucleoid | Silent mutation |
| Nucleases | DNA glycosylases |
|   Endonucleases | AP sites |
|   Exonucleases | Error-prone repair |

SOS response
Homologous genetic recombination
Site-specific recombination
DNA transposition
Holliday intermediate
Heteroduplex DNA
Branch migration
Transposons
Insertion sequences
Complex transposons

# MODULE INTRODUCTION

## DNA Chemistry

Nucleic acids (DNA and RNA) are composed of covalently linked nucleotides that consist of nitrogenous bases (purine and pyrimidine), a pentose (deoxyribose or ribose), and a phosphate group. The bases are attached to deoxyribose in DNA or ribose in RNA through carbon 1 of the monosaccharide. In contrast, the phosphate groups are linked to the monosaccharide at the C-5[1] and C-3[1] positions (Figure 7.1).

The purine bases of both DNA and RNA are always adenine (A) and guanine (G) (Figure 7.2). The pyrimidines in DNA are thymine (T) and cytosine (C)—in RNA—U replaces T.

Although DNA is a double-stranded molecule (a duplex) with the two strands in an antiparallel arrangement, it can be single stranded in some phages and viruses. The integrity of the strands is maintained via repeating 3[1] 5[1] phosphodiester linkages constituting the covalent backbone of the macromolecules with the purines and pyrimidines composing the side chains (Figure 7.3).

Although the base composition of DNA is species-dependent, a base on one strand pairs with a base on the other strand—e.g., A with T and G with C through hydrogen bonding (Figure 7.4).

While the most common conformation of the linear duplex DNA is β-DNA, a right-handed helical structure (Figure 7.5), there are other conformations possible under different conditions. These possibilities are dependent upon nucleotide sequences and degree of hydration as well as interactions with proteins. In addition, circular DNA (Figure 7.6) molecules exist that are topologically confined so that these ends are not capable of rotating but can form supercoils that are either right- (negative) or left- (positive) handed. Whereas the former is characteristic of prokaryotes, eukaryotes form a coiled coil DNA possessing equal weight of histones, basic proteins. These can occur as an octamer, i.e., four histones in pair form about the DNA duplex in a left-handed helix. The octamer is known as a nucleosome, which contains approximately 140 base pairs in a nuclease-resistant nucleosome.

Finally, duplex DNA can be denatured yielding single strands and renatured through regulating temperature. The above discussion is a modification of the summary concluding chapter 25 of Zubay[1] to which the student is referred for thorough coverage of DNA chemistry.

## DNA Synthesis

While there are many contemporary reviews of DNA synthesis, the student is referred to chapter 26 of Zubay[1] for thorough coverage of the topic. Here, we abbreviate the topic by paraphrasing selected information within Zubay's summary to the chapter.

The process of gene cloning utilizes the following steps:

a. The restriction-endonuclease generation of DNA fragments appropriate for cloning.
b. Linkage of fragments to be cloned to a vector plasmid or viral genome via treatment of the DNA fragment/vector mixture with DNA ligase.
c. Introduction of the DNA ligation products into a suitable host cell system by transformation, transfection or infection.
d. Detection of transformed clones containing hybrid DNA species.
e. Isolation and characterization of the hybrid species in those clones.

The student is encouraged to consult the various reviews, monographs, and laboratory protocols at the conclusion of this module for both the theory and technical aspects of cloning. Recently, Broda et al.[2] published a monograph regarding the organization and regulation of the eukaryotic genome. Treatises that specifically concern themselves with the plant genome include those of Ciferri and Dure,[3] Vloten-Doting et al.,[4] Beckman and Osborn,[5] and Anderson and Beardell.[6]

DNA replication is semi-conservative occurring via synthesis of a new strand from each of the pre-existing strands (Figure 7.7). In prokaryotes, DNA synthesis proceeds in a biodirectional fashion for an initiation (primer) site upon the *E. coli* chromosome. Although biodirectional synthesis appears to be the rule in *E. coli*, DNA replication for other prokaryotic chromosomes can be unidirectional (Figure 7.8).

As for DNA replication in eukaryotes, replication can begin at several points along the chromosomes' length. This replication proceeds bidirectionally at each initiation site with termination sites interspersed between the initiation loci. There are two strands in the DNA duplex, i.e., leading and lagging strands. Synthesis occurs continuously on the leading strand as synthesis proceeds from the 5[1] to 3[1] direction by DNA polymerase, and the two strands composing the

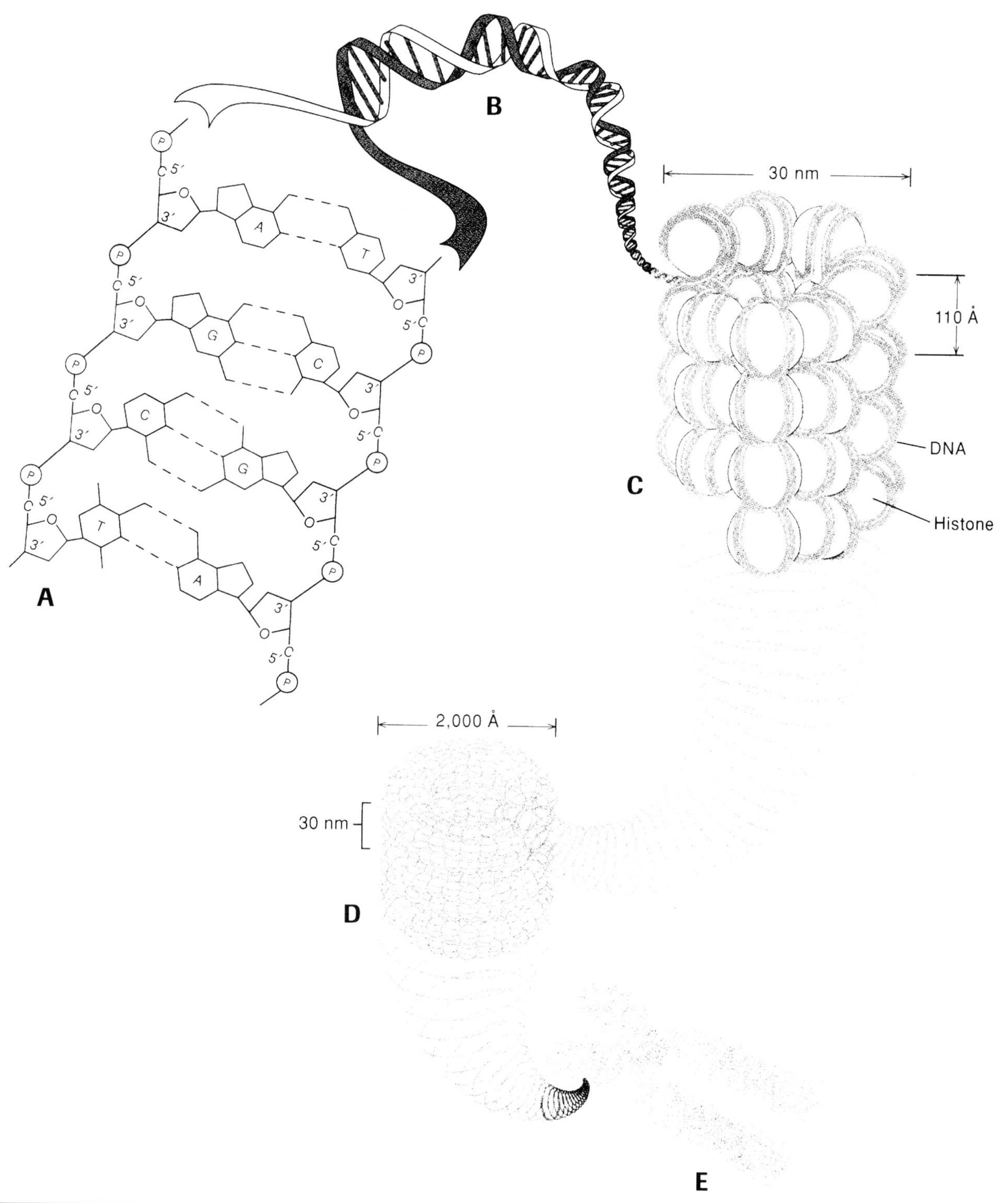



**FIGURE 7.1**  DNA is a long linear polymer of nucleotides. The nucleotides are very similar but they differ in several minor respects. All nucleotides are composed of a purine or a pyrimidine linked to a deoxyribose on one side. On the other side the deoxyribose is attached to a phosphate group. There are two purines, adenine and guanine, and two pyrimidines, cytosine and thymine. The polymeric structure is held together by phosphodiester linkages formed between the 5′ phosphate group of one nucleotide and the 3′OH group of the adjacent nucleotide. This gives the polynucleotide strands a directional sense (5′ to 3′). Two chains come together to form a duplex structure with hydrogen bonds formed between purines and pyrimidines on opposing chains, **(A).** The two-chain structure twists to form a right-handed duplex structure, **(B).** In eukaryotes the DNA wraps around a histone protein octamer, forming a hollow core, **(C).** This structure forms a 200-nm-wide (2,000-Å) fiber by further coiling, **(D).** The chromosome is a coiled-coil structure formed from very long 200-nm filaments that extend over the entire length of the chromosome, **(E).**

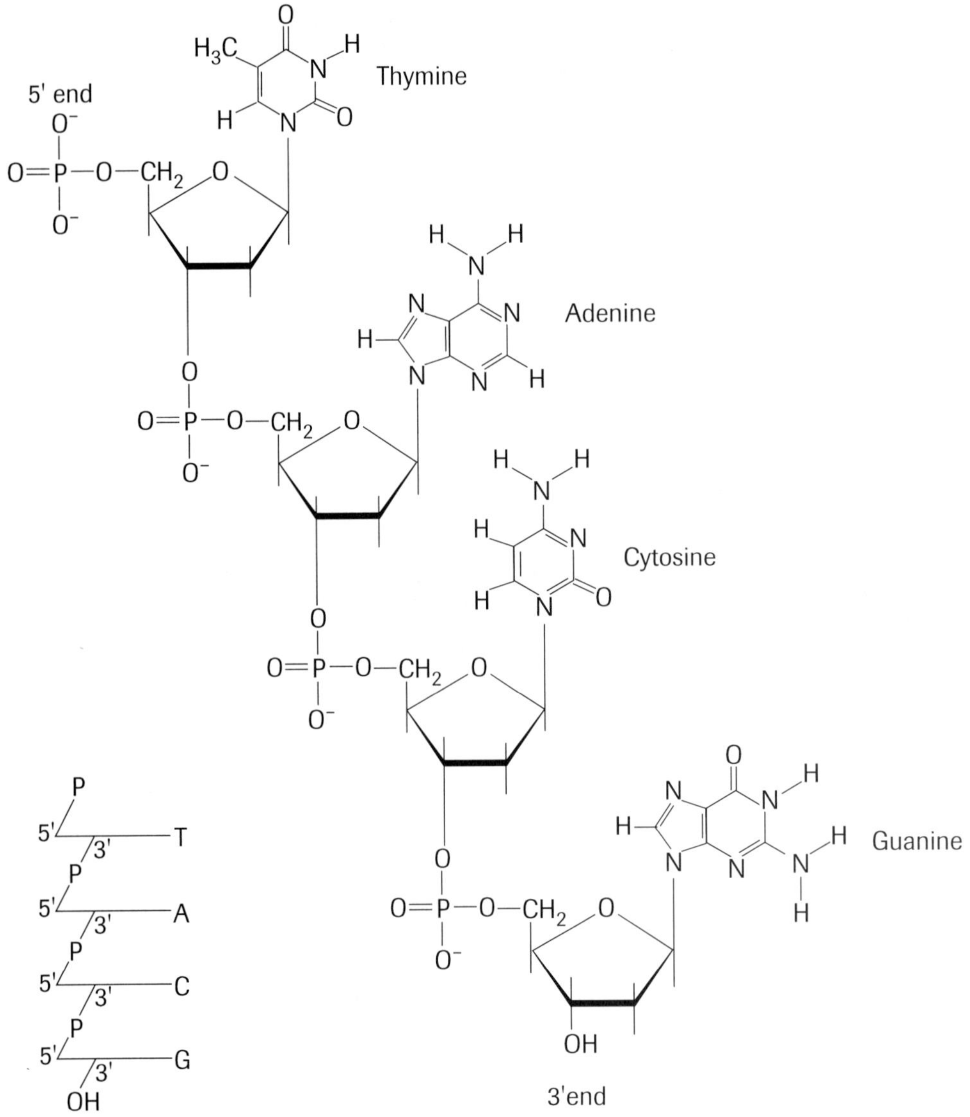

FIGURE 7.2   This structure of a deoxyribonucleotide. Drawn in abbreviated form at lower left, the illustrated structure is written pTpApCpG.



DNA duplex are oriented in opposite directions. In contrast, the lagging strand must unwind and this synthesis on this strand does not occur continuously, but sporadically, instead yielding Okazaki fragments. Synthesis of the lagging strand involves a primasome protein complex including a primase. The fidelity of DNA replication is sustained by: polymerase mediated base selection, $3^1 \to 5^1$ proofreading exonuclease, and a repair system.

DNA synthesis and chromosomal replication require a host of proteins most of which are enzymes (Table 7.1). For example, polymerases (Figure 7.9) catalyze the addition of mononucleotides), helicases (separate the strands of the DNA duplex), type I and II topoisomerase (Figure 7.10, allow rotation and supercoiling) and a DNA ligase (Figure 7.11, a sealing enzyme). In addition, there are enzymes that synthesize RNA primers. Also, there are DNA-binding proteins that stabilize single-stranded regions that are temporarily formed during replication. Lastly, there are enzymes, e.g., recombinases concerned with DNA repair, degradation, and recombination. Refer to chapter 26 of Zubay[1] for an in-depth view of prokaryotic and eukaryotic DNA synthesis as well as replication of single-stranded DNA viruses, animal viruses, adeno-retroviruses, and hepatitis B virus.

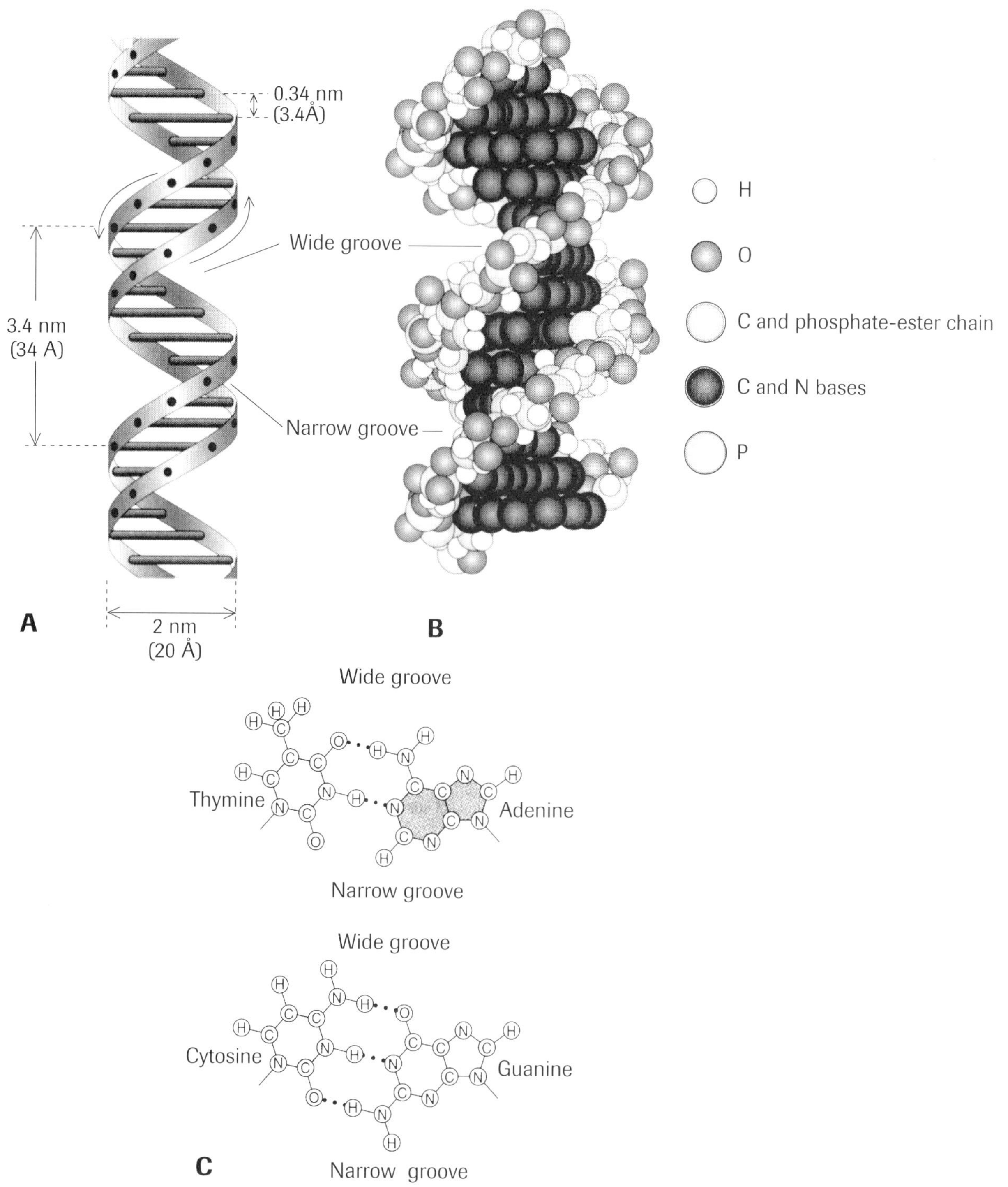



**FIGURE 7.3** The most common form of the double-helix DNA. Base-pairing forms the helix structure shown in (**A**) and (**B**) by a right-handed twist. The two strands are antiparallel as indicated by the curved arrows in (**A**). In (**B**) a space-filling model depicts the sugar-phosphate backbones as strings of mostly gray, red, white, and yellow spheres, while the base pairs are rendered as horizontal flat plates composed of dark blue spheres, In (**C**) the orientation of the groups in the base pairs with respect to the wide and narrow grooves is indicated. The wide and narrow grooves formed in the right-handed duplex structure are sometimes referred to as the major and minor grooves.

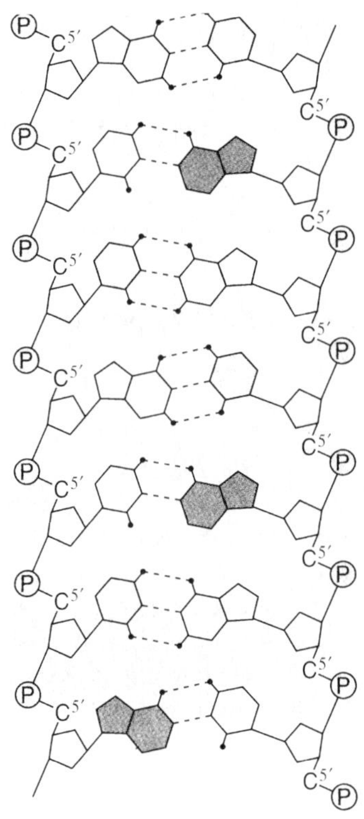

**FIGURE 7.4**   Segment of DNA, drawn to emphasize the hydrogen bonds formed between opposing chains. Each type of base is represented by a different color, with the sugar-phosphate backbones in black. Note the three hydrogen bonds in the G-C pairs and the two in the A-T pairs. (A, red; T, green; G, yellow; C, blue). The two strands are antiparallel: one strand (left side) runs 5' to 3' from top to bottom and the other strand (right side) runs 5' to 3' from bottom to top. The planes of the base pairs are turned 90° to show the hydrogen bonds between the base pairs.

From Zubay, Geoffrey. *Biochemistry,* 3rd edition. Copyright © 1993. Wm. C. Brown Communications, Inc., Dubuque, Iowa. All rights reserved. Reprinted by permission.

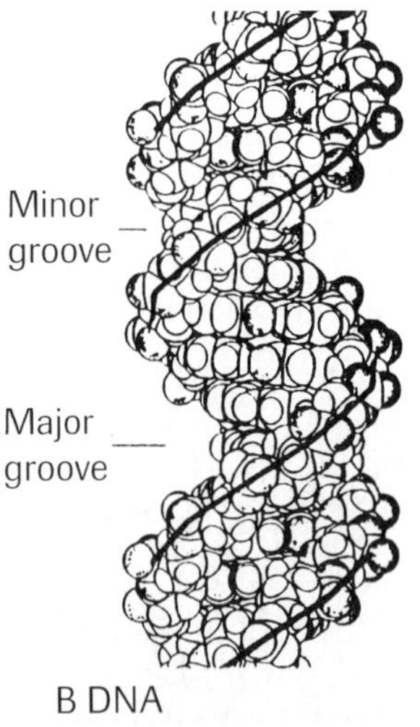


**FIGURE 7.5**   Space-filling models of B DNA.

From Zubay, Geoffrey. *Biochemistry,* 3rd edition. Copyright © 1993. Wm. C. Brown Communications, Inc., Dubuque, Iowa. All rights reserved. Reprinted by permission.

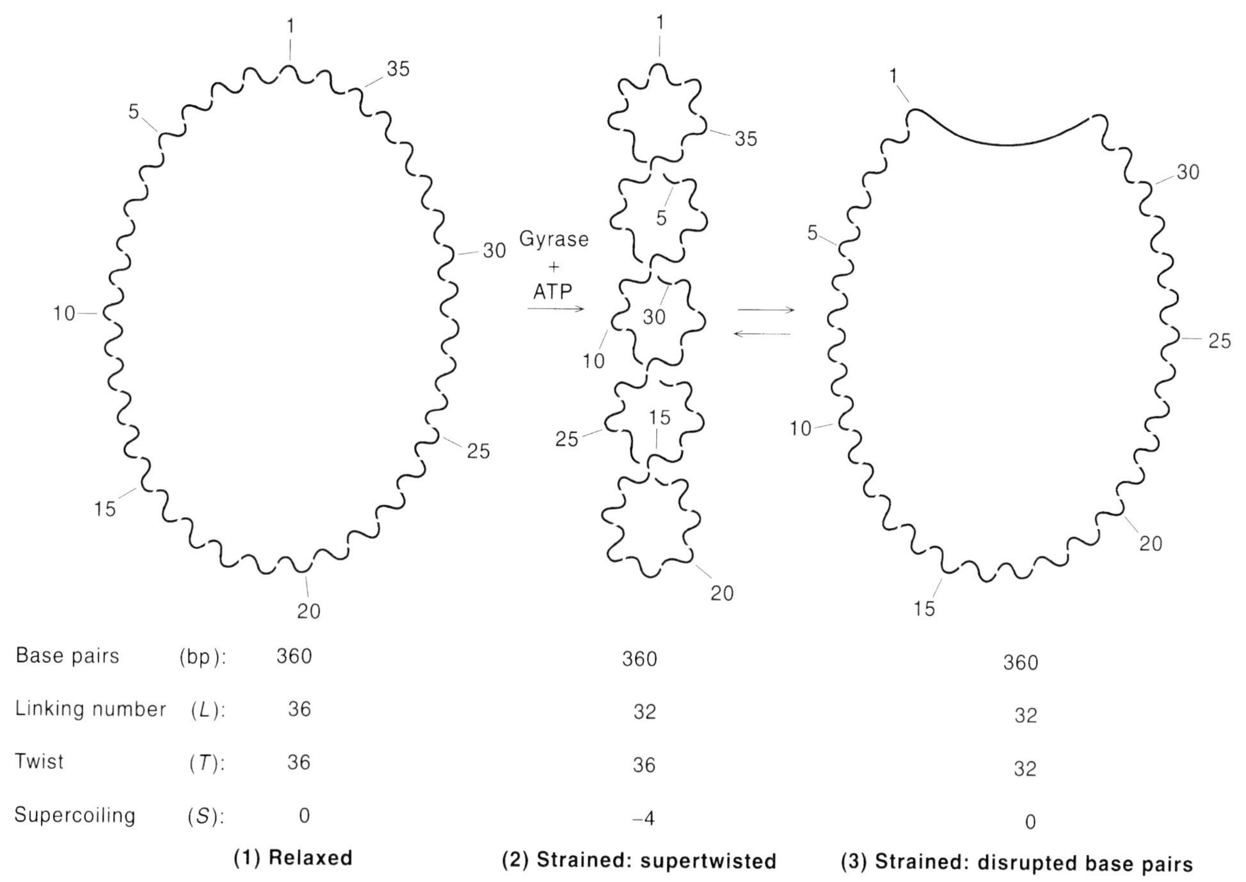



| Base pairs | (bp): | 360 | 360 | 360 |
|---|---|---|---|---|
| Linking number | (L): | 36 | 32 | 32 |
| Twist | (T): | 36 | 36 | 32 |
| Supercoiling | (S): | 0 | −4 | 0 |
| | | **(1) Relaxed** | **(2) Strained: supertwisted** | **(3) Strained: disrupted base pairs** |

**FIGURE 7.6**   A circular duplex molecule in different topological states. The linking number can be changed only by breakage and re-formation of the phosphodiester linkages, as shown in the conversion of the relaxed circular from (1) to the strained negatively supercoiled form (2). The strain in the negatively supercoiled form can be partitioned in different ways between twist ($T$) and supercoiling ($S$) as shown in the interconversion between (2) and (3). No phosphodiester linkages are broken in making this interconversion and consequently there is no change in the linking member ($L$) as indicated.



## Recombinant DNA

Exchange of genetic information between different chromosomes involves genetic recombination. There are two main types of recombination: reciprocal and nonreciprocal exchange. Whereas reciprocal recombination involves exchange of genetic material requiring sequence homology, nonreciprocal recombination centers about insertion of DNA segments, e.g., insertion segments or transposons at chromosomal sites exhibiting little or no sequence homology. Nonreciprocal recombination usually occurs with the insertion of DNA segments emanating within the same organism in which the recombination takes place as originating from a parasite of that organism.

Genetic recombination consists of the breakage and joining of DNA molecules. With the advent of gene cloning procedures, it is possible to selectively "cut" DNA derived from diverse prokaryotes and eukaryotes and to incorporate the fragments into self-replicating genetic elements known as cloning vectors (plasmid or viral genomes).[7,8] The resultant hybrid molecules can be introduced into a suitable host that can be propagated. Extensive coverage of genetic recombination can be found in Wilson[9] and Kucherlapati et al.[10]

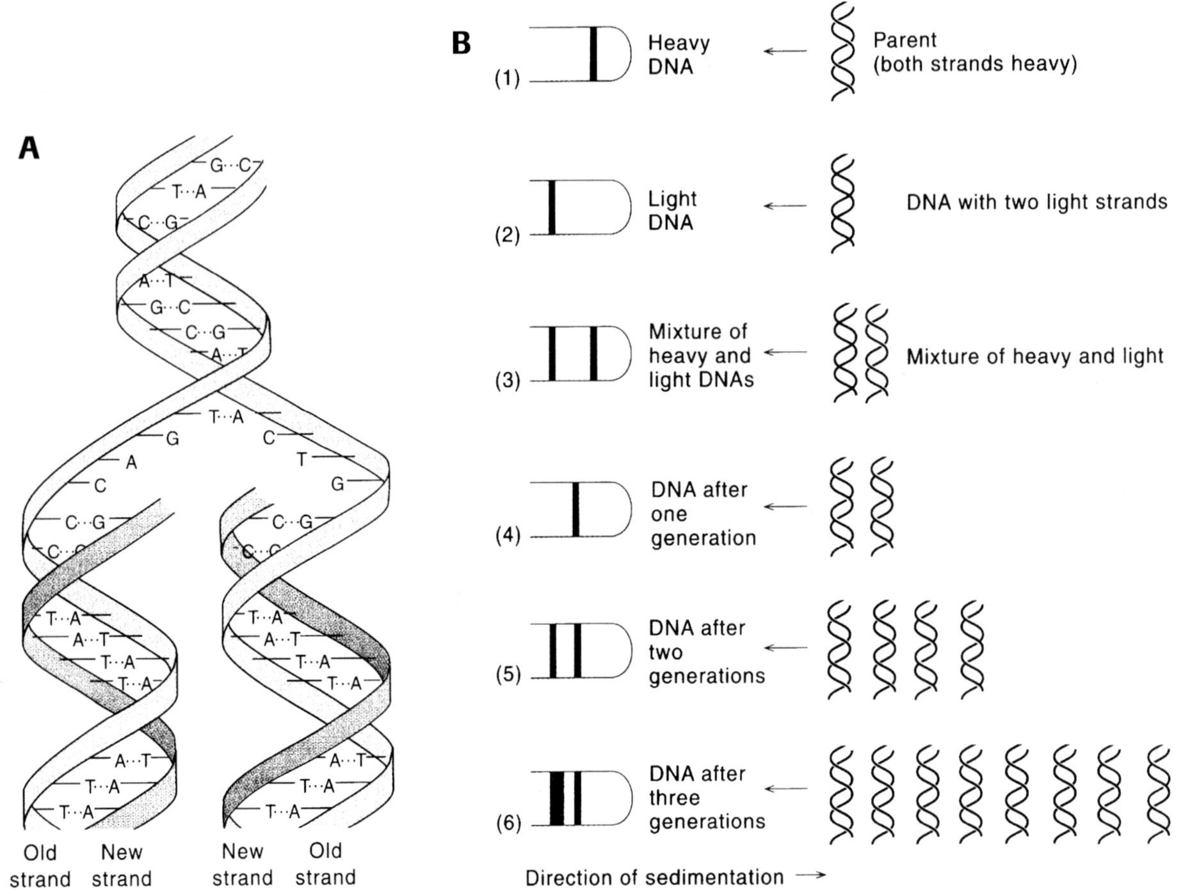


**FIGURE 7.7** **(A)** Watson-Crick model for DNA replication. The double helix unwinds at one end. New strand synthesis begins by absorption of mononucleotides to complementary bases on the old strands. These ordered nucleotides are then covalently linked into a polynucleotide chain, a process resulting ultimately in two daughter DNA duplexes. **(B)** The Meselson-Stahl experiment demonstrating semiconservative replication for *E. coli* chromosomal DNA. CsCl density-gradient centrifugation is used to discriminate between DNAs of different densities. *E. coli* DNA has different densities when cells are grown in $^{14}N$ or $^{15}N$ medium (frames 1 and 2). When cells containing pure heavy DNA ($^{15}N.^{15}N$ DNA) are grown in $^{14}N$ medium for one generation, all of the DNA is of intermediate density ($^{14}N.^{15}N$). After two generations of growth in $^{14}N$ medium, the cells contain equal amounts of light ($^{13}N.^{14}N$) DNA and intermediate density DNA (frame 5). In subsequent generations the hybrid DNA reappears in constant amounts, but the amount of light DNA increases. The results support the model of a semiconservative mode of DNA replication.

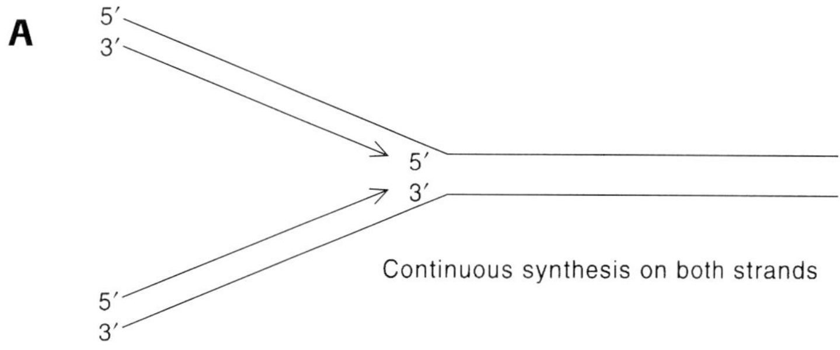


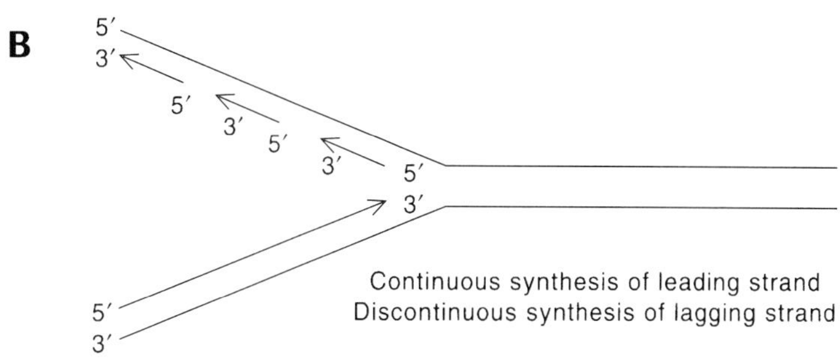


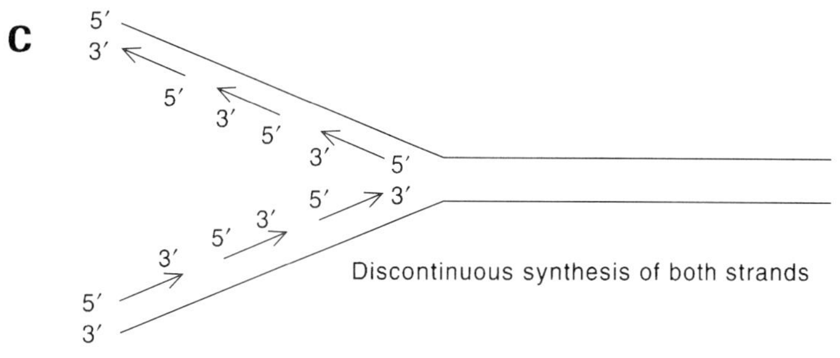


**FIGURE 7.8** Models for synthesis at the replication fork. **(A)** Continuous synthesis on both strands. Note that both growth arrows are pointing in the same direction, which would require growth in the $5' \rightarrow 3'$ direction on one strand and in the $3' \rightarrow 5'$ direction on the other strand. If growth occurs only in the $5' \rightarrow 3'$ direction, synthesis would have to be discontinuous on one strand, as in **(B)**. Alternatively, it could be discontinuous on both strands, **(C)**.

**TABLE 7.1**   Proteins Involved in DNA Replication

| Protein | Gene(s) | Function |
| --- | --- | --- |
| DnaA | *dnaA* | Initiator protein: Binds *oriC*; promotes double-helix opening; DnaB loading |
| DnaC | *dnaC* | Complexes with DnaB; delivers DnaB to DNA |
| DNA polymerase III holoenzyme | | |
| α | *dnaE* | Polymerase |
| ε | *dnaQ* | 3' to 5' exonuclease |
| θ | *holE* | Unknown |
| τ | *dnaX* | |
| γ | *dnaX* | |
| δ | *holA* | |
| δ' | *holB* | |
| χ | *holC* | |
| ψ | *holD* | |
| β | *dnaN* | Processivity factor: "Sliding clamp" |
| PolI | *polA* | Prime removal and gap filling; Initial leading strand synthesis on ColEl |
| Ligase | *lig* | Joins nascent DNA fragments |
| Gyrase | *gyrA* | Type II topoisomerase; replication swivel, DNA |
| | *gyrB* | supercoiling, decatenation |
| DnaB | *dnaB* | 5' to 3' helicase and activator of primase |
| Primase | *dnaG* | Primer synthesis |
| SSB | *ssb* | Binds single-stranded DNA |



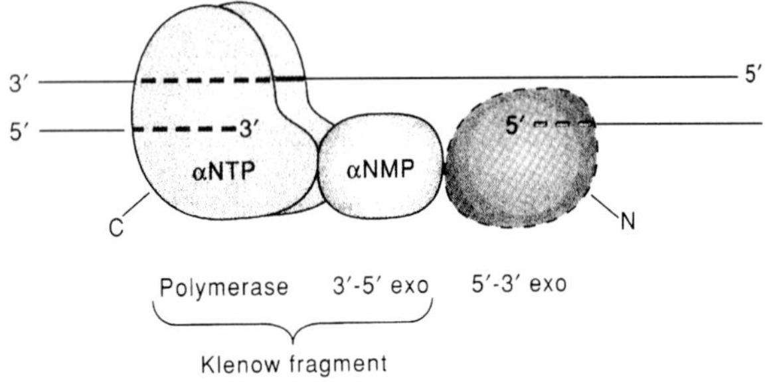


**FIGURE 7.9**   A schematic drawing of the apparent domain structure of *E. coli* DNA polymerase I. The solid line represents the experimentally determined Klenow fragment structure: the dashed lines indicate a possible location for the small fragment produced by proteolytic cleavage of Pol I. The orientation of the nicked DNA substrate is such that the 5' → 3' exonuclease domain would be able to interact with the DNA 5' downstream of the primer terminus.



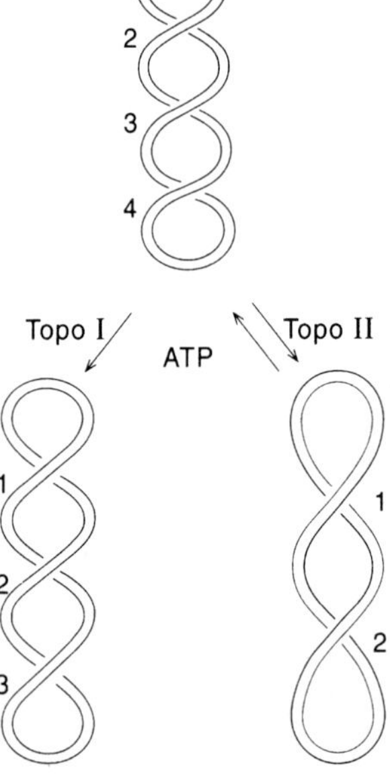


**FIGURE 7.10**   Type I and type II topoisomerases relax negatively supercoiled DNA in steps of one and steps of two, respectively. Type II topoisomerases can also add additional negative supercoils, as indicated by the double reaction arrow. The latter reaction requires energy input, which is encoded by ATP cleavage.

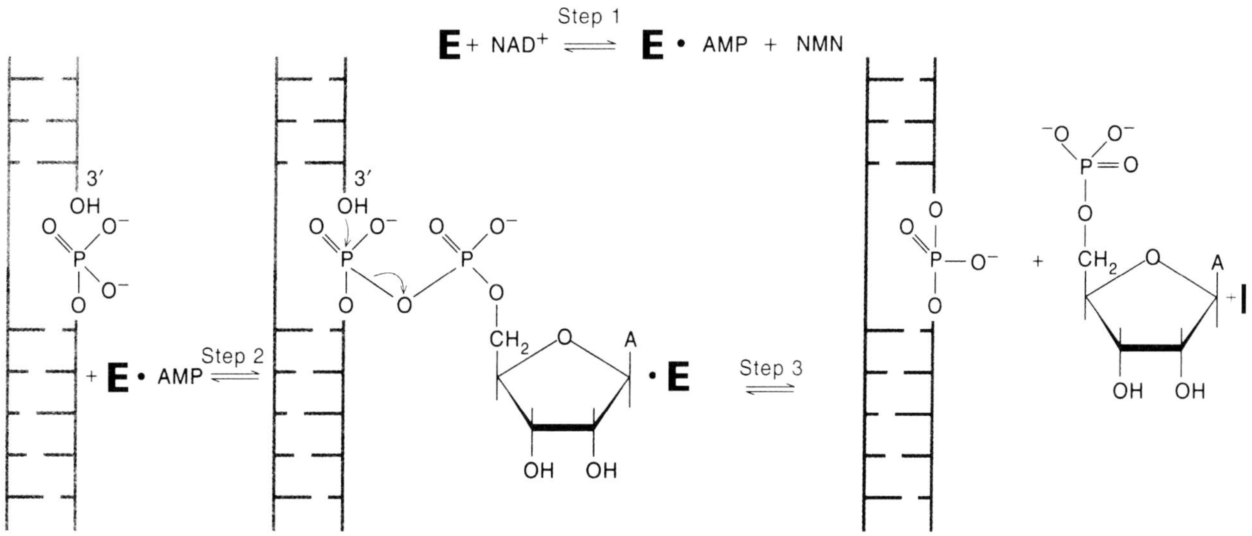

**FIGURE 7.11**   Steps in the sealing of a DNA nick, catalyzed by DNA ligase. The bacterial ligase uses $NAD^+$ to make an enzyme-AMP intermediate. Mammalian DNA ligases and bacteriophate T4 ligase use ATP for the same purpose.



# LABORATORY EXERCISES

## cDNA Cloning of Mono Amine Oxidase*

### ISOLATION OF TOTAL RNA FROM LIVER

#### Supplies and Equipment

Conical centrifuge tube, 50 ml, disposable, sterile
Corex tubes, 15 ml, thick-walled
Gloves, disposable
Microfuge tubes, 1.5 ml, sterile
Micropipette tips, sterile
Pipette, disposable, sterile (10 ml)
Microcentrifuge
Micropipettes
Polytron homogenizer
Sorvall centrifuge with SS34 rotor or equivalent
Speed Vac (Savant Instruments, Inc.)

#### Chemicals

Reagents
Chloroform
Ethanol
GITC solution
  Mix 50 g of guanidinium isothiocyanate,
    58.6 ml of water, 3.52 ml of 0.75 M sodium

citrate (pH 7.0), and 5.28 ml of 10% sarcosyl. Dissolve at 65° C. Store the stock solution at room temperature. Before using, add 0.36 ml of β-mercaptoethanol per 50 ml of the stock solution.
Isopropanol
Lithium chloride, 4 M
Phenol, saturated with 0.1 M sodium citrate, pH 4.5
Sodium acetate, 2 M, pH 4.0
Sodium acetate, 2 M, pH 5.0
TES Solution (10 mM Tris-HCl, pH 7.5, 1 mM EDTA, and 0.5% SDS)

### Introduction

This procedure is an adaptation of the method of Chomczynski and Sacchi[11] as modified by Puissant and Houdebine.[12] All glassware, metal spatulas, razor blades, and so forth should be rendered RNase-free by baking for 4 hours at 250° C. All chemicals should be reserved for RNA work only and handled only with baked spatulas. Water should be treated with 0.1% diethyl pyrocarbonate at 37° C overnight and then autoclaved for 45 minutes. Gloves should be worn at all times.

*Contributed by Dr. Simon Kwok.

# Procedures

a. Cut a piece of frozen liver (1 g–2 g) with a razor blade and a hammer. Quickly add tissue to 5 ml GITC solution in a 50 ml conical centrifuge tube and homogenize with a Polytron homogenizer at medium speed for 1 minute or until tissue is thoroughly dispersed.

b. Add the following reagents, in order; mix after each addition:
0.5 ml of 2 M sodium acetate, pH 4.0
5.0 ml of phenol
1.0 ml of chloroform

c. After thorough mixing, transfer the mixture to 15 ml Corex tubes. Centrifuge at 10,000 $xg$ for 10 minutes at 4° C, using appropriate rubber adaptors.

d. Transfer the upper phase (containing the RNA) to a clean Corex tube, avoiding the DNA and protein at the interphase. Add an equal volume (5.5 ml) of isopropanol; mix well, and store at –20° C for at least 2 hours.

e. Pellet the RNA by centrifugation at 3,000 $xg$ for 10 minutes at 4° C.

f. Resuspend the RNA pellet by vigorous vortexing in 1 ml of 4 M lithium chloride. Transfer the mixture to a microfuge tube. Centrifuge at 3,000 $xg$ for 10 minutes.

g. Dissolve the resulting RNA pellet in 0.5 ml of TES solution. Add 0.5 ml of chloroform and mix well. Centrifuge at 3,000 $xg$ for 10 minutes.

h. Transfer the upper phase to a new microfuge tube and add 50 µl (1/10 volume) of 2 M sodium acetate, pH 5.0 and 550 µl (equal volume) of isopropanol. Mix well and store at –20° C for at least 2 hours.

i. Collect the RNA by centrifugation at 3,000 $xg$ for 10 minutes. Rinse the RNA pellet in 70% ethanol. Dry in Speed Vac for 2–3 minutes.

j. Redissolve the RNA pellet in TES Solution. Determine the concentration by adding 5 µl of the RNA solution to 1 ml of water and reading the absorbance at 260 nm. Multiply this value by 8.04 to get the concentration of RNA in µg/µl. Store RNA at –80° C.

# ISOLATION OF LIVER POLY(A)⁺ RNA

*Supplies and Equipment*
Chromatography columns, 2 ml, disposable, RNase-free
Microfuge tubes
Microcentrifuge
Speed Vac (Savant Instruments, Inc.)

*Chemicals*
Reagents
Elution buffer (10 mM Tris-HCl, pH 7.5, 1 mM EDTA, and 0.05% SDS)
Ethanol
2X loading buffer #1 (40 mM Tris-HCl, pH 7.6, 1.0 M lithium chloride, 2 mM EDTA, and 0.2% SDS)
Loading buffer #2 (20 mM Tris-HCl, pH 7.6, 0.1 M lithium chloride, 1 mM EDTA, and 0.1% SDS)
Oligo(dT)-cellulose (Collaborative Research, Inc.)
Sodium acetate, 3 M, pH 5.2
Washing solution (0.1 M NaOH, and 5 mM EDTA)

# Introduction

Reagent kits for selecting poly(A)⁺ RNA from total RNA are being offered by a number of companies. They employ oligo(dT)-cellulose, oligo(dT)-latex beads, or oligo(dT)-magnetic beads. The procedure described here is an adaptation of the classical oligo(dT)-cellulose chromatography method of Aviv and Leder.[13]

# Procedures

a. Equilibrate oligo(dT)-cellulose in 1X loading buffer #1 (dilute 2X with equal volume of sterile water).

b. Pour into a 2 ml, RNase-free chromatography column.

c. Wash the column with 3 column-volumes each of:
sterile water
washing solution
sterile water

d. Check to make sure the column effluent is less than pH 8.0. If not, wash with additional water.

e. Wash the column with 5 volumes of 1X loading buffer #1.

f. Dissolve the total RNA in sterile water and heat to 65° C for 5 minutes. Add an equal volume of 2X loading buffer #1. Cool the sample to room temperature and apply sample to column.

g. Collect flow-through, heat to 65° C again, cool, and reapply to column.

h. Wash the column with 5–10 column-volumes of 1X loading buffer #1.

i. Wash the column with 4 column-volumes of loading buffer #2.

j. Elute the poly(A)⁺ RNA with 2–3 column-volumes of elution buffer.

k. Add 0.1 volume of 3 M sodium acetate, pH 5.2 and 2.2 volumes of ethanol. Store at –20° C for at least 2 hours.

l. Centrifuge in microfuge for 15 minutes to pellet the poly(A)⁺ RNA. Rinse the pellet in 70% ethanol.

Dry 2–3 minutes in Speed Vac. Dissolve the pellet in sterile water.

m. Determine the concentration as described above for the total RNA.

## Construction of cDNA Libraries

Pre-made liver cDNA libraries of various species are available from American Type Culture Collection, or from companies such as Clontech Laboratories, Inc., and Stratagene Cloning Systems. If custom-made libraries are desired, reagent kits can be purchased from various companies. The protocol accompanying the reagent kit should be followed exactly, since the strategy of cDNA synthesis in each kit may be different. Construction of cDNA library is laborious[14] and time-consuming, and therefore, is not recommended for laboratory exercise. In general, it involves the following steps:

a. Synthesis of first-strand cDNA.
b. Synthesis of second-strand cDNA.
c. Blunt-ending the double-stranded cDNA with T4 DNA polymerase.
d. Methylation of the cDNA with the appropriate methylase (if methylated nucleotide is not used during the first strand synthesis).
e. Ligation of synthetic linkers to the double-stranded cDNA.
f. Digestion of the linkered cDNA with appropriate restriction enzyme to create the sticky ends.
g. Ligation of the linkered cDNA to lambda gt11 vector DNA.
h. Packaging of the recombinant phage DNA into viable phages.
i. Amplification of the cDNA library.

# SCREENING OF cDNA CLONES

### *Supplies and Equipment*
Conical centrifuge tube, 50 ml, disposable, sterile
India ink
Nitrocellulose circles, 137 and 82 mm
Pasteur pipette, autoclaved
Petri dishes, disposable, 150 mm and 90 mm
Whatman 3 MM paper
End-over-end shaker
Incubator
Orbital shaker
Water bath

### *Chemicals*
Antibody against mono amine oxidase (polyclonal, raised in rabbits)

Anti-rabbit IgG-alkaline phosphatase conjugate (Pierce or Sigma)
AP buffer (100 mM Tris-HCl, pH 9.5, 100 mM NaCl, and 5 mM $MgCl_2$)
BCIP substrate (50 mg/ml in 100% dimethylformamide)
Blocking solution (1% bovine serum albumin in TBST)
Isopropyl-B-D-thiogalactopyranoside (IPTG; 10 mM in water)
Liver cDNA library constructed in an expression vector, such as lambda gt11 or lambda Zap vectors (Stratagene Cloning Systems)
NBT substrate (75 mg/ml in 70% dimethylformamide)
NZYM medium (5 g NaCl, 2 g $MgSO_4$ $7H_2O$, 5 g Yeast Extract, and 10 g NZAmine per liter; adjusted to pH 7.5 with NaOH, sterilize by autoclaving)
NZYM agar plates (add 15 g Difico Agar per liter of NZYM medium, autoclave, cool to 50° C, and pour into petri dishes)
NZYM top agarose (add 0.7 g of agarose per 100 ml of NZYM medium; sterilize by autoclaving)
Maltose in water, 20%, sterile-filter
SM (50 mM Tris-HCl, pH 7.5, 100 mM NaCl, 10 mM $MgSO_4$, and 0.01% gelatin; sterilize by autoclaving)
Stop buffer (20 mM Tris-HCl, pH 8.0, and 5 mM EDTA)
TBS (10 mM Tris-HCl, pH 8.0, and 150 mM NaCl)
TBST (10 mM Tris-HCl, pH 8.0, 150 mM NaCl and 0.05% Tween 20)

# Introduction

A number of screening procedures can be used to "fish" out the cDNA encoding the protein of interest. When antibody against the protein of interest is available, the antibody can be used as a probe for screening the cDNA library.[15] However, polyclonal antibody is more preferable than monoclonal antibody for this purpose. When N-terminal or internal amino acid sequence is available, synthetic oligonucleotide can be used as a probe. The oligonucleotide may be designed as suggested by Lathe.[16] If the gene is highly conserved, cDNA from one species can be used as a probe to isolate the cDNA of another species or closely related gene of the same species by the method of Benton and Davis.[17] The procedure described below uses antibody as a probe for the isolation of the desired cDNA clone.

# Procedures

## Primary Screening

a. Grow a 10 ml overnight culture of *E. coli* host in NYZM medium containing 0.2% maltose in a 50-ml conical centrifuge tube.

b. Dilute an aliquot of amplified cDNA library to 300 pfu/µl with SM. Mix 100 µl diluted library (30,000 pfu); 200 µl *E. coli* host (resuspended in equal volume of 10 mM $MgSO_4$). Incubate at 37° C for 20 minutes.

c. Add 7.5 ml NZYM top agarose (kept molten at 50° C), mix, and pour on a 150 mm NZYM agar plate (pre-warmed to 37° C). Allow to cool at room temperature for 20 minutes. Incubate at 42° C for 4–5 hours (until the plaques are clearly visible).

d. Label and soak the 137 mm nitrocellulose circles in 10 mM IPTG, and then air-dry on sterile 3 MM paper.

e. Overlay the nitrocellulose circle on top of the phage plaques. Incubate the plates at 37° C for another 3.5 hours.

f. Remove the plates to room temp. Mark the position of the nitrocellulose circles with India ink. Store at 4° C overnight to harden the top agarose.

g. Remove the nitrocellulose circles from the plates and wash them in 20 ml/circle of TBST in a glass dish for 20 minutes at room temperature, with gentle shaking to remove any remnants of agarose.

h. Incubate the nitrocellulose in 10 ml/circle of blocking solution for 1 hour at room temperature.

i. Incubate the nitrocellulose in 10 ml/circle of antibody solution (1:1,000 dilution in blocking solution) in a petri dish at room temperature for 1 hour with gentle shaking. Note: The primary antibody raised in rabbits should be absorbed with *E. coli* lysate coated on a nitrocellulose circle before use. Otherwise the background may be high.

j. Wash the nitrocellulose in 20 ml/circle of TBST three times (10 minutes each) at room temperature with gentle shaking.

k. Incubate the nitrocellulose in 10 ml/circle of blocking solution containing anti-rabbit IgG-alkaline phosphatase conjugate (1:1,000 dilution) in a petri dish for 1 hour at room temperature with gentle shaking.

l. Wash the nitrocellulose three times in TBST as in step j, and once in TBS to remove Tween 20.

m. Prepare fresh a color development substrate solution by mixing 10 ml of AP buffer and 40 µl NBT substrate, followed by mixing in 30 µl BCIP substrate. Protect the solution from light.

n. Transfer the nitrocellulose circle (with the phage side up) individually to the color development substrate solution. Shake gently on an end-over-end shaker. Positive clones will appear as purple dots on the circle on a light pink background. When the color has developed to the desired intensity, stop the reaction by washing the circles in stop solution.

o. Blot the circles dry on blotting paper. Mark the positions of positive clones and orientation marks on transparency and use it as template to isolate the positive clones from the original agar plates.

## Secondary/Tertiary Screening

The positive clones obtained above are impure and must go through two more cycles of "plaque purification" as described below:

a. Place the template (step o previously) underneath the agar plate. Align the orientation marks. Use a sterile pasteur pipette to stab the agar plate at the position of the positive signal. Transfer the phages as an agar plug to 1 ml of SM in a microfuge tube. Add two drops of chloroform, mix, and allow phages to diffuse out at room temperature for at least 1 hour.

b. Dilute the phage stock by mixing 1 µl of the phage stock with 1 ml of SM in a microfuge tube (1:1,000 dilution).

c. Mix 100 µl of phage dilution and 100 µl of overnight culture of *E. coli* host. Incubate at 37° C for 20 minutes. Then add 2.5 ml of Top Agarose, mix, and pour on a prewarmed 90 mm NZYM agar plate. Incubate at 42° C for 4–5 hours.

d. Overlay a 82 mm nitrocellulose circle (presoaked with IPTG) on top of the phage plaques. Incubate the plates at 37° C for another 3.5 hours.

e. Repeat steps f–o as described previously, using half of the volume of all reagents due to the smaller size of the nitrocellulose circle used.

# ISOLATION OF PHAGE DNA

*Supplies and Equipment*
    Centrifuge bottles, 250 ml
    Conical centrifuge tubes, 50 ml
    Culture flasks, 2-liter
    Dialysis tubing
    SW40 nitrocellulose tubes (for ultracentrifuge)
    Beckman ultracentrifuge with SW40 rotor or
        equivalent
    Shaker/incubator
    Sorvall centrifuge with GSA rotor or equivalent

**Chemicals**

Chloroform

CsCl

DNase, RNase, proteinase K (Sigma)

Ethanol

EDTA (500 mM in water)

Gelatin (1%, sterilize by autoclaving)

$MgSO_4$ (10 mM)

NZYM medium (5 g NaCl, 2 g $MgSO_4$ $7H_2O$, 5 g yeast extract, and 10 g NZAmine per liter; adjusted to pH 7.5 with NaOH, sterilize by autoclaving)

Phage Dialysis Buffer (10 mM Tris-HCl, 25 mM NaCl, and 1 mM $MgSO_4$ pH 8.0

Phenol-chloroform-isoamyl alcohol (Sigma)

Polyethylene glycol (PEG) 6000

SM without gelatin (50 mM Tris-HCl, pH 7.5, 100 mM NaCl, and 10 mM $MgSO_4$, sterilize by autoclaving)

Sodium dodecyl sulfate (SDS, 20% in water)

TSE buffer (10 mM Tris-HCl, 5 mM NaCl and 1 mM EDTA, pH 8.0, sterilize by autoclaving)

# Introduction

Once the positive clones have been purified, phage DNA can be isolated. The cDNA insert can be then isolated from the recombinant phage DNA and subcloned into a plasmid vector for sequencing. For cDNA libraries constructed with the lambda Zap vector (Strategene Cloning Systems), the positive clones can be converted to a phagemid by the "*in vivo* excision" procedure using a helper phage. No subcloning is necessary. Thus, this step can be skipped.

# Procedures

a. From the positive plaques of the tertiary screening, pick a single, well-isolated plaque with a pasteur pipette and transfer into 1 ml of SM. Add 2 drops of chloroform. Mix and allow phages to diffuse at room temperature for at least 1 hour.

b. Mix 500 µl of the phage suspension (step a) and 100 µl of overnight culture of a suitable *E. coli* host (resuspended in equal volume of 10 $MgSO_4$) in a sterile, 50-ml centrifuge tube. Incubate at 37° C for 20 minutes.

c. Add 4 ml of NZYM medium; shake at 37° C overnight.

d. Add 40 µl of chloroform to the culture; shake at 37° C for another 15 minutes to complete the lysis.

e. Transfer the culture to two 2-ml microfuge tubes, remove cell debris by centrifugation for 2 minutes.

Transfer the supernatant to fresh tubes, add sterile gelatin to 0.01% and 2 drops of chloroform. This is the phage stock. Determine the titer and store at 4° C until use.

f. In a glass 16 × 100 mm tube, add 20 µl of $10^{10}$ pfu/ml phage stock and 2 ml of *E. coli* host overnight culture (resuspended in equal volume of 10 mM $MgSO_4$). Incubate at 37° C for 20 minutes.

g. Pour the phage/bacterial mixture into a 2-liter flask containing 400 ml NZYM medium. Incubate at 37° C for 7–16 hours with shaking.

h. When lysis is evident, add 1.5 ml of chloroform to each flask; shake for another 20 minutes at 37° C.

i. Divide the phage culture into two 250-ml centrifuge bottles. Centrifuge at 7,500 rpm for 30 minutes (GSA rotor) to remove cell debris. Combine supernatant in a 2-liter flask.

j. Add 400 µl each of the DNase and RNase (1 mg/ml in 10 mM Tris-HCl, pH 7.5). Mix well, and incubate at room temperature for 2 hours (or at 4° C overnight).

k. Add 11.69 g of NaCl (to make 0.5 M) and 40 g of PEG 6000 (to make 10%) slowly; shake to dissolve. Leave at 4° C overnight or in ice water for at least 1 hour to precipitate the phages.

l. Collect the phages by centrifugation in two 250-ml centrifuge bottles (GSA rotor, 7,500 rpm, 30 minutes).

m. Resuspend phages in a total of 8 ml of SM, transfer to a 50-ml centrifuge tube.

n. Extract with equal volume of chloroform to remove PEG. Centrifuge to separate phases (PEG precipitates at interphase).

o. Transfer phage suspension (top phase) to a clean 50-ml centrifuge tube. Add 0.75 g of CsCl per ml of phage suspension. Shake gently to dissolve the CsCl. Then transfer the solution to a SW40 nitrocellulose tube and centrifuge in a SW40 rotor at 35,000 rpm for 24 hours at 4° C.

p. Place the tube in front of a black background; pipette out the phage band (can be seen as a white band in the middle of the tube) from the top.

q. Dialyze phages against 3 × 2 liters of Phage Dialysis Buffer at 4° C (to remove CsCl).

r. Transfer the dialyzed phages to a 15-ml tube. Add SDS to 0.2% and EDTA to 10 mM. Lyse phage at 65° C for 15 minutes and digest with 50 µg/ml proteinase K at 37° C for 1 hour.

s. Extract with equal volume of phenol-chloroform-isoamyl alcohol. Centrifuge briefly to separate the phases. Transfer the aqueous DNA solution (top phase) to a clean 15-ml tube.

t. Extract once with chloroform (to remove trace of phenol). Centrifuge briefly to separate the phases.

u. Transfer the aqueous DNA solution (top phase) to dialysis tubing and dialyze against TSE buffer at 4° C overnight.

v. Read absorbance at 260 nm to determine the yield (1 $^A$260 mn unit = 50 µg/ml).

# SUBCLONING

### *Supplies and Equipment*

Conical centrifuge tubes, 50-ml, disposable
Falcon 2059 polypropylene tubes
Microfuge tubes
Petri dishes, 90 mm
Clinical centrifuge
Microcentrifuge
Microcooler or a 15° C water bath
Water bath

### *Chemicals*

Calcium buffer (50 mM $CaCl_2$ and 10 mM Tris-HCl, pH 8.0, filter-sterilize)
Chloroform
EDTA (500 mM, pH 8.0)
Ethanol
Indicator plates
  Dissolve 8 g of tryptone, 5 g of yeast extract, and 5 g of NaCl in 1 liter of water; adjust to pH 7.5 with NaOH. Add 15 g of agar and autoclave for 20 minutes. After cooling to 50° C in water batch, add sterile ampicillin (20 mg in 1 ml of water), sterile IPTG (40 mg in 1.5 ml of water), and X-gal (40 mg in 1.5 ml of dimethyl formamide). Mix and pour in 90 mm petri dishes.
LB medium (10 g of tryptone, 5 g of yeast extract, and 5 g of NaCl in 1 liter of water); adjust to pH 7.5 with 1 N NaOH; autoclave for 20 minutes.
10X ligation buffer (500 mM Tris-HCL, pH 7.6, 100 mM $MgCl_2$, 100 µM DTT, and 500 µg/ml of bovine serum albumin)
Phenol-chloroform-isoamyl alcohol (Sigma)
Pre-cut and dephosphorylated pUC18 vector
Restriction endonuclease EcoRI
Sodium acetate (3 M, pH 5.2)
$T_4$ DNA ligase

# Introduction

Since the yield of phage DNA is low, the cDNA insert is usually subcloned into a plasmid vector, such that a large amount of DNA can be isolated for restriction endonuclease mapping and sequencing. Furthermore, large cDNA may be cut into manageable pieces. Each DNA fragment will be subcloned for sequence analysis. To facilitate subcloning, pre-cut and dephospho-

rylated pUC18 vectors are available from Pharmacia. Competent cells are also available from a number of companies. The following is a general subcloning procedure using $CaCl_2$-treated competent cells.

# Procedures

a. For cDNA clones constructed with lambda gt11 vector, release the cDNA insert by digesting 5 µg of recombinant phage DNA with EcoRI in 50 µl of EcoRI buffer at 37° C for 2 hours.

b. Add EDTA to 15 mM. Extract with equal volume of phenol-chloroform-isoamyl alcohol. Centrifuge briefly to separate the phases.

c. Transfer the aqueous (top) phase to a fresh tube. Extract with equal volume of chloroform. Centrifuge briefly to separate the phases.

d. Transfer the aqueous (top) phase to a fresh tube. Precipitate with 0.1 volume of 3 M sodium, acetate and equal volume of ethanol on dry ice for 15 minutes. Centrifuge for 15 minutes.

e. Redissolve the DNA pellet in 7 µl of water; add 1 µl of EcoRI-cut and dephosphorylated pUC 18 DNA (100 ng/ml). Heat at 45° C for 5 minutes and chill on ice. Note: It is not necessary to isolate the cDNA insert on an agarose gel. The lambda DNA present in the mixture will not interfere with the subcloning.

f. To the DNA mixture, add 1 µl of 10X ligation buffer, 0.5 µl of 10 mM ATP, and 0.5 µl of T4 DNA ligase (1 Weiss unit/µl). Mix and incubate at 15° C overnight. For negative control, set up identical tube except without phage DNA (ligated vector only).

g. Grow an overnight culture of XL1-Blue (Stratagene) in 2 ml of LB medium containing 12.5 mg/ml tetracycline at 37° C (other suitable *E. coli* host may be used, but appropriate antibiotic should be used).

h. Next morning, heat inactivate the ligated DNA (from step f) at 70° C for 10 minutes. Chill on ice. Centrifuge for 1 second.

i. Inoculate 50 ml of LB medium in a 250-ml culture flask with 0.5 ml of the overnight culture. Shake at 37° C until absorbance at 550 nm reaches 0.3 (approximately 2 hours).

j. Chill the bacterial cells on ice for 10 minutes. Transfer to a 50-ml disposable centrifuge tube and centrifuge in a clinical centrifuge at maximal speed for 5 minutes in the cold room.

k. Decant the medium carefully and blot on paper towel. Resuspend bacterial pellet in 25 ml (1/2 of the growth volume) of ice-cold calcium buffer by swirling the tubes gently. Stand on ice for 20 minutes. Note: Handle the bacteria very gently from this step on. Avoid vortexing. It is easier to resuspend cells in small amount of buffer and then mix with the remaining portion of buffer.

l. Centrifuge in clinical centrifuge at maximal speed for 5 minutes in cold room.

m. Decant the buffer. Resuspend pellet in 3.33 ml (1/15 of growth volume) of ice-cold calcium buffer.

n. Pipet 200 µl aliquots of the competent cells (from step m) with big mouth tips into Falcon 2059 polypropylene tubes.

o. To the competent cells, add 2 µl of ligated DNA (from step h; 20 ng) or 2 µl of intact vector DNA (2 ng, positive control) or 2 µl of ligated vector DNA alone (20 ng, negative control). Incubate on ice for 40 minutes. Mix occasionally by tapping the tubes.

p. Heat shock the tubes for 2 minutes at 42° C.

q. Add 1 ml of LB medium (prewarmed to 37° C) to each tube. Incubate at 37° C for 1 hour with gentle shaking.

r. Mix the bacteria gently. Spread 0.2 ml on each indicator plate. Incubate at 37° C overnight.

s. Pick white colonies (those containing DNA inserts) and prepare "miniprep" DNA for further analysis.

# MINI-PREPARATION OF PLASMID DNA

### *Supplies and Equipment*
Culture tubes, 16 × 100 mm
Microfuge tubes
Micropipette tips
Microcentrifuge
Shaker/incubator
Water bath

### *Chemicals*
Ampicillin
Cetyltrimethylammonium bromide (CTMAB, 5%)
Ethanol
LB medium (F10 g of tryptone, 5 g of yeast extract, and 5 g of NaCl in 1 liter of water; adjust to pH 7.5 with 1 N NaOH, autoclave for 20 minutes)
Lysozyme
NaCl, 1.2 M
STET buffer (8% sucrose, 50 mM Tris-HCl, pH 8.0, 50 mM EDTA, and 0.1% Triton X-100)
Stock RNase (DNase-free, Boehringer Mannheim, Catalog No. 1119-915)

## Introduction

This procedure is an adaptation of the method of De Sal et al[18] which is a fast and economical method.

The resulting DNA is clean enough for DNA sequencing. A number of companies also sell "miniprep" DNA isolation reagent kits; they work well but are more expensive.

## Procedures

a. Pick a single colony and grow a 2-ml overnight culture in LB medium containing 50 mg/ml ampicillin at 37° C.

b. Chill the culture on ice for 5 minutes. Pipette 1.5 ml of the culture to a microfuge tube and centrifuge for 2 minutes. Remove the supernatant with a pasteur pipette as completely as possible. Save the remaining culture as glycerol stock at –20° C.

c. To the cell pellet, add 200 µl of STET buffer and resuspend the cells by vortexing.

d. Add 4 µl of a freshly prepared lysozyme solution (50 mg/ml in 10 mM Tris-HCl, pH 7.5), mix, and incubate for 5 minutes at room temperature.

e. Heat the tubes in boiling water bath for 45 seconds.

f. Centrifuge at 4° C for 10 minutes. Remove the pellet with a toothpick.

g. Add 0.5 µl of stock RNase (30 µg–50 µg) each tube. Incubate at 68° C for 10 minutes.

h. Add 10 µl of 5% CTMAB, mix, and centrifuge at 4° C for 5 minutes.

i. Decant the supernatant. Resuspend the pellet in 300 µl of 1.2 M NaCl by vortexing. Add 750 µl of ethanol, mix, and centrifuge for 5 minutes.

j. Decant the supernatant. Add 1 ml of ice-cold 70% ethanol and recentrifuge for 5 minutes.

k. Decant and dry in Speed Vac for 15 minutes.

l. Redissolve the DNA pellet in 20 µl of TE buffer.

# DNA SEQUENCING

### *Supplies and Equipment*
Microfuge tubes
Micropipette tips
Micropipettes
Power supply, 2000 volt
Sequencing pipette
Sequencing gel electrophoresis apparatus
Water bath
Whatman 3MM paper

### *Chemicals*
Acetic acid, glacial
Acrylamide:bis-acrylamide (19:1), 40% solution
Ammonium persulfate
M13 Universal Primer
M13 Reverse Primer
N, N, N', N'-Tetramethylenediamine (TEMED)

Sequenase Version 2.0 DNA Sequencing Kit (US Biochemical, Catalog No. 70770)

[$^{35}$S]dATP

5X TBE Buffer (0.45 M Tris base, 0.45 M boric acid, and 12.5 mM EDTA, pH 8.3)

TDMN Buffer (0.28 M TES, 0.12 M HCl, 0.05 M DTT, 0.08 M MgCl$_2$, and 0.2 M NaCl)

Urea, solid

# Introduction

The following procedure is an adaptation of the method of Sanger et al.[18] as modified by De Sal et al.[19] The Sequenase DNA sequencing reagent kit (US Biochemical) can be used with this procedure without adjustment of reagent volumes.

# Procedures

## Sequencing Reaction

a. Mix 2 µl of plasmid DNA (about 1 µg) 5 µl of water, 2 µl of primer (10 ng/µl), and 1 µl of a freshly prepared solution of 1 N NaOH. Denature the DNA by incubating for 10 minutes at 68° C.

b. Neutralize the DNA solution by adding 4 µl of TDMN buffer. Allow to anneal at room temperature for 10 minutes.

c. While the annealing mixes are incubating, prepare 4 tubes per DNA sample, each containing 2.5 µl of *one* of the four termination mixes.

d. To each annealing mix (step b), add the following: 3 µl of 1X Labeling Mix (dilute the 5X stock with water); 1 µl of [$^{35}$S]dATP (1000 Ci/mmol)

e. Transfer the tubes to microfuge, then pipette 2 µl of T7 DNA polymerase (Sequenase, diluted to 1.5 units/µl with enzyme dilution buffer) to the top of each tube; start the reactions all at once by spinning for 1 second. Mix and incubate at room temperature for 5 minutes.

f. Before the 5-minute incubation is up, transfer 4 µl of each labeling reaction (step e) to the top of each of the four termination mix tubes (step c) in the same group. When the labeling reaction is done, spin the tubes for 1 second to start the termination reactions all at once. Close the caps, mix, and transfer the tubes to a 37° C water bath. Incubate for 10 minutes.

g. Place the tubes in microfuge. Add 5 µl of sequence stop to the top of each tube. Terminate all reactions by spinning for 1 second. Heat the DNA samples at 90° C for 3 minutes and quickly chill on ice.

## Electrophoresis on 6% Sequencing Gel

a. Assemble the gel plates between two spacers as per instruction of the manufacturer.

b. Combine the following in a suction flask (final volume = 90 ml):

| | |
|---|---|
| solid urea | 45.0 g |
| 40% acrylamide solution | 13.5 ml |
| 5X TBE Buffer | 18.0 ml |
| water | 25.5 ml |

Swirl the flask in a 37° C water bath until urea dissolves. (Do not overheat!)

c. Add 50 mg of ammonium persulfate; mix to dissolve.

d. Degas for 5 minutes (rewarm the solution if crystals appear). Filter through Whatman filter paper.

e. Add 35 µl of TEMED to the acrylamide solution, mix, and pour into the gel mold. Insert a straight edge comb. Allow to polymerize for 30 minutes.

f. When ready for electrophoresis, remove the straight edge comb and rinse the gel surface with deionized water. Insert a shark tooth comb with the teeth barely touching the gel surface. Set up the gel, fill up the buffer tank with 1X TBE Buffer, and pre-electrophorese at 1800 V until the gel gets warm (about 30 minutes).

g. Turn off power and rinse the sample slots with 1X TBE Buffer using a 60 ml syringe. Load 1.5 µl of the DNA sample per slot on the sequencing gel. Run the gel at 1800 V. (Note: Lower the voltage if the gel becomes too hot or it will crack!)

h. Make the second loading when the first xylene cyanol (green dye) has migrated to the bottom of the gel.

i. Terminate the electrophoresis when the second bromophenol blue (blue dye) reaches the gel bottom. Remove one glass plate. Soak the gel on the other glass plate in 2.5 liters of 10% acetic acid for 20 minutes.

j. Carefully siphon out the acetic acid. Soak the gel in distilled water for 5 minutes.

k. Carefully siphon out the water. Place the gel right side up on the bench. Transfer the gel onto Whatman 3MM paper by placing the paper on top of the gel and lifting the paper carefully while squirting a little bit of water between the gel and glass plate. Cover the other gel surface with plastic wrap, place on slab gel dryer with the paper side down, and dry at 80° C for 30 minutes.

l. Remove the plastic wrap and expose the gel to X-ray film at room temperature for 16–18 hours.

m. Develop the X-ray film and read the sequence.

# cDNA CLONING OF POLYPHENOL OXIDASE

## *Supplies and Equipment*

Messenger affinity paper

Nitrocellulose

Miracloth

Waring blender
Preparative centrifuge

### Chemicals

Chloroplast preparation
50 mM Hepes/KOH, pH 7.6 supplemented with
2 mM EDTA, 1 mM $MgCl_2$, 1 mM $MnCl_2$ 330 mM
sorbitol and 5mM ascorbate
RNA isolation Phenol/chloroform/isoamyl alcohol
(24:24:1,v,v,v)
2M Tris/HCl, pH 8.0
0.1 mM Tris/HCl, pH 8.0
20 mM phosphate buffer, pH 7.0
Chloramphenicol (50 µg, ml$^{-1}$)
Guanidine buffer 8 M guanidine hydrochloride,
20 mM Mes (4-morpholineethane sulfonic acid)
20 mM EDTA, 50 mM mercaptoethanol, pH 7.0
Ethanol
Acetic acid          Agarose
Sodium acetate        50mM NaOH
Formaldehyde          4 µg ethidium bromide
Pancreatic DNase (RNase free)
0.05 M sodium acetate, pH 7.0, 0.01 M $MgCl_2$
2 mM $CaCl_2$
cDNA construction, screening and DNA
sequencing
mRNA purification kit

Tween 20
Stratagene ZAP—cDNA Synthesis Kit*
Stratagene Uni-ZAP XR Cloning Vector*
Stratagene Gigapack II Gold Packaging Extracts*
Stratagene Pico Blue Immunoscreening Kit*
USB Sequenase Version 2.0 DNA Sequencing Kit*

# Procedures

## Chloroplast Preparation

The preparation of chloroplasts from *Vicia faba* is presented in Figure 7.12.

## Isolation of RNA

Various procedures exist for the isolation of poly(A)-mRNA including oligo (dt) chromatography. In recent years, scientific companies have developed kits for the rapid isolation of this RNA species. Instructors are encouraged to divide the students into groups and use the oligo (dt) chromatographic method presented in Module 9 as well as the Pharmacia and Promega kits.

---

*These kits are available from Stratagene Cloning Systems (11011 North Torrey Pines Road, La Jolla, CA 92037)

---

Grow *Vicia faba* (broad bean) in the greenhouse in a soil/sand/peat mixture (1:1:1, v/v).
↓
Isolate chloroplasts from leaves of 2–4 week old seedlings by washing leaves in ice-cold  water.
↓
Mince leaves into ice-cold homogenization buffer (50 mM Hepes/KOH, pH 7.6, supplemented with 2mN EDTA, 1 mM $MgCl_2$, 1mM $MnCl_2$ 330 mM sorbitol, and 5mM ascorbate).
↓
Homogenize using a Tekman homogenizer or equivalent at a setting of 6 for 1 minute.
↓
Filter brei through eight layers of cheesecloth and one layer of Miracloth.
↓
Concentrate chloroplasts by centrifugation at 1,000 *xg* for 3 minutes.
↓Resuspend chloroplasts in 0.5M sorbitol, 50 mM HEPES and 5mM $MgCl_2$, pH 7.6,  containing 1% (v/v) Tween 20; the Suspension should be diluted to yield 0.1 mg  chlorophyll ml$^1$.
↓
Ice in the dark for 30 minutes.
↓
Centrifuge 100,000 *xg*, 2 hours.

| Pellet | Supernatant |
|---|---|
| Tween 20 treated thylakoids | Tween 20 solubilized components including PPO |

**FIGURE 7.12**   Procedure for solubilizing polyphenol oxidase from isolated chloroplasts.

Lax, A. R., and K. C. Vaughn. 1991. "Colonization of polyphenol oxidase and photosystem II proteins." *Plant Physiol.* 96:26–31.

The Promega company manufactures a poly(A)-mRNA isolation kit that eliminates the requirement for oligo (dt) chromatography. The Promega technology utilizes a biointylated olig (dt) primer to hybridize (at high efficiency in solution) to the poly(A) region in most eukaryotic mRNA species. The resulting hybrids are then captured and washed at high stringency employing streptavidin coupled to paramagnetic particles and a magnetic separation strand. The poly(A)-mRNA is eluted from the solid phase via the addition of ribonuclease-free deionized $H_2O$ (Promega Technical Manual, poly(A)-mRNA isolation system). The isolation of bulk and mRNA from plant tissues is shown in Figure 7.13.

## Construction of cDNA libraries

This exercise is concerned with the construction of a cDNA duplex from a poly(A)-mRNA.

### *cDNA Construction*

The formation of a cDNA duplex from a poly(A)-mRNA consists of four main steps as summarized by Zubay (Figure 7.14).[1]

Step 1.  The production of a DNA-RNA hybrid duplex from a poly(A)-mRNA via reverse transcriptase and the 4 deoxynucleotide triphosphates. Poly(A)-mRNA serves as a template for the synthesis for the first strand of the cDNA.

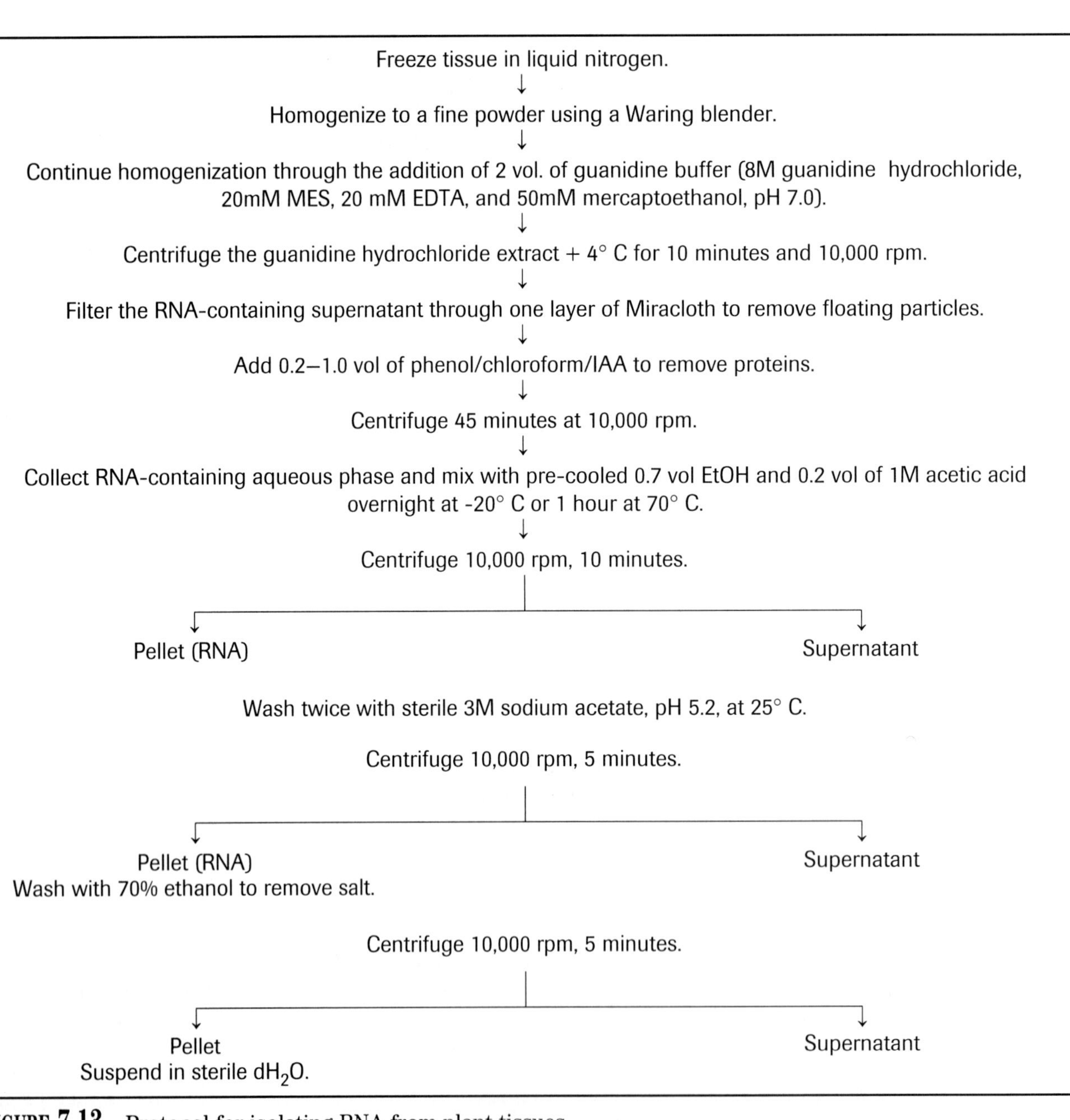


**FIGURE 7.13** Protocol for isolating RNA from plant tissues.
Longeman, I., I. Schell, and J. Willimitzer. 1987. "Improved method for the isolation of RNA from plant tissues." *Analytical Biochemistry.* 163: 16–20.

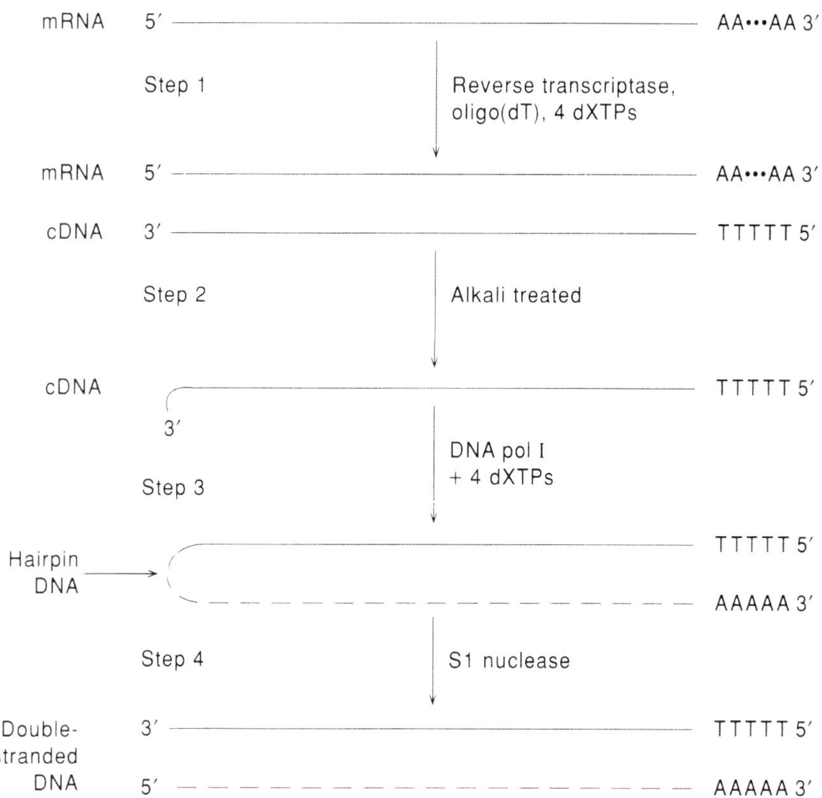



**FIGURE 7.14** Formation of a cDNA duplex from a poly(A)-mRNA. Step 1: To the poly(A)-mRNA are added reverse transcriptase, oligo(dT), and the four-deoxynucleotide triphosphates. In favorable circumstances this step results in a full-length DNA-RNA hybrid duplex. Step 2: The RNA is removed and degraded by alkali treatment. Step 3: The remaining single-stranded DNA serves as both template and primer for the synthesis of a complementary DNA strand. The newly synthesized DNA is linked to the primer template DNA by a hairpin turn. Step 4: The hairpin is cleaved with S1 nuclease.



Step 2.  Removal and degradation of the RNA.

Step 3.  The remaining single stranded DNA serves as both template and primer for the synthesis of a complementary DNA strand in the presence of DNA polymerase one and the 4 deoxynucleotide diphosphates. Note the newly synthesized DNA is linked to the primer template DNA by a hairpin turn.

Step 4.  The hairpin DNA is cleaved with a $S_1$ nuclease.

The construction of cDNA libraries (Figure 7.15) can be facilitated via the employment of commercial kits such as those manufactured by Stratagene (Uni-Zap-cDNA synthesis/Gigapack II Packaging Extract)[20,21] and a combination of Stratagene Uni-Zap-cDNA Synthesis Kit and Epicentre Technologies Max Plax Packaging Extract. MaxPlax™ packaging extract is a convenient economical and simple-to-use reagent to perform *in vitro* lambda packaging reactions. MaxPlax packaging extract is supplied as pre-dispensed single-tube reactions that have been optimized for high-efficiency packaging of methylated and unmethylated DNA. The packaging extract routinely yields packaging efficiencies of $1 - 3 \times 10^9$ pfu/μg of control λ DNA. The extract can be used in the construction of representative cDNA libraries, genomic cloning of highly modified (methylated) DNA into λ-phase or cosmid vectors, and in the rescue of λ-shuttle vectors from transgenic animal systems. The technical bulletins for the use of these kits are available through Stratagbe Cloning Systems (LaJolla, CA) and Epicentre Technologies (Madison, WI). With regard to screening of transcribed cDNA, immunoscreening[22] is often employed (Figure 7.16). A Pico Blue Immunoscreening Kit is available from Stratagene.

## Synthesis of cDNA

1. Prepare $\geq$ 10µg µl⁻¹ poly (A)-RNA (see *in vitro* synthesis protein synthesis module). Heat poly(A)-RNA in tightly sealed microfuge tube at 65° C for 5 minutes, and place on ice.
2. Add the following to a separate tube (180 µl final volume):
   20 µl 5 mM dNTPs (500 µM final, each)
   40 µl 5x RT buffer (1 x final)
   10 µl 200mM DTT (10 mM final)
   20 µl 0.5 mg/ml oligio (dT) 12–18 (50 µg/ml final)
   60 µl H₂O
   10 µl (10 U) RNAsin (50 U/ml final)
   Vortex, microcentrifuge, and add to 10 µl RNA. Add 20 µl (200 U) AMV reverse transcriptase. Mix and transfer 10 µl to tube containing 1 µl of [α-³²P]dCTP. Maintain both tubes 5 minutes at 25° C, then incubate 1.5 at 42° C.
3. Add 1 µl of 0.5 M EDTA, pH 8.0, to the radioactive tube and place at –20° C. To the main reaction, add 4 µl of 0.5 M EDTA, pH 8.0, and 200 µl buffered phenol. Vortex, microcentrifuge 1 minute at 25° C, and transfer upper (aqueous) phase to a new tube.
4. Add 100 µl TE buffer to phenol layer, mix, microcentrifuge, and combine aqueous phases. Add 1 ml diethyl ether, vortex, microcentrifuge, and discard upper layer. Repeat diethyl ether extraction.
5. Add 125 µl of 7.5 M ammonium acetate and 950 µl of 95% ethanol. Place 15 minutes in dry ice/ethanol bath, warm to 4° C. Discard supernatant, fill tube with ice-cold 70% ethanol, and microcentrifuge 3 minutes at 4° C. Remove supernatant and dry in vacuum desiccator.
   *Thaw tube containing radioactive aliquot from step 3 and spot onto nitrocellulose membrane filter. Wash membrane with ice-cold 10% TCA Count bound radioactivity and calculate amount of cDNA synthesized.*

### Conversion of cDNA into double-stranded cDNA
6. Resuspend RNA pellet (step 5) in 284 µl water and add in order (final volume 400 µl):
   4 µl 5mM dNTPs (50 µM final, each)
   80 µl 5X SS buffer (1 x final)
   12 µl 5 mM β-NAD⁺ (150 µM final)
   2 µl [α-³²P]dCTP (50 µCi/ml, final)
7. Vortex, microcentrifuge, and add:
   4 µl (4 U) RNase H (10U/ml final)
   4 µl (20 U) *E. coli* DNA ligase (50 U/ml final)
   10 µl (100 U) *E. coli* DNA polymerase I (250 U/ml final)
8. Vortex, microcentrifuge, and add:
   4 µl (4 U)RNase H (10 U/ml final)
   4 µl (20 U) *E. coli* DNA ligase (50 U/ml final)
   10 µl (100 U) *E. coli* DNA polymerase I (250 U/ml final)
9. Vortex, microcentrifuge, and incubate 12–16 hours at 14° C.
10. Remove 4 µl to new tube and freeze at –20° C for later determination of incorporation of radioactivity as in step 5. Extract remaining reaction with 400 µl buffered phenol and back extract phase with 200 µl TE buffer.
    *Expect incorporation of 1–10 $\times$ 10⁶ cpm.*
11. Pool aqueous phases and extract twice with 900 µl ether (step 4). Divide between two tubes, add ammonium acetate, and ethanol precipitate (step 5).

---

**FIGURE 7.15**  Summary of procedure for cDNA library construction.
Modified from Ausubel et al. 1992. *Short Protocols in Molecular Biology.* New York: John Wiley and Sons.

## DNA Sequencing

The sequencing of DNA by the chain-termination method involves the synthesis of a DNA strand via a DNA polymerase employing a single-stranded DNA template *in vitro*. This synthesis is initiated at the site where an oligonucleotide primer anneals to the template. Synthesis is terminated by the incorporation of nucleotide analogs (2, 3¹-dideoxynucleotide-5¹-triphosphates) that will not support continued DNA elongation. The ddNTPs analogs do not possess the 3¹-OH group requisite for DNA chain elongation (see DNA synthesis). Employing mixtures of dNTPs and

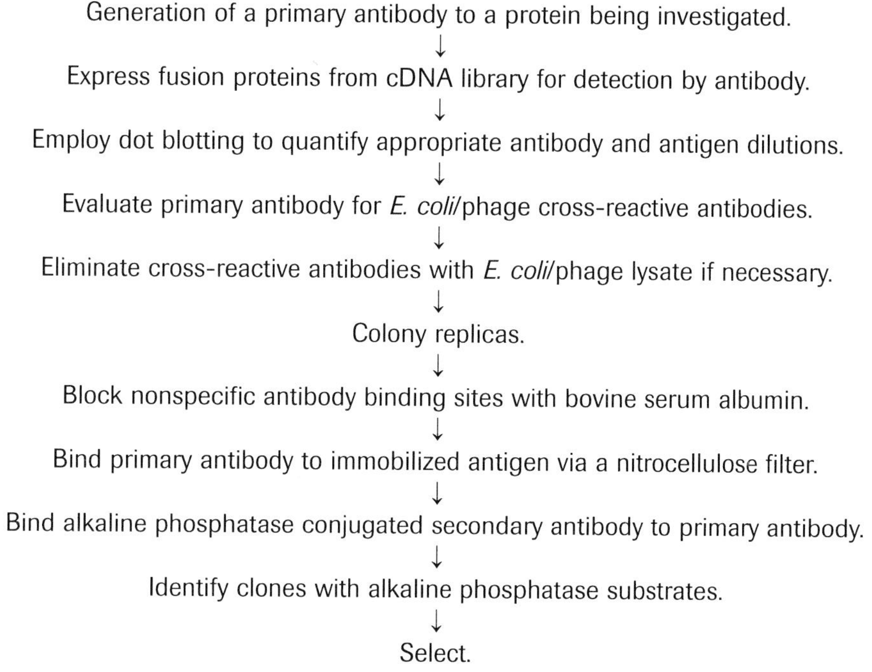



**FIGURE 7.16**  Generalized procedures for immunoscreening of transcribed cDNA. Adapted from *Stratagene Pico Blue Immunocreening Kit Instructional Manual.* 1993. La Jolla, CA: Stratagene.

one of the four ddNTPs results in the termination of enzyme-catalyzed polymerization at each site where the ddNTPs can be incorporated. Complete sequence information can be obtained utilizing four separate reactions each with a different ddNTP. To visualize the labeled chains of various lengths by autoradiography following separation by high resolution electrophoresis, a radioactively labeled nucleotide is included in the kit (USB Technical Bulletin, Sequenase Version 2.0 DNA Sequencing Kit). The brief protocol for the DNA sequencing sections is presented in Figure 7.17. Each of the steps is expanded upon in "detailed notes or protocols" in the aforementioned USB Technical Bulletin. The 28-page bulletin, which is available from United States Biochemical (PO Box 4 Cleveland, Ohio, 44122), should be read critically in order to understand all of the subtleties involved in each of the steps mentioned in the brief protocol.

## *DNA Sequencing Gels*

The preparation, loading, running, autoradiography, and "reading" of a sequencing gel require considerable time as sequencing gels are rather long (40 cm–50 cm). Since the preparation of the gel is rather time-consuming, it is recommended that the gel be prepared 25 hours in advance of the laboratory, as denaturing polyacrylamide gels are stable up to 24 hours at room temperature. The details for preparing these gels are presented on pages 13.45–13.53 Sambrook et al.[23] Pages 13.54–13.58 present protocols for loading and running gels (13.54–13.55), autoradiography (13.56– 13.57), and reading gel sequences (13.58). Here we abbreviate these procedures in sufficient detail to perform each of the steps. The major variables relevant to denaturing polyacrylamide sequencing gels are depicted in Table 7.2.

As mentioned, pages 13.47–13.53 of Sambrook et al.[23] are concerned with the preparation of buffer-gradient polyacrylamide gels (Figure 7.18). The loading and running of the gels are as follows:

a. Following polymerization of the gel, remove any dried polyacrylamide/urea external to the gel mold. Remove the shark's tooth comb from the gel's top, and remove the electrical tape from the bottom of the gel mold. Sambrook et al.[23] recommend that tape removal can be accomplished with a scalpel cutting the tape into several segments.

b. Attach the gel mold to the electrophoresis apparatus with bulldog binder clips or plastic-coated laboratory clamps. Fill the top reservoir with 0.5X TBE and the bottom reservoir with TBE.

c. Remove the cover from the microliter plate containing the completed sequencing reactions. Isolate the open plate in an incubator or float it in a 80° C for 10 minutes.

d. While the plate is incubating, fill a 10-ml hypodermic syringe with attached 22-gauge needle with 1X TBE. To remove any fragments of urea and polyacrylamide, squirt the TBE across the submerged flat loading surface.

## Sequencing Reaction Protocols

All sequencing reactions are run in small plastic centrifuge tubes (typically 0.5 ml), these should be kept capped to minimize evaporation of the small volumes employed. Additions should be made with disposable-tip micropipettes and care should be taken not to contaminate stock solutions. The solutions must be thoroughly mixed after each addition, typically by "pumping" the solution two or three times with the micropipette, avoiding the creation of air bubbles. At any stage where the possibility exists for some solution to cling to the walls of the tube, it should be centrifuged. With care and experience these reactions can be completed in 10–15 minutes.

### Brief Protocols

1. Denature double-stranded templates.
2. Annealing mixture:

   | | |
   |---|---|
   | DNA | _μl (Up to 7 μl) |
   | $H_2O$ | _μl (To adjust total volume) |
   | Reaction Buffer | 2 μl |
   | Primer | 1 μl |
   | | ___ |
   | Total | 10 μl |

   Anneal by heating 2 min, 65° C then cool slowly to <35° C over 15–30 minutes. Centrifuge briefly and chill on ice for use in step 6.
3. While cooling, label, fill and cap tubes with 2.5 μl of each Termination Mixture (G,A,T and C). Use mixtures from red-capped tubes for dGTP or orange-capped tubes for dITP. Keep covered at room temperature for steps 5 and 7.
4. Dilute Labeling Mix 5-fold to working concentration if needed, dGTP (green capped tube) or dITP (yellow capped tube). Retain for use in step 6.

   Labeling Mix _____ μl (Typically 2 μl)

   $H_2O$ _____ μl (Typically 8 μl)
5. Pre-warm 4 termination tubes from step 3 (G,A,T and C) in 35° C bath.
6. LABELING REACTION

   | | |
   |---|---|
   | To ice-cold Annealed DNA Mixture | (10 μl) |
   | Add: DTT, 0.1M | 1 μl |
   | Diluted Labeling Mix | 2 μl |
   | [35S] dATP | 0.5 μl |
   | Diluted Sequenase Polymerase | 2 μl |
   | TOTAL | 15.5 μl |

   Mix and incubate at room temperature 2–5 minutes.
7. TERMINATION REACTIONS

   Transfer 3.5 μl of labeling reaction to each termination tube (G,A,T and C), mix and continue incubation of the termination reactions at 37° C for 5 minutes.
8. Stop the reactions by adding 4 μl of Stop Solution.
9. Heat samples to 75° C, 2 minutes. immediately before loading onto sequencing gel. Load 2 – μl in each lane.

**FIGURE 7.17**  Summary of dideoxy chain termination from protocol for sequencing. Reprinted from *Sequenase™ Version 2.0 DNA Sequencing Kit, Step-by-Step Protocols for DNA Sequencing with Sequenase Version 2.0 T7 DNA Polymerase,* 8th edition, with permission from United States Biochemical Corporation, Cleveland, Ohio.

**TABLE 7.2**  Major Variables Relevant to Denaturing Polyacrylamide Sequencing Gels

| Variable | Remarks |
|---|---|
| Gel length | 40 cm–50 cm is standard, but apparatuses exist for 100 cm long gels. Note: Long gels are difficult to process requiring large containers for fixation and exposure to X-ray film. |
| Gel width | 20 cm gels (narrow) contain up to 40 lanes and are easier to bundle than 40 cm gels. |
| Gel thickness | Gel thickness is defined by the thickness of the thin plastic strips employed as spacers between the front and back glass plates. The thicknesses of gels are standard sequencing 0.3 mm–0.4 mm, thin 0.2 mm and thick 0.6 mm. Thin gels yield excellent resolution and are fragile; thick gels accept longer volumes and are difficult to fix and dry. |
| Cross-sectional stage of the gel | Tapered or wedge-shaped gels are thicker at the bottom (0.6 mm–0.75 mm) than at the top (0.25 mm). These gels increase resolution by compressing the spacing between adjacent DNA bands through lowering electrical resistance. Disadvantages to wedge-shaped gels—longer to dry and often crack. Also, resolution of bands at the bottom is sometimes inadequate. |
| Loading slots | Two types: conventional, employing a plastic template; and slots, inserting a shark's tooth comb.<br>a) Conventional slots—insert a plastic template into the top of the acrylamide prior to polymerization. Wash slots with buffer after removing the template, before loading the samples.<br>b) Shark's tooth comb—create slots by adding a shark's tooth comb into the top of a gel after polymerization. |
| Gel temperature gradient | Temperature gradient occurs as a gel "heats-up" during electrophoresis, results in curved DNA bands. Simplest temperature control systems are plastic-coated metal plates clamped to the glass plates holding the gel. In some apparatuses, the electrophoresis apparatus cools the back of the gel. |

Adapted from Sambrook, J., E. F. Fitsch, and T. Maniatis. 1989. *Molecular Cloning: A Laboratory Manual,* 2d ed. Cold Spring Harbor, NY: Cold Spring Harbor Laboratory Press.

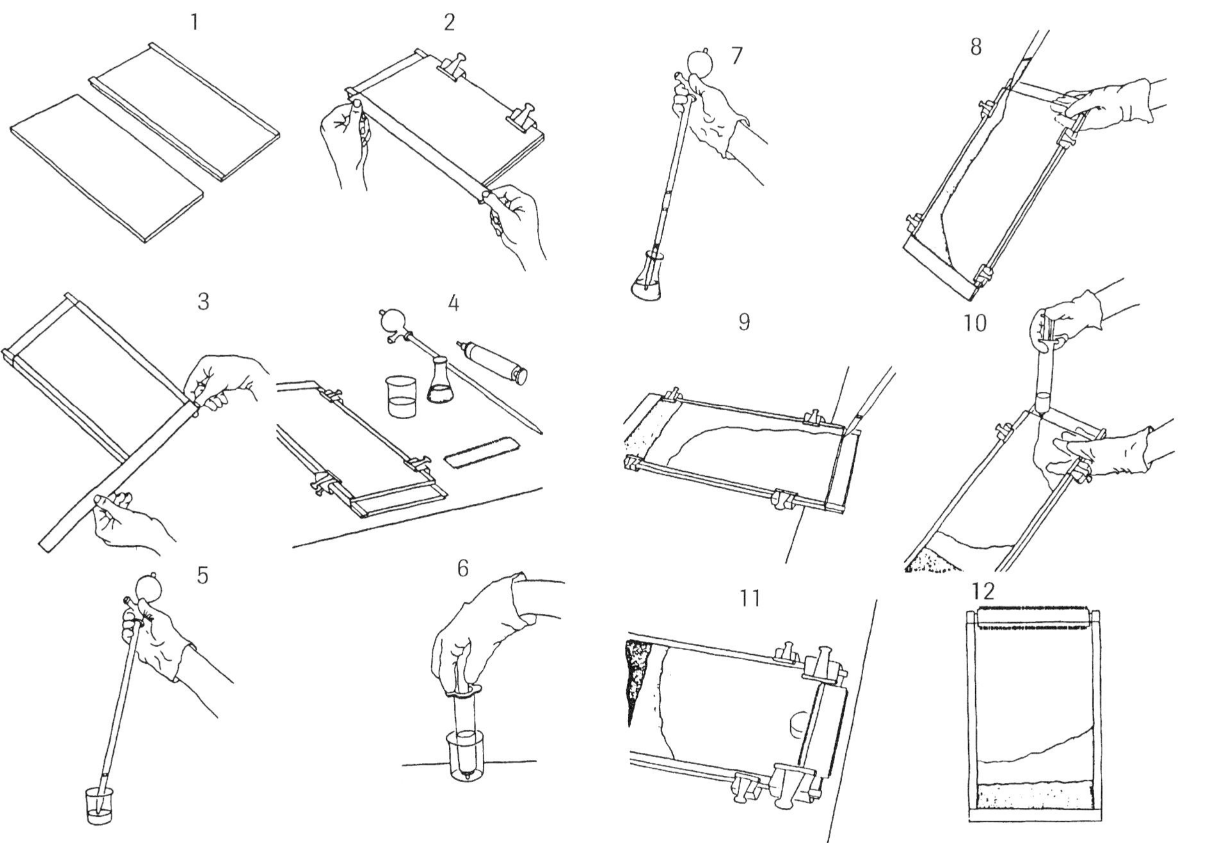

**FIGURE 7.18**  Preparation of sequencing gels.

Adopted from Sambrook, J., E. F. Fitsch, and T. Maniatis. 1989. *Molecular Cloning: A Laboratory Manual,* 2d ed. Cold Spring Harbor, NY: Cold Spring Harbor Laboratory Press.

e. Reinsert the shark's tooth comb with its teeth protruding into the loading surface of the gel. Remove the comb and wash the slots of the gel with 1X TBE.

f. Remove the microtiter plate from the 80° C oven or water bath and transfer it immediately into packed ice. Keep the plate at 80° C until the samples have been loaded onto the gel.

g. Load 1 ml–2 ml of each sequencing reaction into adjacent gel slots maintaining a record of the templates[14] order. Note: load the samples in every reaction set in the same order. TCGA is the best order. Sambrook et al.[23] discuss the methods for loading the samples into the gel's wells. Please include a Hamilton syringe equipped with a 30-gauge needle, and automatic micropipettor (20µl) or a drawn-out glass capillary.

h. Following loading of the samples, connect the electrodes to the power pack and the electrophoresis apparatus. Run the gel at constant power (35–40 W for a 20 cm × 40 cm gel, ~1700) until optimal resolution. Estimate the time by the migration of the marker dyes in the formamide/EDTA/XC/ BPB gel-loading buffer (Table 7.3).

## Autoradiography of Sequencing Gels

a. Upon completion of the electrophoresis, turn off the power and disconnect the sequencing apparatus from the power pack. Remove the gel from the apparatus and dispose of the electrophoresis buffer.

b. Place the gel mold flat onto plastic-backed protective bench paper with the "notched" plate uppermost. Following the removal of any electrical tape, slowly pry apart the mold's plates with a metal spatula.

c. After separating the plates, "cut-off" the bottom cover of the side of the gel that was "loaded" first, thereby orienting the gel for subsequent manipulation.

d. Transfer the gel (together with its supporting plate) to a shallow bath containing 10% methanol and 10% acetic acid in $dH_2O$ in order to fix the gel and remove urea and sucrose. (Note: agitation not requested.) The agents prohibited the gel from drying completely causing it to adhere to the autoradiographic film. The time needed for fixation depends upon the thickness of the gel—e.g. 5, 15, and 40 minutes for 0.2, 0.4 and 0.6 mm gel thickness, respectively.

e. Pickup the supporting plate by its edges and lift the gel from the fixation fluid being careful that the gel does not slide from the plate after the plate is clear of the fixation fluid; secure the gel with a gloved hand and tilt the plate permitting excess fixation fluid to drain.

f. Place the plate (gel uppermost) onto a piece of protective paper removing any folds or disturbances by gently "kneading" the gel with gloved fingers. Remove excess fixation fluid from the glass plate with kimwipes without touching the gel's surface.

g. Place a piece of Whatman 3MM paper over the gel. Cut the paper to be slightly larger (2 cm–3 cm) than the gel in both length and width, and center over the gel. Apply slight pressure so that the gel firmly attaches to the paper.

h. Hold the paper in place with one hand and pickup the supporting glass plate with the other. Turn the plate over and lay it on a dry piece of protective paper.

i. Slide the protective paper to the bench's edge. Grasp the leading edge of the 3MM paper and pull it downward while moving the glass plate slowly toward the bench's edge. The gel should adhere to the paper as it is peeled from the glass plate.

j. Place the 3MM paper (gel uppermost) onto two pieces of 3MM paper of the same size. Cut a piece of plastic wrap to be slightly larger than the gel and place it on top of the gel avoiding creases and bubbles. (Use another person.) Hold the covers of the wrap and pull outward stretching it. Lower the stretched warp onto the gel's surface. (Don't attempt to remove the wrap; this will cause the gel to tear.)

k. Trim all three pieces of 3MM paper and the plastic wrap with a paper cutter so that they approximate the size of the gel.

l. Dry the gel 30–40 minutes under vacuum on a commercial gel dryer at 80° C.

**TABLE 7.3** Migration Rates of Marker Dyes through Denaturing Polyacrylamide Gels

| % Polyacrylamide | Bromophenol Blue[a] | Xylene Cyanol FF[a] |
| --- | --- | --- |
| 5 | 35 | 130 |
| 6 | 26 | 106 |
| 8 | 19 | 76 |
| 10 | 12 | 55 |
| 20 | 8 | 28 |

[a]The numbers are the approximate sizes of DNA (in nucleotides) with which the marker dyes will comigrate.
Adapted from Sambrook, J., E. F. Fitsch, and T. Maniatis. 1989. *Molecular Cloning: A Laboratory Manual,* 2d ed. Cold Spring Harbor, NY: Cold Spring Harbor Laboratory Press.

m. Remove the gel from the dryer and peel off the wrap. The dried gel should feel smooth but not sticky. To orient the autoradiography, attach a small adhesive label marked with radioactive ink to the 3MM paper in the space created by cutting the bottom corner of the gel.

n. Construct an autoradiograph by exposing the gel to X-ray film (Kodak XAR-2, XAR-5, or equivalent) for 16–24 hours at room temperature; if possible, use spring-loaded notch cassettes to ensure direct contact between the entire surface of the dried gel and the film emulsion.

o. Reading the gel: After the autoradiograph has been developed, label it with the date and the names of the templates clearly identifying each set of sequencing reactions.

p. Distinguish the left and the right sides of the gel. The radioactive ink image should be at the bottom of the sequencing reaction that was loaded on the gel's first track.

q. When looking for correspondence between the new sequence and the one already known, search for obvious "signatures" such as homopolymeric ones (consecutive T residues) or alternating purine and pyrimidines (GTGTGT). Once observed, these signatures can locate the sequence of interest.

r. Read and record an unknown sequence at least twice by more than one investigator. Next, compare the two readings and resolve discrepancy for additional sequence if required.

s. Develop the autoradiograph and read the DNA sequence.

t. If the gel was loaded in the order TCGA, read the complementary strand's (3'–5') sequence by turning the autoradiograph over and reading the gel from the bottom. Record the tracks on the inverted autoradiograph from left to right, and their order is presumed to be TCGA.

The above is paraphrased from Sambrook et al.,[23] who suggest guidelines for reading the gels. These are:

- single C bands are generally weaker than single bands of the three other nucleotides
- the first A in a homopolymeric run of A's is generally stronger than the rest
- the first C in a homopolymeric run of C's is usually much weaker than the second
- G bands are weak when they are preceded by a T

Figures 7.19 and 7.20 assist in reading gels.

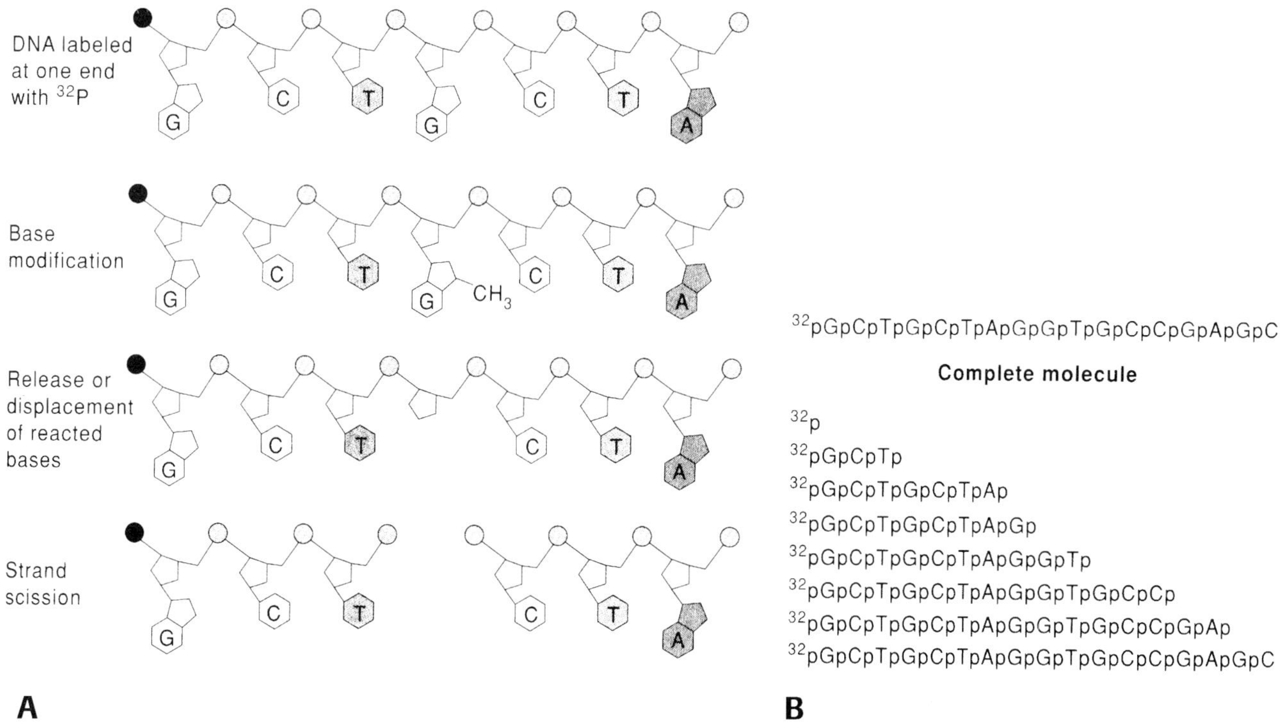



**FIGURE 7.19**  Sequencing end-labeled DNA by limited, base-specific chemical cleavage. **(A)** A sequence of three reactions leading to strand cleavage at a guanine. If the entire DNA is subjected to this reaction sequence for a limited time the result is **(B)**, a family of labeled fragments that will be cleaved at different guanine sites.

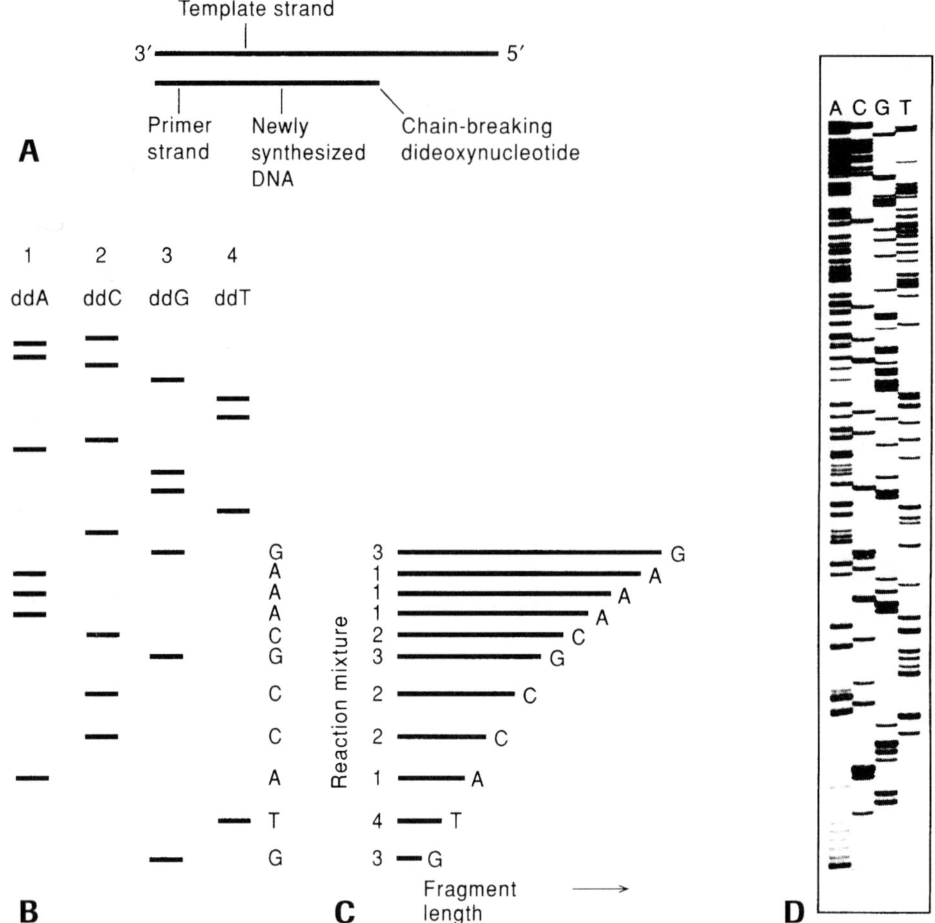



**FIGURE 7.20**  The Sanger dideoxynucleoside method of sequencing DNA. **(A)** A suitable template is chosen, and the primer is chosen so that DNA synthesis begins at the point of interest. In addition to the template-primer complex the reaction mixture contains all four radioactive deoxyribonucleoside triphosphates and small amounts of a single dideoxynucleoside triphosphate. The dideoxy compound serves as a chain terminator. **(B)** After synthesis in the presence of DNA polymerase I, the products of the reaction mixture are separated by gel electrophoresis and analyzed by autoradiography. For a given dideoxy compound all fragments terminating with that particular base should give rise to bands on the gel. The interpretation of the gel pattern is given in **(C)**. The smallest labeled fragment moves the fastest and appears at the bottom of the gel. **(D)** A typical sequencing film. The sequence begins CAAAAAACGG.



# Oligonucleotide Synthesis and Protein Sequencing

## Oligonucleotide Synthesis

There are two main methods for chemical synthesis of oligonucleotides. Table 7.4 compares procedures for the two methods.

The solid phase synthesis of oligodeoxyribonucleotides has been made easier because of modern machine-aided methods. The history of the solid phase method as well as the methods including automation of the solid phase deoxyribonucleotide synthesis have been thoroughly described by Brown and Brown.[24] The Schematic diagram of solid phase oligodeoxynucleotide synthesis is presented in Figure 7.21. Should instructors wish to acquaint the student with such synthesis, students should read the above reference in Eckstein.[25] In addition to the chapter regarding modern machine-aided methods of oligo-dexoyribonucleotide synthesis, the Eckstein volume contains chapters concerned with oligoribonucleotide synthesis, $2^1$-O-methyloligoribonucleotides:

**TABLE 7.4**  Comparison of Two Procedures for the Synthesis of Oligonucleotides

| Features Shared in Common by Procedures for Synthetic Oligonucleotide Synthesis | Phosphodiester | Merrifield Solid-Phase |
| --- | --- | --- |
| $3^l$ activation is involved instead of $5^l$ activation protecting groups as required in intermediate steps to prevent other groups from reacting | A. Sequential treatment of the $5^l$-0-protected nucleoside with:<br>1) RoPC1$_2$<br>2) a $3^l$-0-protected nucleoside<br>3) iodine water<br>B. Repetition of cycle<br>C. Remove all phosphoryl protecting groups and $3^l$-0-terminal protecting group (see Figure 7.21) | Couple $5^l$-OH group of the initial nucleotide to a resin support facilitates stepwise addition of nucleotides |

Summarized from Zubay, G. 1993. *Biochemistry.* Wm. C. Brown Communications, Inc., Dubuque, Iowa. 763–65.

synthesis and applications, phosphorodithioate derivatives, synthesis of oligonucleotide phosphorodithioates, and oligo-$2^l$-deoxyribonucleotide methylphosphonates, nucleotides containing modified bases, oligonucleotides with reporter groups attached to the $5^l$-terminus site specific attachment of labels to the DNA backbone, oligonucleotides for affinity chromatography, oligodeoxynucleotides with reporter groups attached to the base and oligionucleotides attached to interactive, photoreactive and cleavage agents.

### *Protein Sequencing*

Contemporary methods for sequencing proteins can be found in Wittmann-Liebold et al.[26] These are:

1. separation and amino acid analysis of proteins and peptides for microsequencing studies
2. manual and solid-phase microsequencing methods
3 gas-phase and radiosequence analysis.

The volume discusses contemporary methodologies for sensitive amino acid analysis, N-terminal sequence analysis of protein and peptide purification. In addition, recent mass spectrometric approaches are described as an alternative to the stepwise degradative sequence analysis.

Other useful references concerning protein sequencing are a three-volume treatise consisting of review articles[27] and a monograph on the topic by Allen.[28] Finally, there is a variety of protein sequencing references pre-dating 1983, but these are not presented here.

## Review Questions

1. Beginning with sugar, bases, and phosphate, construct a DNA molecule.
2. Compare and contrast the various procedures for isolating genomic DNA. How can you verify its purity?
3. Compare and contrast the isolation of genomic DNA and plasmid DNA.
4. Design an experiment to isolate and purify total RNA.
5. Design an experiment to isolate and purify poly(A)-RNA.
6. What are restriction endonucleases? Suppose that you discovered a new endonuclease; how could you determine its recognition site if you had available to you a variety of synthetic oligonucleotides?
7. Distinguish between cosmids, plasmids, phage, and expression vectors. What factors dictate which you would employ in an evolving cloning protocol?
8. Explain how you would construct a cDNA library. How does it differ form a genomic library? How can cDNA libraries be used?
9. Compare and contrast the Maxam Gilbert and Sanger et al. methods for sequencing DNA.
10. By means of diagrams, demonstrate how a piece of genomic DNA generated by a restriction endonuclease can be inserted into the *E. coli* genome. In addition, detail an experiment that would provide evidence that *E. coli* harbors a DNA insert.

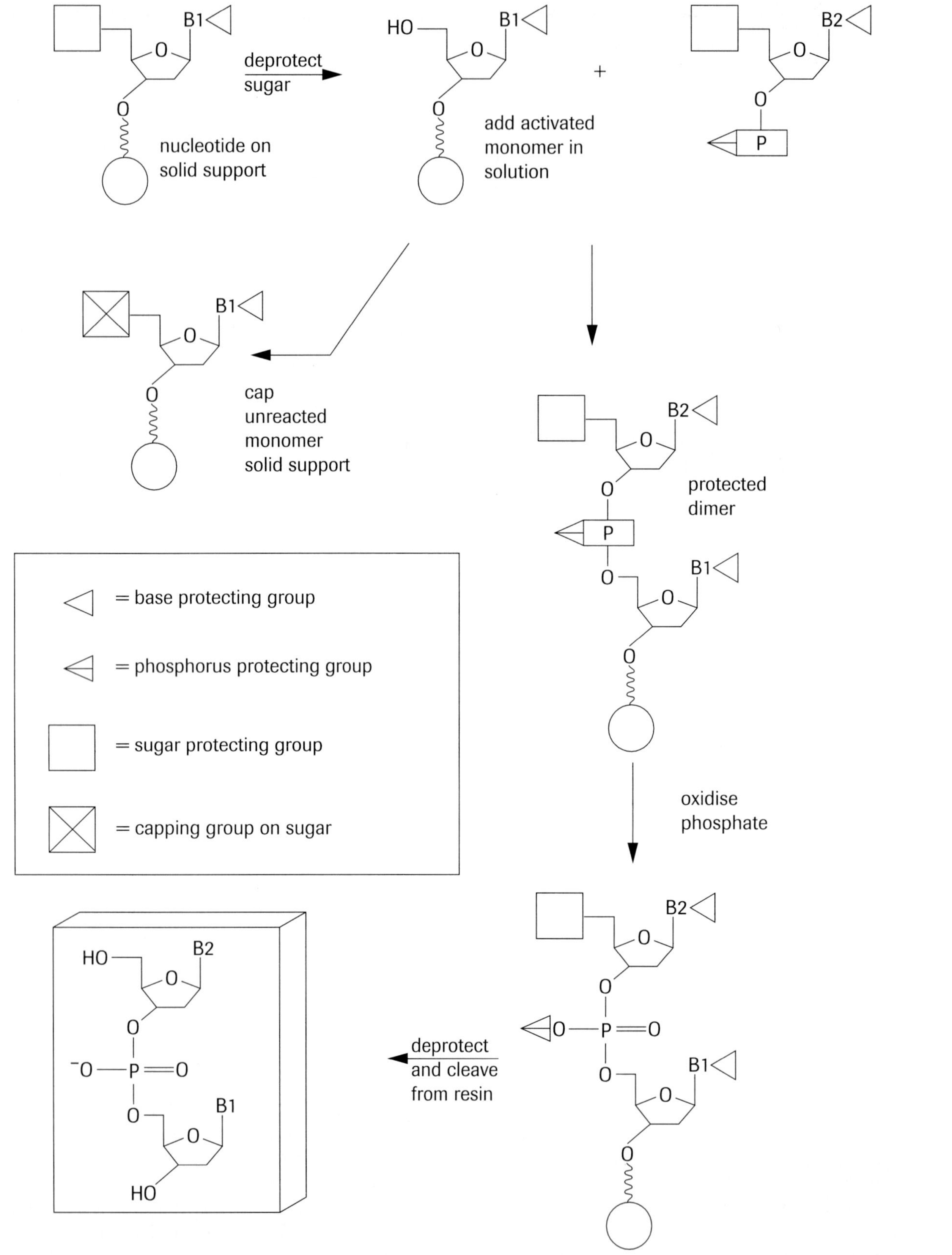



**FIGURE 7.21**   A schematic diagram of solid phase oligonucleotide synthesis.
From Brown, T., and D. J. S. Brown. 1991. "Modern machine-aided methods of oligodeoxyribonucleotide synthesis." *Oligonucleotides and Analogues. A Practical Approach.* Oxford: IRI Press.

11. Design an electrophoretic experiment to verify that a genomic DNA fragment generated by a restriction endonuclease has been inserted into plasmid DNA.

12. Suppose that you incorporated a DNA fragment bearing the gene/genes for PPO into the *E. coli* chromosome; how could you employ antibodies to establish that *E. coli* was expressing the gene(s) for PPO?

# REFERENCES


1. Zubay, G. 1993. *Biochemistry.* 3rd ed. Dubuque, IA: Wm. C. Brown Communications, Inc.

2. Broda, P. M. A., S. G. Oliver, and P. F. G. Sims. 1993. *The Eukaryotic Genome. Organization and Regulation.* Great Britain: Cambridge Univ. Press.

3. Ciferri, O., and L. Dure. 1983. *Structure and Function of Plant Genomes.* New York: Plenum Press.

4. Vloten-Doting, L., G. S. P. Groot, and T. C. Hall. 1985. *Molecular Form and Function of the Plant Genome.* New York: Plenum Press.

5. Beckman, J. S. and T. C. Osborn. 1992. *Plant Genome Methods for Genetic and Physical Mapping.* Netherlands: Kluwer Academic Publishers.

6. Anderson, J. W., and J. Beardell. 1991. *Molecular Activities of Plant Cells: An Introduction to Plant Biochemistry.* Oxford: Blackwell Scientific Publications.

7. Powels, P. H., B. C. Enger-Valk, and W. J. Branman. 1985. *Cloning Vectors: A Laboratory Manual.* Amsterdam: Elsevier.

8. Armitage, P., R. Walden, and J. Draper. 1988. "Vectors for the transformation of plant cells using *Argobacterium.*" *Plant Genetic Transformation and Gene Expression: A Laboratory Manual.* Oxford: Blackwell Scientific Publications.

9. Wilson, J. H. 1985. *Genetic Recombination.* Menlo Park, CA: Benjamin Cummings.

10. Kucherlapati, R., and G. R. Smith. 1988. *Genetic Recombination.* Washington, D.C.: *American Society for Microbiology.*

11. Chomczynski, P., and N. Sacchi. 1987. "Single-step method of RNA isolation by guanidinum thiocyanate-phenol-chloroform extraction." *Anal. Biochem.* 162:156–59.

12. Puissant, C., and M. Houdebine. 1990. "An improvement of the single step method of RNA extraction by acid guanidinum-phenol-chloroform extraction." *BioTechniques.* 8:148–49.

13. Aviv, H., and P. Leder. 1972. "Purification of biologically active globin messenger RNA by chromatography on oligothymidylic acid-cellulose." *Proc. Natl. Acad. Sci. USA* 69:1408–12.

14. Gubler, U., and B. J. Hoffman. 1983. "A simple and very efficient method for generating cDNA libraries." *Gene.* 25:263–69.

15. Young, R. A., and R. W. Davis. 1983. "Yeast RNA polymerase II genes: Isolation with antibody probes." *Science.* 222:778–82.

16. Lathe, R. 1985. "Synthetic oligonucleotide probes deduced from amino acid sequence data: Theoretical and considerations." *J. Mol. Biol.* 183:1–12.

17. Benton, W. D., and R. W. Davis. 1977. "Screening lambda gt recombinant clones by hybridization to single plaques in situ." *Science.* 196:180–82.

18. De Sal, G., G. Manfioletti, and C. Schneider. 1989. "A one-tube plasmid DNA mini-preparation suitable for sequencing." *BioTechniques.* 7:514–20.

19. Sanger, F., S. Nicklen, and A. R. Coluson. 1977. "DNA sequencing with chain-terminating inhibitors." *Proc. Natl. Acad. Sci. USA* 74:5463–67.

20. *Stratagene Uni-Zap™ XR Cloning Kit Instructional Manual.* 1993. La Jolla, CA: Stratagene.

21. *Stratagene Gigapack® Packaging Extract Instructional Manual.* 1993. La Jolla, CA: Stratagene.

22. *Stratagene Pico Blue Immunoscreening Kit Instructional Manual.* 1993. La Jolla, CA: Stratagene.

23. Sambrook, J., E. F. Fitsch, and T. Maniatis. 1989. *Molecular Cloning: A Laboratory Manual,* 2d ed. Cold Spring Harbor, NY: Cold Spring Harbor Laboratory Press.

24. Brown, T., and D. J. S. Brown. 1991. "Modern machine-aided methods of oligodeoxyribonucleotide synthesis." *Oligonucleotides and Analogues. A Practical Approach.* Oxford: IRI Press.

25. Eckstein, F. 1991. *Oligonucleotides and Analogues: A Practical Approach.* Oxford: IRL Press.

26. Wittmann-Liebold, B., 1989. *Methods in Protein Sequence Analysis.* New York: Springer-Verlag.

27. Tschesche, H. *Modern Methods in Protein Chemistry,* vols 1–3. Berlin: de Guyter.

28. Allen, G. 1981. *Sequencing of Proteins and Peptides.* Amsterdam: Elsevier.

29. Lax, A. R., and K. C. Vaughn. 1991. "Colonization of polyphenol oxidase and photosystem II proteins." *Plant Physiol.* 96:26–31.

30. Longeman, I., I. Schell, and J. Willimitzer. 1987. "Improved method for the isolation of RNA from plant tissues." *Analytical Biochemistry.* 163:16–20.


## Supplementary References—Monographs


Beckman, J. S., and T. C. Osborn. 1992. *Plant Genome Methods for Genetic and Physical Mapping.* Netherlands: Kluwer Academic Publishers.

Berger, S. L., and A. R. Kimel. 1987. *Guide to Molecular Cloning Techniques*. Orlando: Academic Press.

Brown, T. A. 1990. *Gene Cloning: An Introduction*. London: Chapman and Hall.

Circy, M. 1993. "Function of the genetic molecule: Transposable elements in lower eukaryotes." *Progress in Botany*. 54:306–17.

Darbre, A. 1986. *Practical Protein Chemistry: A Handbook*. Chichester, NY: John Wiley and Sons.

Dennis, D. T., and D. H. Turpin. 1990. *Plant Physiology, Biochemistry and Molecular Biology*. New York: Longman Scientific & Technical and John Wiley and Sons.

Dennis, E. S., and D. J. Llewellyn. 1991. *Molecular Approaches to Crop Improvement*. New York: Springer-Verlag.

Draper, J., R. Scott, P. Armitage, and R. Walden. 1988. *Plant Genetic Transformation and Gene Expression: A Laboratory Manual*. Oxford: Blackwell Scientific Publication.

Draper, J., and R. Scott. 1988. "The isolation of plant nucleic acids." *Plant Genetic Transformation and Gene Expression: A Laboratory Manual*. Oxford: Blackwell Scientific Publications.

Ehrlich, H. A. 1989. *PCR Technology: Principles and Applications for DNA Amplification*. New York: Stockton Press.

Fini, C. 1990. *Laboratory Methodology in Biochemistry: Amino Acid Analysis and Protein Sequencing*. Boca Raton, FL: CRC Press.

Gelvin, S. B., and R. A. Schilperoort. 1988. *Plant Molecular Biology Manual*. Boston: Kluwer Academic Press.

Glover, D. M. 1985. *DNA Cloning: A Practical Approach*. Oxford: IRL Press.

Gresshoff, P. M. 1992. *Plant Biotechnology*. Boca Raton, FL: CRC Press.

Grierson, D. 1991. *Plant Genetic Engineering*. Glasgow: Blackie.

Griffeths, A. J. F., J. H. Miller, D. T. Suzuki, R. C. Lewontin, and W. M. Gelbart. 1993. *An Introduction to Genetic Analysis*. New York: W. H. Freeman & Co.

Herman, R. G., and B. A. Larkins. 1990. *Plant Molecular Biology*. New York: Plenum Press.

Huynh, T. V., R. A. Young, and R. W. Davis. 1985. In *DNA Cloning: A Practical Approach*. Oxford: IRL Press.

Kurg, Shan-dow, and C. J. Arntzh. 1989. *Plant Biotechnology*. Boston: Dittenvortts

Lunneburg, A., and C. Janson. 1993. "Isolation of plant genes by T-DNA and Transpose Mitogenesis: Gene Mapping." *Progress in Botany*. 54:295–305.

Maniatis, T., E. F. Fritsch, and J. Sanbrook. 1982. *Molecular Cloning: A Laboratory Manual*. Cold Spring Harbor, NY: Cold Spring Harbor Laboratory Press.

Mantell, S. H. 1985. *Principles of Plant Biotechnology: An Introduction to Genetic Engineering in Plants*. Oxford: Blackwell Scientific Publications.

McPherson, M. J., G. R. Taylor, and P. Querke. 1991. *PCR: A Practical Approach*. Oxford: Oxford University Press.

Puhler, A., and K. N. Timmis. 1984. *Advanced Molecular Genetics*. Berlin: Springer-Verlag.

Schlesinger, D. H. 1988. *Macromolecular Sequencing and Synthesis Selected Methods and Applications*. New York: A. R. Liss.

Scott, N., J. Drape, R. Jefferson, G. Cuey, and L. Jacob. 1988. "Analysis of gene organization and expression in plants." *Plant Genetic Transformation and Gene Expression: A Laboratoary Manual*. Oxford: Blackwell Scientific Publications.

Williams, J. 1988. *Genetic Engineering*. Oxford: IRL Press.

## Supplementary References—Journals

Back, A. W., N. C. Lan, D. L. Johnson, C. W. Abel, M. E. Blembenek, S-W Kwan, P. H. Sweeburg, and J. C. Shih. 1988. "cDNA cloning of human liver monoamine oxidase A and B: Molecular basis of differences in enzymatic properties." *Proc. Natl. Acad. Sci.* 85:4934–38.

Cary, J. W., A. R. Lax, and W. H. Flurkey. 1992. "Cloning and characterization of cDNAs coding for *Vicia faba* polyphenol oxidase." *Plant Molecular Biology*. 20:245–53.

Chargwin, J. M., A. E. Przybylar, R. J. MacDonald, and W. J. Ritter. 1979. "Isolation of biologically active ribonucleic acid from sources enriched in ribonuclease." *Biochemistry*. 18:5794–99.

Chen, S., J. C. Shih, and Q-P Xu. 1985. "Inhibition of monoamine oxidase by phenyl azides." *Journal of Neurochemistry*. 45:940–45.

Feinberg, A. P., and B. Vogelstein. 1983. "A technique for radiolabeling DNA restriction fragments to a high specific activity." *Anal Biochem*. 132:6–13.

Ganal, M. W., and S. Tanksley. 1989. "Analysis of tomato DNA by pulsed field gel electrophoresis." *Plant Molecular Biology Report*. 7:17.

Gubler, U., and B. J. Hoffman. 1983. "A simple and very efficient method for generating cDNA libraries." *Gene*. 25:263–69.

Huber, M., G. Hinderman, and K. Lerch. 1985. "Primary structure of tyrosinase for *Streptomyces glauceseens*." *Biochemistry*. 24:6038–44.

Hunt, M. D., N. T. Eannetta, H. C. Yu, S. M. Newman, and J. C. Steffers. 1993. "cDNA cloning and expression of potato polyphenol oxidase." *Plant Molecular Biol*. 21:59–68.

Huynh, T. V., R. A. Young, and R. W. Davis. 1984. "Constructing and screening cDNA libraries in λg10 and λg11." *DNA Cloning Techniques, A Practical Approach.* vol. 1. Oxford: IRL Press.

Lax, A. R. 1991. "Comparison of thykaloid membrane proteins stabilized using Tween 20 or Triton X-114." *Plant Physiol.* 96:122.

Lax, A. R., and K. C. Vaughn. 1991. "Colonization of polyphenol oxidase and photosystem II proteins." *Plant Physiol.* 96:26–31.

Massing, J., R. Crea, and P. H. Seeburgt 1981. "A system for shotgun DNA sequencing." *Nucleic Acids Res.* 9:309–21.

Newman, S. M., N. T. Eannetta, H. Yu, J. P. Prince, D. C. Vicente, S. D. Tanksley, and J. C. Steffens. 1993. "Organization of the tomato polyphenol oxidase gene family." *Plant Molecular Biology.* 21:1035–51.

Proudfoot, N. J., and G. G. Brownlee. 1976. "3[l] Noncoding region sequences in eukaryotic messenger RNA." *Nature.* 263:211–14.

Sanger, F., S. Nicklen, and A. R. Cousen. 1977. "Sequencing by chain termination with dideoxynucleotides." *Proc. Natl. Acad. Sci. USA* 74:5463–67.

Shahar, T., N. Hening, T. Guttfinger, D. Harever, and E. Lifschitz. 1992. "The starch 66.3 KD polyphenol oxidase gene molecular identification and developmental expression." *Plant Cell.* 4:135–47.

Stephen, D., C. Jones, and J. P. Schofield. 1990. "A rapid method for isolating high quality plasmid DNA suitable for DNA sequencing." *Nucl. Acids. Res.* 18:7463–64.

Upcroft, P., and A. Healey. 1993. "PCR priming for the restriction endonuclease site 3[l] extension." *Nucleic Acids Res.* 21:4854.

Webber, A. N., K. A. Platt Altoria, R. I. Heath, and W. W. Thomson. "The marginal regions of thylakoid membranes: A partial characterization by polyethylene sorbitan monolaurate (Tween 20) solubilization by spirach thykaloids." *Physiol. Plant.* 72:288–97.

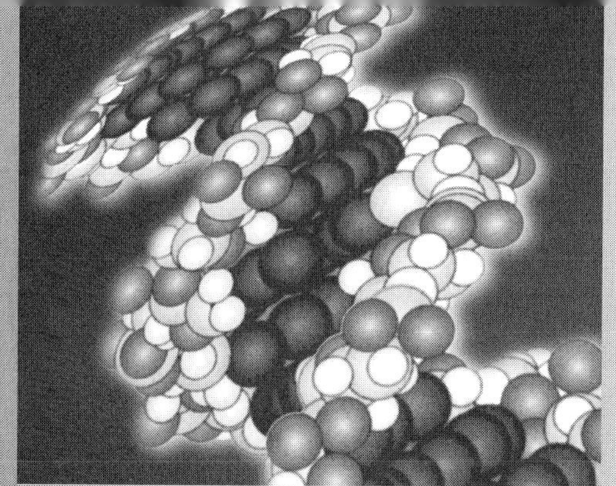



# DNA Amplification Using the Polymerase Chain Reaction

## Outline of Module

ISOLATION OF TOMATO DNA

*Duration:* 3 hours

AMPLIFICATION OF POLYPHENOL OXIDASE FROM TOMATO

*Duration:* PCR setup 1 hour, thermocycler operation 2–3 hours

ESTIMATION OF DNA FRAGMENT SIZES AND DNA QUANTIFICATION

*Duration:* 2 hours

# MODULE INTRODUCTION

The polymerase chain reaction (PCR) is a commonly used method to amplify DNA *in vitro* for molecular analysis. This reaction can be thought of as a modification of DNA replication, since only small portions of the entire DNA strand are being replicated instead of the entire genome. Reactions are conducted in microfuge tubes containing a buffer, dATP, dCTP, dGTP, dTTP, single-stranded oligonucleotide primers, a thermostable DNA polymerase, and the target DNA to be amplified (Figure 8.1). Target DNA can be from any source, including eukaryotes, prokaryotes, and viral cells. In addition, for some PCR techniques, whole cells can be used without prior DNA extrac-

tion.[1] Each primer sequence is chosen from the region of the gene to be amplified; one sequence from each end of the DNA. PCR reactions are conducted using a thermocycler that heats and cools the contents of the tube to the desired temperature for each cycle of the reaction. A cycle of the PCR includes denaturation, annealing, and extension steps (see Figure 8.1). The thermocycler heats the reaction mix to separate the target DNA into single strands and then cools to allow the oligonucleotide primers to attach to their complimentary sequences followed by completion of the synthesis of the DNA fragment. This sequence represents one cycle of the polymerase chain reaction (see

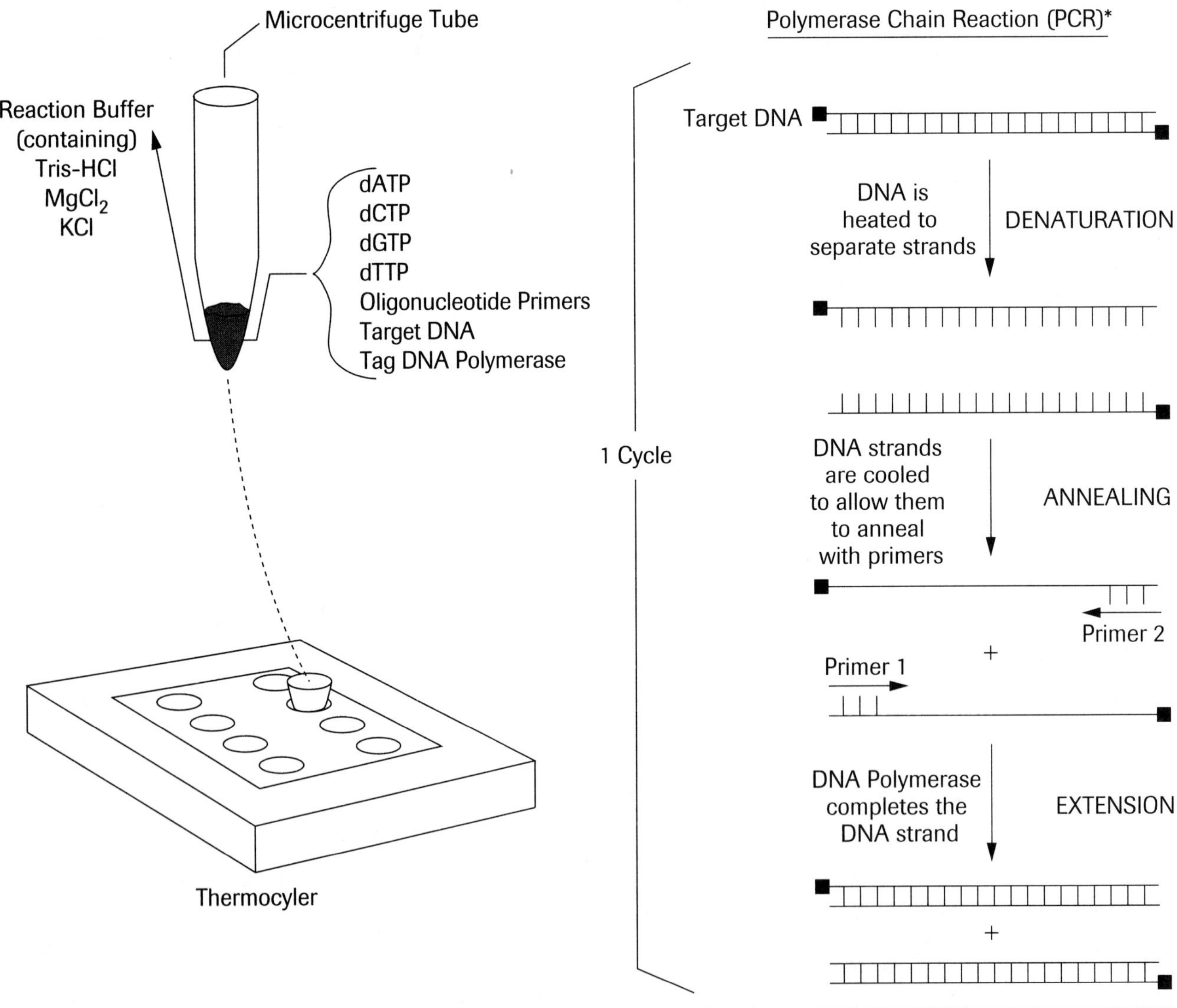


**FIGURE 8.1**   DNA amplification.

Figure 8.1). The total number of cycles typically ranges from 15–45. Also the denaturation, annealing, extension times, and temperature vary for each DNA fragment to be amplified. In general, the annealing temperature is 3°–5° C below the $T_m$ of the primers used for the DNA amplification, while the extension times are often set at one minute per 1000 nucleotides to be amplified.[2,3] However, denaturation, annealing, and extension times also depend on the type of thermocycler used for conducting PCR.

The polymerase chain reaction has had a dramatic influence on molecular biology in a short time. This technique is being used for numerous applications because it is so sensitive (picogram quantities of DNA can be detected easily) and because the yield of amplified product can be so high. A few of the applications include the modification of the DNA fragment to aid in cloning or molecular analysis, gene expression studies, genome characterization, and mutation detection.

Ends of primers used in the PCR reactions can be modified so they contain restriction sites (site recognized by a restriction enzyme) at their 5' end.[4] Newly amplified DNA would then contain a restriction site at the end. A different restriction site could be added to each end of the PCR fragment by modification of both primers. These restriction sites could facilitate putting this piece of DNA into a cloning vector.

Another way PCR can be used as a tool for recombinant DNA techniques, is the use of mixed (degenerate) primers. Typical PCR reactions contain primers derived from a known DNA sequence. However, it is possible to use the amino acid sequence to design primers.[5] Since many amino acids are specified by more than one codon, primers derived from the amino acid sequence are usually a mixture of possible sequences. These mixed (degenerate) primers can be used to clone a DNA fragment or the entire gene.[5,6] PCR procedures can also be used to characterize regions of the genome without prior DNA or amino acid sequence information.

Regions that reside beyond a known DNA sequence can be amplified using inverse PCR.[7,8] In this procedure, primers are chosen that will amplify away from the known sequence. The target DNA is cut with a restriction enzyme and then circularized. When the PCR reaction is conducted, the region outside the known sequence is amplified.

Different regions of a gene or segments from completely different genes can be combined into unique combinations using recombinant PCR.[9,10] This procedure can be used to combine any two DNA segments. For example, a promoter sequence can be added to a coding sequence to produce a unique promoter-gene

combination. Joining fragments together may help contribute to an understanding of the function of the gene segment. For instance, recombinant PCR could help define the region of the gene coding for the active site of a protein.

Regulation of gene expression is a vital part of developmental biology and molecular genetics. One of the ways to measure the activity of a gene is to quantitate the level of transcription of the gene. The polymerase chain reaction is being used to quantitate mRNA levels for different proteins in the cell,[11–15] or the mRNA levels of multiple forms of the same enzyme.[16] DNA can be amplified for use in *in vitro* transcription and translation systems by inserting promoter sequences at the 5' end of one of the primers. The amplified DNA is transcribed in an *in vitro* transcription system with RNA polymerase to produce mRNA and put into an *in vitro* translation system to produce the protein product.[17,18]

Genome characterization can also be studied using PCR techniques. For instance, the Alu class of repetitive sequences is commonly found in mammals. This class of repeats has conserved regions (like TL65) from which a primer can be made and used to amplify chromosome specific regions between two Alu sequences that are in opposite directions.[19–21] Alu PCR and other techniques, such as vectorette PCR, which is used for genomic walking,[22] are being used by scientists working on the human genome project to locate human genes of interest. Alu PCR is unique because only one primer is used; other techniques are also modifications of the basic PCR protocol.

Random amplified polymorphic DNA (RAPD) are generated using another modification of the basic PCR procedure. This technique and other similar methods use one short primer of about one-half the length of a normal primer. Since the primer is short, it will recognize and amplify a number of regions.[23–25] The RAPD procedure is one of the most widely used techniques since it requires no prior sequence information. This procedure has been used to amplify a number of bands that are then used for linkage experiments. The RAPD generated markers have been used to locate disease resistant genes,[26–28] to measure genetic diversity,[29] and for DNA fingerprinting.[30]

One of the most important applications of PCR is the detection of mutations and genes causing genetic disorders.[31–35] In fact, PCR can be used to detect changes that occur in certain genes in cancer patients.[36,37] PCR is one of the most widely used techniques used in molecular biology. The purpose of this laboratory exercise is to acquaint the student with this procedure and related techniques in molecular biology.

# Laboratory Exercises*

## ISOLATION OF TOMATO DNA

### Supplies and Equipment
Sterile microfuge tubes
Sterile pipette tips
Microcentrifuge
65° C water bath
Kontes disposable pellet pestle with tubes
Pipetter

### Chemicals
Liquid nitrogen
Tomato leaf tissue
70% ethanol
95% ethanol
Phenol/chloroform/isoamyl alcohol
24:1 chloroform/isoamyl alcohol
CTAB extraction buffer: [4% CTAB (hexade-
cyltrimethyl ammonium bromide), 5 M NaCl,
2 M Trizma base (pH 8.0), 0.5 M EDTA][40]
TE buffer: 0.01M Tris (pH 7.5), 0.001 M EDTA

## Introduction

Before a DNA fragment can be amplified using the polymerase chain reaction, the DNA to be used as a target must be purified. A number of techniques are available to isolate DNA.[38–41] Exercise 1 demonstrates the isolation of high molecular weight tomato DNA.

## Procedures

Target DNA for the PCR procedure will be isolated from tomato leaf tissue purchased from a garden center or stored in a –70° C freezer.

a. Before starting the isolation procedure, set a water bath at 65° C and heat the CTAB buffer in the bath. A tomato leaf weighing 200 mg–300 mg should be used for the isolation procedure. After weighing the tissue, gently wipe the leaf off with alcohol.

b. Put the tissue in a sterile 1.5-ml Kontes microfuge tube and immediately add liquid nitrogen into the tube.

c. Use a sterile Kontes disposable pellet pestle to crush the frozen tissue.

d. Immediately add 600 µl of the CTAB extraction buffer, vortex the tube for 15 seconds, and then put the micro-centrifuge tube into the 65° C water bath for 45 minutes.

e. Remove the tube from the water bath and add 600 µl of phenol/chloroform/isoamyl alcohol to the tube and mix by gentle inversion. Note: If there is a fume hood available this step should be conducted in the hood. The experiment could be done using chloroform/isoamyl alcohol at this step instead of phenol/chloroform/isoamyl alcohol.

f. Centrifuge the tube for 4 minutes at 14,000 rpm in a microcentrifuge.

g. Remove the tube from the microcentrifuge, carefully remove the top layer and put it in a fresh 1.5-ml microcentrifuge tube (any sterile microcentrifuge tube). Be careful not to transfer any of the bottom layer. *Be sure to draw the liquid up slowly!*

h. Add 600 µl chloroform/isoamyl alcohol to the tube and mix it using gentle inversion.

i. Centrifuge the tube at 14,000 rpm for 5 minutes and then transfer the upper layer to a new microcentrifuge tube.

j. Add 900 µl–1000 µl of 95% ethanol and invert gently several times to ensure mixing. Put the tube in a –20° C freezer for 30 minutes to 1 hour. The longer that you keep the solution in the freezer, the greater the yield of DNA that will be obtained.

k. Centrifuge the tube for 20 minutes at 14,000 rpm to pellet the DNA.

l. Pour off the ethanol and wash the pellet in 70% ethanol and centrifuge for 3 minutes at 14,000 rpm.

m. Pour off the 70% ethanol and invert the tubes to allow the alcohol to drain out while the DNA pellets air dry.

n. Redissolve the pellet in 50 µl TE buffer.

o. Quantitate the concentration of the DNA sample using a miniflourometer or spectrophotometer, or using a gel method. A quick method to estimate the quantity of your DNA sample is to determine the OD260 reading using a spectrophotometer. Take a small aliquot of the DNA stock and dilute it with water in a 1ml cuvette. An OD 260 reading of 1 corresponds to 50 µg/ml of double stranded DNA.[42] As an alternative to using the spectrophotometer, the DNA can be quantitated using gel electrophoresis (see exercise 3).

## AMPLIFICATION OF POLYPHENOL OXIDASE FROM TOMATO

### Supplies and Equipment
Sterile microfuge tube (0.6 ml)
Pippeter, sterile tips

---

*Contributed by Dr. D. E. McMillin

Gloves
Thermocycler

***Chemicals***
Sterile distilled water
PPO specific primers
10X reaction buffer (with 1.5 mM Mg)
Mineral oil
dNTP's
Target DNA

NOTE: A typical PCR reaction mix contains a total volume of 100, 50, or 25 microliters. Since the reaction volume is so small, *pipeting errors are magnified;* therefore, it is important that the student try to pipette as accurately as possible.

# Introduction

In this exercise you will amplify a portion of the polyphenol oxidase gene from tomato. The primers used for this experiment were chosen from the published gene sequence.[43]

# Procedures

a. Remove the microcentrifuge tubes from the freezer labeled dATP, dCTP, dGTP, dTTP, and 10X reaction buffer and thaw each tube; then put on ice. Also remove the PPO primers (PPO1 and PPO2) and isolated tomato DNA from the refrigerator.

b. Take a sterile microcentrifuge tube, label it, and add the following components being sure to vortex each component before you add it to your tube:

    36.5 µl $H_2O$
    5.0 µl 10X reaction buffer
    1.0 µl dATP
    1.0 µl dCTP
    1.0 µl dGTP
    1.0 µl dTTP
    1.0 µl PPO1 primer (10pMole)
    1.0 µl PPO2 primer (10pMole)
    1.0 µl target DNA (50 ng)
    1.0 µl BSA (200 µg/ml)

c. Put the tube on ice until all groups are finished.

d. Heat the tubes in a thermocycler for 5 minutes at 95° C.

e. Remove the tubes (carefully) from the thermocycler and put back on ice. Your teacher will add 0.5 µl DNA polymerase to each of the tubes, being careful to mix the enzyme before the aliquot is removed.

f. Take the tube to the thermocycler and add 2 drops mineral oil into the tube and 1 drop into the well of the thermocycler.

g. The PCR protocol that will be run will be denaturation at 94° C for 1 minute, annealing at 55° C for 1 minute and 30 seconds, and extension at 72° C for 2 minutes for 37 cycles. The 37 cycles will be followed by a 7 minute period at 72° C to allow all DNA fragments to complete extension.

h. The product of the PCR reaction will be analyzed using agarose gel electrophoresis.

# ESTIMATION OF DNA FRAGMENT SIZES AND DNA QUANTIFICATION

***Supplies and Equipment***
Sterile pipette tips
Small electrophoresis gel chamber
Power supply
Semi-log paper
Microwave
250-ml flask
Balance
Pipetter
Parafilm or foil
Film
Transilluminator/UV shield or glasses
Trays for staining the gels
Kitchen spatula

***Chemicals***
High melting agarose
pGEM DNA marker
TAE buffer (0.04 M Tris-acetate, 0.001 M EDTA)
60% sucrose
Dye solution
Known DNA
Ethidium bromide (10 mg/ml)

# Procedures

Gel electrophoresis is a commonly used technique for numerous types of molecular biology techniques. After PCR amplification, the size of the DNA fragments are estimated using gel electrophoresis. This helps to ensure that the proper fragment was amplified.

a. Weigh out 0.5 g of high-melting agarose, put it in a flask, and mix it with 50 ml of TAE buffer to make a 1% solution. Heat the solution in a microwave or melt the agarose in a water bath. If you use a microwave, be sure to watch the gel mixture to ensure that the solution will not overflow.

b. While the agarose is cooling, assemble the gel tray with a thick comb for the wells.

c. Pour the agarose solution into the gel tray and allow the gel to harden.

d. When the gel is completely formed, remove the comb and cover the gel with 1X TAE buffer.

e. Cut a piece of parafilm 1 inch in length and put it near the gel.

f. Mix 4 µl of loading dye with 0.6 µl of the pGEM marker (your teacher may have diluted the marker before the class used it).

g. Mix 4 µl loading + 6 µl 60% sucrose with the PCR-amplified DNA fragment from the last experiment in a microfuge tube. (Usually the student should load as much of the sample as possible to account for errors.)

h. Carefully load the marker first and then all the samples. Load the samples by *slowly* expelling the sample into the well through the buffer. Load as much as possible without causing the sample to run into the next well.

i. When all the samples are loaded, close the lid of the gel chamber and attach the electrodes to the power supply. Be sure that the red electrode is at the opposite end from where you loaded your samples.

j. Turn on the power supply and run the gel at 100 volts until the dye marker migrates 75% of the way down the gel.

k. Turn off the power supply.

l. Your instructor should set up an isolated area with trays for the straining of your gel. To each tray the instructor will add enough water to cover the gel and approximately 25 µl of ethidium bromide stain stock and stain for 20–25 minutes. *Ethidium bromide is a mutagen!*

m. After the gels have stained, remove them using a kitchen spatula, and put on the transilluminator to be viewed or photographed. Always use protective eyeware!

n. Using the photograph, measure the distance in centimeters from the end of the well to each DNA band of the pGEM marker and your amplified fragment.

o. A linear relationship exists for the log of the size in base pairs of a DNA fragment and electrophoretic mobility of the fragment. The smaller a DNA fragment is, the further it migrates down the gel. On semilog paper plot the electrophoretic mobilities (distance from the well to the DNA band) for all the DNA standards as a function of the fragment length. The fragment length of each fragment for the pGEM marker is shown in Figure 8.2. From the graph, estimate the fragment size of the PCR fragment you amplified. If your sample did not amplify, use a classmate's lane. Compare your gel with the gel shown in Figure 8.2; use lane 2 to estimate the DNA fragment size shown in the figure (top band should be tomato PPO) and a unknown (lane 3). Note: gel electrophoresis can also be used to quantitate DNA.[42] Known quanti-

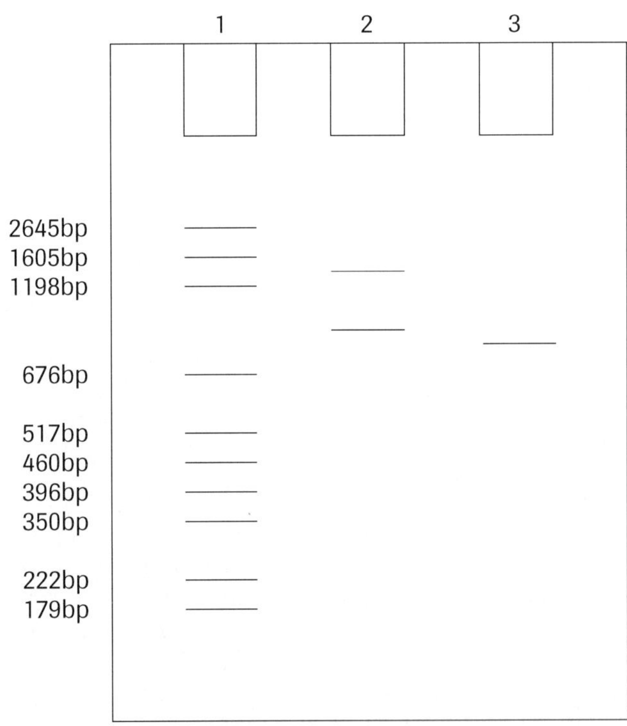



**FIGURE 8.2** Agarose gel electrophoresis. Lane 1— pGEM DNA markers; Lane 2—typical amplification of tomato PPO; Lane 3—unknown DNA fragment.

ties of DNA are loaded onto the gel with an unknown sample; after electrophoresis and staining the intensity of the unknown DNA is compared with the known standards.

## Review Questions

1. How would the length of the primer influence the Tm of that primer?
2. What changes would you make to the protocol listed if you wanted to amplify a DNA fragment 5,000 bp in length?
3. Why would contamination be a particular concern for a PCR application involving human disease diagnosis?
4. Why was the use of a thermostable DNA polymerase so important for DNA amplification?
5. What is the purpose of the denaturation temperature? What would happen if this temperature was too low?
6. What factors must be considered if one wanted to use the PPO1 and PPO2 primers from tomato to amplify the region from the polyphenol oxidase gene of a different organism?
7. What are the estimated sizes of the amplified DNA fragments?
8. What is the approximate size of the DNA bands in lane 2 and 3 of Figure 8.2?

# REFERENCES


1. Mercier, B., C. Gaucher, O. Feugeas, and C. Mazurier. 1990. "Direct PCR from whole blood, without DNA extraction." *Nucleic Acids Res.* 18: 5908.
2. Innis, M. A., and D. H. Gelfand. 1990. "Optimization of PCR's." *PCR Protocols: A Guide to Methods and Applications.* New York: Academic Press. 3–12.
3. Saiki, R. K. 1989. "The design and optimization of the PCR." *PCR Technology: Principles and Applications for DNA Amplification.* New York: Stockton Press.
4. Higuchi, R. 1990. "Using PCR to engineer DNA." *PCR Technology: Principles and Applications for DNA Amplification.* New York: Stockton Press. 61–70.
5. Compton, T. 1990. "Degenerate primers for DNA amplification." *PCR Protocols: A Guide to Methods and Applications.* New York: Academic Press. 39–45.
6. McMillin, D. E., L. L. Muldrow, S. J. Leggette, Y. Abdulahi, and U. M. Ekanemesang. 1991. "Molecular screening of *Clostridium difficile* toxins A & B genetic determinants & identification of mutant strains." *FEMS Microbiol. Lett.* 78:75–80.
7. Huang, S. H., C. H. Wu, B. Cai, and J. Holcenberg. 1993. "cDNA cloning by inverse polymerase chain reaction." *PCR Protocols: Current Methods and Applications.* Totowa, NJ: Human Press. 349–56.
8. Triglia, T., M. G. Peterson, and D. J. Kemp. 1988. "A procedure for *in vitro* amplification of DNA segments that lie outside the boundaries of known sequences." *Nucl. Acids Res.* 16:8186.
9. Horton, R. M, H. D. Hunt, S. N. Ho, J. K. Pullen, and L. R. Pease. 1989. "Engineering hybrid genes without the use of restriction enzymes: gene splicing by overlap extension." *Gene.* 77:61–68.
10. Horton, R. M. 1993. "*In vitro* recombination and mutagenesis of DNA: SOEing together tailor-made genes. *PCR Protocols: Current Methods and Applications.* Totowa, NJ: Human Press. 241–50.
11. Owczarek, C. M., P. Enriquez-Harris, and N. J. Proudfoot. 1992. "The primary transcription unit of the human-2-globin gene defined by quantitative RT/PCR." *Nucleic Acids Res.* 20:851–58.
12. Murphy, L. D., C. F. Herzog, J. B. Rudick, A. Y. Fojop, and S. E. Bates. 1990. "Use of the polymerase chain reaction in the quantitation of mdr-1 gene expression." *Biochemistry.* 29:10351–56.
13. Robinson, J. M., and M. I. Simon. 1991. "Determining transcript number using the polymerase chain reaction: Pgk-2, mP2 and PGK-2 transgene mRNA levels during spermatogenesis." *Nucleic Acids Res.* 19:1557–62.
14. Sam-Singer, J., M. O. Robinson, A. R. Bellvue, M. I. Simon, and A. D. Riggs. 1990. "Measurement by quantitative PRC of changes in HPRT, PGK-1, PGK-2, APRT, MTase and Zfy gene transcripts during mouse spermatogenesis." *Nucleic Acids Res.* 18:1255.
15. Tan, S. S., and J. H. Weis. 1992. "Development of a sensitive reverse transcriptase PCR assay, RT-RPCR utilizing rapid cycle times." *PCR Methods Applic.* 2:137–43.
16. Brooks, P., P. Sims, and P. Broad. 1993. "Isozyme specific polymerase chain reaction analysis of differential gene expression: A general method applied to lignin peroxidase genes of *Phanerochaete chrysosporium.*" *Biotechnology.* 11:830–34.
17. Mackow, E. R., M. Y. Yamanaka, M. N. Dang, and H. B. Greenberg. 1990. "DNA amplification-restricted transcription-translation: Rapid analysis of rhesus rotavirus neutralization sites." *Proc. Natl. Acad. Sci.* 87:518–22.
18. Sarkar, G., and S. S. Sommer. 1989. "Access to a messenger RNA sequence or its protein product is not limited by tissue or species specificity." *Science.* 244:331–34.

19. Fors, L., R. A. Saavedra, and L. Hood. 1990. "Cloning of the shark Po promoter using a genomic walking technique based on the polymerase chain reaction." *Nucleic Acids Res.* 18:2793–99.

20. Licther, P., S. A. Ledbetter, D. H. Ledbetter, and D. C. Ward. 1990. "Fluorescence *in situ* hybridization with Alu L1 polymerase chain reaction probes for rapid characterization of human chromosomes in hybrid cell lines." *Proc. Natl. Acad. Sc.* 87:6634–38.

21. Nelson, D. L., S. A. Ledbetter, L. Corbo, M. F. Victoria, R. Ramirez-Solis, T. D. Webster, D. H. Ledbetter, and C. T. Caskey. 1989. "Alu polymerase chain reaction: A method for rapid isolation of human-specific sequences from complex DNA sources." *Proc. Natl. Acad. Sci.* 86:6686–90.

22. Arnold, C., and I. J. Hodgson. 1991. "Vectorette PCR: A novel approach to genomic walking." *PCR Methods and Applications.* 1:39–42.

23. Caetano-Anolles, G., B. J. Bassam, and P. M. Gresshoff. 1991. "DNA amplification fingerprinting using very short arbitrary oligonucleotide primers. *Bio/technology.* 9:533–57.

24. Welsh, J., K. Chada, S. S. Dalal, R. Cheng, D. Ralph, and M. McClelland. 1992 "Arbitrarily primed PCR fingerprinting of RNA." *Nucleic Acids Res.* 20:4965–70.

25. Williams, J. G. K., A. R. Kubelik, K. J. Livak, J. A. Rafalski, and S. V. Tingey. 1990. "DNA polymorphisms amplified by arbitrary primers are useful as genetic markers." *Nucleic Acids Res.* 18:6531–35.

26. Haley, S. D., P. N. Miklas, J. R. Stavely, J. Byrum, J. D. Kelly. 1993. "Identification of RAPD markers linked to a major rust resistance gene block in common bean." *Theor. Appl. Genet.* 86:505–12.

27. Martin, G. B., J. G. K. Williams, and S. D. Tanksley. 1991. "Rapid identification of markers linked to a *Pseudomonas* resistance gene in tomato by using random primers and near-isogenic lines." *Proc. Natl. Acad. Sci.* 88:2336–40.

28. Michelmore, R. M., I. Paran, and R. V. Kesseli. 1991. "Identification of markers linked to disease-resistance genes by bulked segregant analysis: A rapid method to detect markers in specific genomic regions by using segregating populations." *Proc. Natl. Acad. Sci.* 88:9828–32.

29. Goffreda, J. C., W. B. Burnquist, S. C. Beer, S. D. Tanksley, and M. E. Sorrells. 1992. "Application of molecular markers to assess genetic relationships among accessions of wild oat, *Avena stervilis.*" *Theor. Appl. Genet.* 85:146–51.

30. McMillin, D. E., L. L. Muldrow. 1992. "Typing of toxic strains of *Clostridium difficile* using DNA fingerprints generated with arbitrary polymerase chain reaction primers." *FEMS Microbiol. Lett.* 92:5–10.

31. Fortina, P., G. Dotti, R. Conant, G. Monokian, T. Parrella, W. Hitchcock, E. Rappaport, E. Schwartz, and S. Surrey. 1992. Detection of the most common mutations causing Thalassemia in mediterraneans using a multiplex amplification refractory mutation system (MARMS)." *PCR Methods and Applications.* 2:163–66.

32. Hayashi, K. 1991. "PCR-SSCRP: A simple and sensitive method for detection of mutations in the genomic DNA." *PCR Methods and Applications.* 1:34–38.

33. Highsmith, W. E. 1993. "Carrier screening for cystic fibrosis." *Clinical Chemistry.* 39:706–07.

34. Prior, T. W., A. C. Papp, P. J. Snyder, A. H. M. Burghes, M. S. Sedra, L. M. Western, C. Bartello, and J. R. Mendell. 1993. "Identification of two point mutations and a one base deletion in exon 19 of the dystrophin gene by heteroduplex formation." *Human Molecular Genetics.* 2:311–13.

35. Raskin, S., J. A. Phillips, G. Kaplan, M. McClure, and C. Vnencak-Jones. 1992. "Cystic fibrosis genotyping by direct PCR analysis of Guthrie blood spots." *PCR Methods and Applications.* 2:154–56.

36. Soto, D. S. and S. Sukumar. 1992. "Improved detection of mutations in the p53 gene in human tumors as single-stranded conformation polymorphs and double-stranded heteroduplex DNA." *PCR Methods and Applications.* 2:96–98.

37. Van Mansfield, A. D. M., and J. Bos. 1992. "PCR-based approaches for detection of mutated ras genes." *PCR Methods and Applications.* 1:211–16.

38. Stewart, C. N., and L. E. Via. 1993. "A rapid CTAB isolation technique used for RAPD fingerprinting and other PCR applications." *BioTechniques.* 14:748–50.

39. Edwards, K., C. Johnstone, and C. Thompson. 1991. "A simple and rapid method for the preparation of plant genomic DNA for PCR analysis." *Nucleic Acids Res.* 19:1349.

40. Hillis, D. M., and C. Moritz (eds.). 1990. *Molecular Systematics.* Sunderland, MA: Sinauer Associates Inc.

41. Cheung, W. Y., N. Hubert, and B. S. Landry. 1993. "A simple and rapid DNA microextraction method for plant, animals, and insect suitable for RAPD and other PCR analyses." *PCR Methods and Applications.* 3:69–70.

42. Sambrook, J., E. F. Fritsch, T. Maniatis (eds.). 1989. *Molecular Cloning.* New York: Cold Spring Harbor Laboratory Press.

43. Newman, S. M., N. T. Eannetta, J. Yu, J. P. Prince, M. Carmen de Vicente, S. D. Tanksley, and J. C. Steffens. 1993. "Organization of the tomato polyphenol oxidase gene family." *Plant Mol. Biol. Rep.* 21:1035–51.

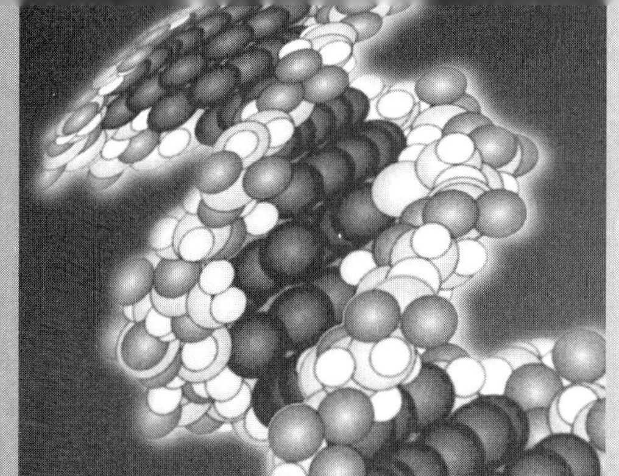

# MODULE
# 9

# Restriction Fragment Length Polymorphism (RFLP)

## Outline of Module

The RFLP module consists of eleven laboratory periods. However, if the material is carefully organized, the module can be accomplished in six laboratory periods.

DNA isolation from a suitable set of plants
*Duration:*   One laboratory period

Digestion of DNA with a restriction enzyme
*Duration:*   One laboratory period and an overnight digestion; following morning, add stop reaction and store at 4° C

Separation of DNA fragments by agarose gel electrophoresis
*Duration:*   One laboratory period and an overnight run

Transfer of the separated restriction enzyme-generated fragments to a nylon filter by Southern blotting
*Duration:*   One laboratory period and overnight run

Detection of individual restriction fragments by nucleic acid and hybridization with a radioactively labeled probe
*Duration:*   Transformation with recombinant plasmids; one laboratory period, and allow the plates to grow overnight

Overnight culture of transformed bacteria
*Duration:*   One laboratory and an overnight culture

Isolation of bacterial plasmids, measuring DNA and checking the quality of DNA
*Duration:*   One laboratory and an overnight agarose gel electrophoresis

Preparation of $^{32}$P-labeled probes by the Random Primer Method
*Duration:*   One laboratory period

149

DNA hybridization
*Duration:*   One laboratory period; scanning of RFLPs by direct observation of radioautograms

Washing filters and autoradiography
*Duration:*   One laboratory period and overnight

Scoring sample filters and calculations
*Duration:*   One laboratory period

# LABORATORY EXERCISES

## RESTRICTION FRAGMENT LENGTH POLYMORPHISM

### *Supplies and Equipment*
Agarose gel electrophoresis apparatus
Autoradiographic film
Eppendorf μl pipettes and tips
Falcon tubes
Genescreen
Gloves
Gyrotory shaker
$H_2O$ bath
Hybridization boxes
Incubator
Lyophilyzer
Magnetic stirrer and stirring bars
Microfuge
Miracloth
Mortar and pestle
Paper towels
Pasteur and regular pipettes
Polytron
Plastic wrap
Sephadex G-50
Spectrophotometer
Vacuum oven
Vortex
Waring blender
Whatman 3MM paper

### *Chemicals*
### Solutions Other than Buffers for RFLP Analysis
Ethidium bromide 10 mg ml$^{-1}$ $dH_2O$
2.5 N HCl (10X) stock—833 ml concentrated HCl
                         3167 ml $dH_2O$
1.0 M NaOH/1.0 M NaCl (2X) stock
1.0 M Tris 12.0 M NaCl, pH 7.5 (2X) stock
20X SSC

175.3 g NaCl
88.2 g $Na_3$—citrate dihydrate
pH to 7.0
$dH_2O$ to 1 l
10% sarcosyl—N-lauroylsarcosine
LB broth
  10 g bacto-tryptone
  5 g yeast extract
  10 g NaCl
  $dH_2O$ to 1 l, autoclave
  for plates, add 15 g Difco agar
Blue juice
  350 ml glycerol
  25 ml 10X neutral electrophoresis buffer
  40 ml 0.25 M EDTA
  50 ml $dH_2O$
  5 ml 20% SDS
  30 ml 10% bromphenol blue

### LS for Random Hexamer RX
TM
  250 mM Tris-HCl pH 8.0
  25 mM $MgCl_2$
  50 mM β-mercaptoethanol
OL
  90 units/ml hexadeoxynucleotides
    (Pharmacia-PL) in TE
  Store at –20° C
DTM
  100mM dATP, dTTP, dGTP in TM
  Store at –20° C
Hepes
  1M HEPES buffer titrated to pH 6.6 with NaOH
  Store at 4° C
Use above solutions to make up:
LS
  25 Hepes: 25 DTM: 7 OL
  Store at –70° C in aliquots

Stop solution
  10% SDS
  125 mM EDTA
  Blue dextran 200 mg 50 ml$^{-1}$
  Bromphenol blue 10 mg ml$^{-1}$ stock 1 ml
Denhardt's solution
  5 g ficoll
  5 g polyvinylpyrrolidone
  5 g bovine serum albumin
  250 ml dH$_2$O
  Prefilter through Whatman 3MM
  Filter through disposable millipore
  Freeze in 50 ml aliquots
St DNA
  1 g salmon sperm DNA
  200 ml dH$_2$O
  Stir at room temperature until DNA is in solution
  Shear by sonication

*Other Chemicals*
  Ampicillin
  Cetyl-trimethyl-ammonium bromide
  Chloroform
  EDTA
  EtOH
  Isoamyl alcohol
  Klenow fragment
  Liquid scintillation cocktail
  Lysozyme
  Phenol
  Radionuclide $^{32}$PdCTP
  Restriction endonucleases
  Sodium hydroxide
  Spermidine
  Tris-HCl

**DNA Extraction Buffer**
  420 g Urea
  70 ml 5M NaCl
  50 ml 1.0M Tris, pH 8.0
  80 ml 0.25M EDTA
  200 ml 10% sarcosyl
  50 ml phenol reagent
  dH$_2$O to 1 liter

**DNA Storage Buffer**
  800 ml 95% ethanol
  200 ml 1.0 M Na acetate pH 7.0

**TE Buffer, pH 8.0**
  10 ml 1.0M Tris, pH 8.0
  4 ml 0.25M EDTA, pH 7.0
  986 ml dH$_2$O

**LiCl Buffer**
  5 ml 1.0M Tris, pH 7.5
  2.33 g EDTA

  0.4 ml Triton X-100
  10.6 g LiCl
  dH$_2$O to 100 ml

**Neutral Electrophoresis Buffer (10X Stock)**
  1210 g Tris base
  33.6 g EDTA (Na$_2$)
  170.1 g NaAc-3H$_2$O
  Adjust pH to 8.1 with concentrated acetic acid
  dH$_2$O to 10 liters

**Hybridization Buffer**
  1000 ml 20XSSC
  200 ml 1M NaPO$_4$
  200 ml Denhardt's solution
  40 ml 0.25M EDTA
  120 ml 20% SDS
  400 ml 50% dextran sulfate
  2040 ml dH$_2$O
  pH to 7.3 with HCl, if necessary

# Introduction

Restriction fragment length polymorphisms (RFLPs) are restriction fragments from a given chromosomal locus that vary in size (or length) in different individuals of the same or different species.[1] The RFLPs, which originate in base sequence changes or in DNA rearrangements, "are naturally occurring, simply inherited, Mendelian characters."[1] The RFLPs are markers and stable attributes of the DNA itself, which can be employed to construct genetic maps. With regard to the uses of RFLPs, Kochert[2] has thoroughly reviewed these and RFLP maps. These involve DNA fingerprinting, employment as a phenetic character, comparative RFLP mapping, plant breeding, tagging genes, introgression, analysis of quantitative traits, and cloning genes. A summary of these is presented in Table 9.1. Students are encouraged to read the reviews and papers of Kochert and his associates (see References).

The RFLP analysis method, consists of the six steps depicted in Figure 9.1. In this module, students will learn a variety of molecular biological skills through their performances. In addition, students will enhance their computer skills.

# Procedures

## DNA Extraction Procedure for RFLP Analysis Beginning with Young Leaves[1]

a. Starting material: young leaves—fresh, from tomato or potato plants. Lyophilized dried in an oven or dried at room temperature.

b. Fresh leaves—homogenize into a powder with a mortar and pestle in liquid nitrogen.

**TABLE 9.1**   Summary of RFLP Uses

| RFLP Use | Beneficial Result |
|---|---|
| DNA fingerprinting | Identification of a cultivar, clone or an individual plant |
| Phenetic character | Indication of change at the DNA sequence level |
| | Comparison of whole chloroplast genomes from one plant to another |
| Comparative RFLP mapping | Taxonomic relationship and chromosomes' evolution |
| Plant bleeding | Follow the chromosome segments of two parent plants through a cross and into the progeny |
| Tagging genes | Segregation of genes can be followed simultaneously, and plants containing multiple genes for resistance can be selected for progeny |
| Introgression | Use of RFLP analysis to locate chromosome segments derived from each parent to estimate their size |
| Analysis of quantitative traits | Follow every chromosome segment of both parents through a cross and correlate with the quantitative trait being studied |
| Cloning genes | Cloning of genes for RFLP markers |

Adapted from Kochert, G. 1991. "Restriction fragment length polymorphism in plants and its implications." *Subcellular Biochemistry: Plant Genetic Engineering*, vol. 17. New York: Plenum Press. 167–90.

c. Lyophilized or dried—use a mechanical homogenizing mill.

d. Scrape powder into an extraction buffer, pH 8.0, a salt such as NaCl to aid in dissociating proteins from the DNA, detergents such as SDS or Sarkosyl to solubilize plant membranes, an agent to inactivate the DNase such as EDTA, phenol, chloroform, or urea.

e. Other plants such as legumes—i.e., peanuts or alfalfa—purify DNA from carbohydrates or glycoproteins.

f. Rupture cells of fresh tissue in a blender or Polytron tissue homogenizer.

g. Obtain a "crude" nuclear preparation.[2,3]

h. DNA extraction:[4,5] mix with urea-phenol extraction buffer.

i. Add SDS to 0.6% and incubate at 65° C. Invert gently at 5 minute intervals for 15 minutes.

j. Extract with chloroform: Isoamyl alcohol (24:1). Centrifuge 2,000 $xg$, 15 minutes.

l. Pipette off the upper phase containing the DNA through a layer of Miracloth.

m. Precipitate DNA with 2 volumes 95% EtOH, rinse in 70% EtOH, dissolve in TE, and recentrifuge.

n. Treat supernatant with 10 µg ml$^{-1}$ RNase.

o. Reprecipitate with 0.1 volume of 6 M LiCl and 2 volumes 95% EtOH.

p. Redissolve in TE.

q. Quantify DNA by spectrophotometry.

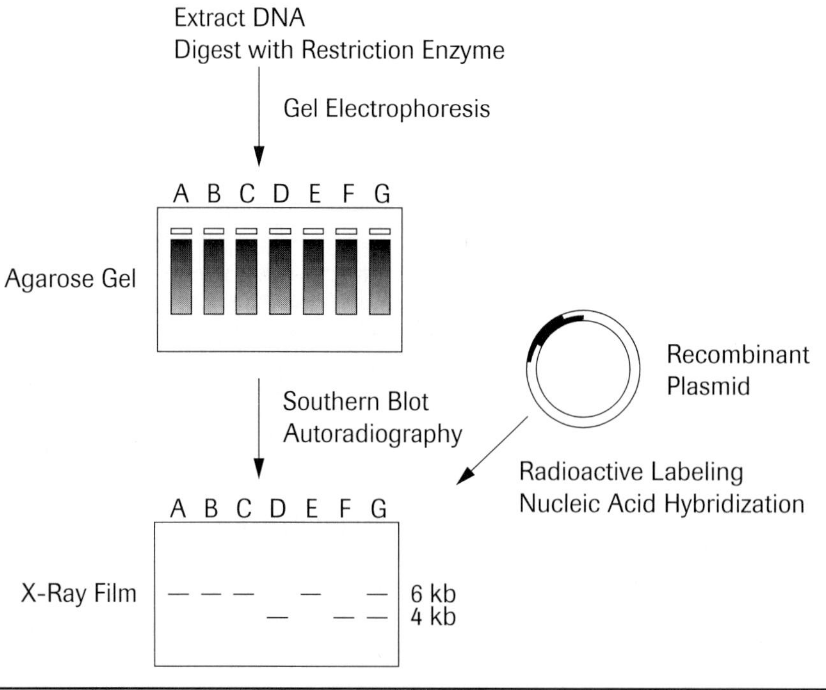


**FIGURE 9.1**   Detection of RFLP.

Adapted from Kochert, G., S. Tanksley, and J. P. Prince. 1989. RFLP *Training Course Laboratory Manual*. New York: The Rockefeller Foundation Program on Rice Biotechnology.

## Digestion of DNA with Restriction Endonucleases[6,7]

a. Following isolation of genomic DNA from tomato and assessment of purity by agarose gel electrophoresis.

b. Digest DNA with various restriction endonuclease in small volumes (~30 µl).

c. *Eco*RI, *Eco*RII, *Dra*I, *Taq*I, *Hae*III, *Hin*fl, *Hin*dIII, *Msp*I, *Xho*I.*Blu*I, *Pst*I (NE Biolabs) (Table 9.2).[8]

d. Add 4 mM spermidine to promote complete digestion. (Use 2 units enzyme/µg DNA; 1 unit defined is the amount that can cut 1 µg DNA in 1 hour).

General protocol for restriction endonuclease cleavage:

a. 1 µg of DNA and 1 U enzyme are mixed in an appropriate incubation buffer to give a final volume of 20 µ–25 µl.

b. Incubate at recommended temperature for 60 minutes.

c. Stop reaction by heating the sample to 70° C for 5 minutes (in the case of λ *Eco*RI/*Hin*dIII digestion, transfer the heated sample subsequently onto ice!).

d. Before loading onto the gel, add 1/10 vol. of a solution containing 50% glycerol and 0.25% bromophenol blue.

Note: Choice of restriction enzyme for RFLPs performed on the basis of cost and efficiency in detecting variability.

**TABLE 9.2**   Restriction Endonucleases

| Restriction Enzyme | Sequence Recognized | Fragment End Structure after Cleavage | Mean Fragment Length (bp) |
|---|---|---|---|
| *Hin*fl | GANTC | $_p$ANTCNN<br>GNN | 256 |
| *Hpa*II | CCGG | $_p$CGGNN<br>CNN | 256 |
| *Sau*I | GATC | $_p$GATCNN<br>NN | 256 |
| *Taq*I | TCGA | $_p$CGANN<br>TNN | 256 |
| *Eco*RII | CC$^A_T$GG | $_p$CCXGGNN<br>NN | 512 |
| *Ava*II | GG$^A_T$CC | $_p$GXCCNN<br>GNN | 512 |
| *Ava*I | CYCGRG | $_p$YCGRGNN<br>CNN | 1024 |
| *Bam*HI | GGATCC | $_p$GATCCNN<br>GNN | 4096 |
| *Bgl*II | AGATCT | $_p$GATCCNN<br>ANN | 4096 |
| *Eco*RI | GAATTC | $_p$AATTCNN<br>GNN | 4096 |
| *Hin*dIII | AAGCTT | $_p$AGCTTNN<br>ANN | 4096 |
| *Sal*I | GTCGAC | $_p$TCGACNN<br>GNN | 4096 |
| *Xba*I | TCTAGA | $_p$CTAGANN<br>TNN | 4096 |
| *Xho*I.*Blu*I | CTCGAG | $_p$TCGAGNN<br>CNN | 4096 |
| *Xma*L | CCCGGG | $_p$CCGGGNN<br>CNN | 4096 |

## Protocol for Agarose Gel Electrophoresis[6]

a. Add 3 g agarose to 300 ml neutral electrophoresis buffer to yield a 1.0% gel (NEB 10X stock, 1210 g Tris base, 33.6 g NaEDTA, 170.1 g NaAC, $3H_2O$; adjust pH to 8.1 with conc. acetic acid, $dH_2O$ to 10 l).

b. Bring to boiling, stir vigorously, remove from heat and cool to 60° C.

c. Seal ends of the agarose plastic gel frame with tape and pour agarose into the gel mold (Figure 9.2).

d. Position a "comb" in the gel; cool at least 30 minutes. Pour about 2 l of 1X NEB buffer into the gel buffer reservoir.

e. Carefully pull the tape off your gel and place the gel in the gel box; allow to remain 5 minutes.

f. Remove the comb; make certain that the buffer covers gel to a depth of 5 mm.

g. Load the digested samples into the wells (2 µg–3 µg for plants with a small genome to 15 µg for those with a large genome).

h. Attach the lid and the electrodes.

i. Adjust voltage to 20–30 volts.

j. Run gel overnight.

## Protocol for Southern Blotting

a. Remove the gel frame containing gel from the performed agarose gel electrophoresis.

b. Slide the gel out of the frame and place it upside down onto a plexiglass sheet in a tray.

c. Stain the gel by covering it with $dH_2O$ and adding one drop of ethidium bromide (10 mg ml⁻¹); shake the gel gently 15 minutes.

d. Rinse the gel with $dH_2O$ and destain by shaking the gel in $H_2O$ 15 minutes.

e. Photograph the gel in UV light.

f. Soak the gel in 1 l 0.25 N HCl 10 minutes, with shaking to depurinate the DNA.

g. Rinse the gel with $dH_2O$ and soak in 1 l 0.5N NaOH/0.5 M NaCl 60 minutes, with shaking to denature the DNA.

h. For each gel to be blotted prepare: 3 pieces of Whatman 3MM; chromatography paper—21 cm × 14 cm; 1 piece of Genescreen Plus—20 cm × 13.5 cm.

i. Dry Genescreen is curled; use permanent ball point to write the number of gel in the lower left-hand corner of the convex side of the Genescreen.

j. Wet 2 of the Whatman squares and the Genescreen in $dH_2O$; then in 0.5 M NaOH/0.5 M NaCl GeneScreen *Plus* (New England Nuclear) hybridization membrane.

k. Place two sheets of Whatman 3MM on tip of the sponge in the blotting set up (Figure 9.3). Genescreen curls the opposite way when wet.

l. Slide the gel onto the blotting apparatus (be sure there is adequate 0.5 M NaOH/0.5 M NaCl in the plastic tray containing the sponges) when the hour has been completed.

m. Use the dry pieces of Whatman paper to blot off excess buffer on top of gel.

n. Place the Genescreen plus onto the gel (label-side up with the writing at the low molecular weight's gel end) so that the entire gel is covered.

o. If the gel protrudes, cut off the excess with a razor blade to prevent the blot from short-cutting.

p. Roll a glass pipette over the top of the filter to eliminate air bubbles.

q. Place the two moist Whatman squares on top of the Genescreen (match edges as closely as possible); eliminate air bubbles.

r. Put the stack of paper towels on top, covering the Whatman squares completely.

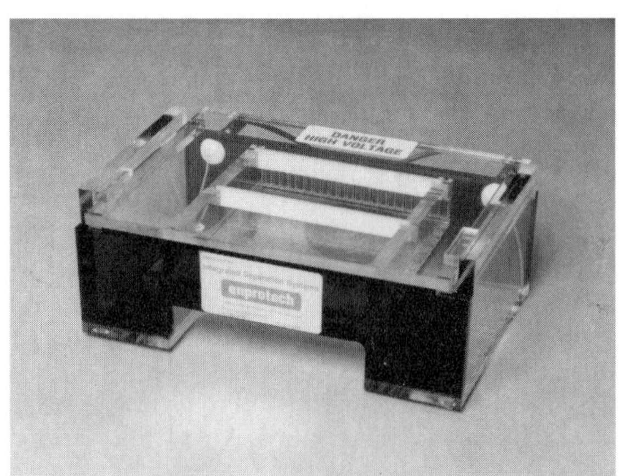

**FIGURE 9.2** **(A)** Mini and standard gel casting stands. **(B)** Standard horizontal device.

Adapted from FMC Corporation, Rockland, Maine, and Integrated Separation Systems, Natick, MA.

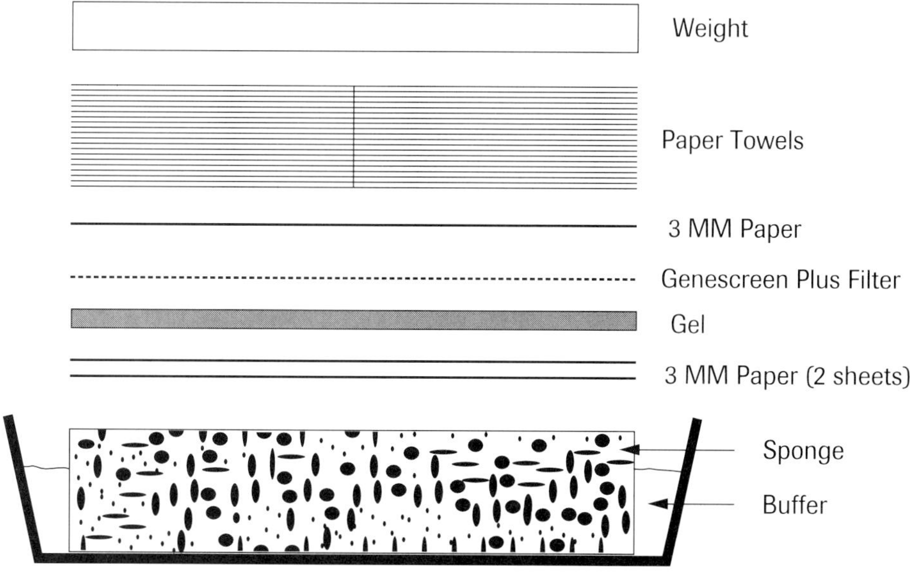


**FIGURE 9.3**   Summary of protocol for southern blotting.
Adapted from Kochert, G., S. Tanksley, and J. P. Prince. 1989. *RFLP Training Course Laboratory Manual.* New York: The Rockefeller Foundation Program on Rice Biotechnology.

s. Add a 1 kg weight over the paper towels.

t. Allow the blot to "pull" the moisture out of the gel and transfer the DNA onto the filter for 16 hours.

## Summary of Protocols for RFLP Probe Libraries; Transformation with Recombinant Probes

a. Make *E. coli* cells of the strain DH 5-α competent (able to take up plasmids) (Tables 9.3, 9.4; Figures 9.4, 9.5).

b. Remove competent cells from the –70° C freezer and thaw on ice.

c. Mix gently, then aliquot 50 µl competent cells into a chilled 1.5 ml microfuge tube. Add 1 ng plasmid DNA by moving the pipette tip through the cells while dispersing; gently shake cells ~5s immediately after addition.

d. Incubate cells on ice 30 minutes.

e. Heat shock cells 35s in a 42° C water bath without shaking.

f. Place on ice 2 minutes.

g. Add 1 ml sterile LB (minus ampicillin) at room temperature.

h. Shake at 225 rpm (37° C) 1 hour.

i. Pour 100 µl onto a plate containing LB with 50 µg ml$^{-1}$ ampicillin.

j. Spread cells gently.

k. Incubate upside down at 37° C overnight; check immediately next morning so that the colonies are not confluent.

## Overnight Culture of Transformed Bacteria

a. Examine the plate from overnight incubation for small white colonies; each colony should be a clone arising from a single transformed bacterium.

b. Store plate at 4° C until late afternoon.

c. Pinch off a single colony with a sterile toothpick and incubate it into a 5 ml culture of LB + ampicillin in a 15 ml Falcon tube (save plate at 4° C).

d. Shake the culture at ~225 rpm for ~12–16 hours at 37° C.

## Isolation of Bacterial Plasmids[6]

a. Centrifuge the bacteria from the overnight culture in an SM24 rotor at 2,000 rpm 5 minutes.

b. Aspirate off the supernatant.

c. Resuspend the pellet of bacterial cells in 500 µl LiCl buffer and transfer to a 1.5 ml microfuge tube.

d. Add 50 µl fresh lysozyme 10 ml dH$_2$O$^{-1}$, mix, and incubate 5 minutes at room temperature.

e. Place in 100° C dry bath 3 minutes; cool on ice 5 minutes.

f. Centrifuge 10,000 rmp in a microfuge (8 minutes at room temperature).

g. Following centrifugation, remove and discard the pellet by "stabbing" it and sliding it out of the tube with a Pasteur pipette.

h. Recentrifuge and transfer the supernatant to a fresh tube (removes more debris).

**TABLE 9.3**  Plasmid Vector Systems

| Vector | Original Host | Host Range |
| --- | --- | --- |
| ***Gram-negative bacteria*** | | |
| Col E1 | *Escherichia coli* | *E. coli* |
| pBR322 | *E. coli* | *E. coli* |
| RSF1010 | *E. coli* | *E. coli*, some *Pseudomonas* spp. |
| pRK290 | *Klebsiella aerogenes* | Most gram-negative bacteria |
| RP1 | *Pseudomonas aeruginosa* | Most gram-negative bacteria |
| ***Gram-positive bacteria*** | | |
| pUB110 | *Bacillus* | *Bacillus* spp. |
| SCP-2 | *Streptomyces* | Some *Streptomyces* spp. |
| Fungi | | |
| YEp; YRp | *Saccaromyces* | Yeast |
| Plant Cells | | |
| Ti | *Agrobacterium tumefaciens* | Dicotyledenous plant species |
| ***Animal cells*** | | |
| Defective SV40[a] | Mammalian cells | Mammalian cell lines |

From Karl Kammermeyer and Virginia L. Clark. 1989. *Genetic Engineering Fundamentals: An Introduction to Principles and Applications.* New York: Marcel Dekker, Inc.
[a]Control of transmittance is obtained in part by use of defective viral forms.

**TABLE 9.4**  Properties of Some Common Plasmids

| Source: *E. coli* | Size Kilobases | Marker Genes | Single Restriction Enzyme Sites |
| --- | --- | --- | --- |
| pBR322 | 4.3 | Amp$^r$<br>Tet$^r$ | *Eco*RI, etc. |
| pBR313 | | Ampt$^r$<br>Tet$^r$<br>Kan$^r$ | *Eco*RI, *Bam*HI, *Sal*I<br>*Hind*III, *Sma*I, *Hpa*I<br>*Xho*I, *Hind*III, *Pst*I |
| pACYC 177 | 3.7 | Amp$^r$ | *Bam*HI, *Sma*I, *Hind*II |
| pMB9 | 5.2 | Tet$^r$<br>ColEI imm. | *Bam*Hi, *Hind*III, *Sal*I<br>*Eco*RI |
| pUR | 2.7 | Amp$^r$<br>Lac Op.<br>Z gene | *Pst*I, *Pvu*I<br>*Eco*RI |
| PSC101 | 5.8 | Tet$^r$ | *Eco*RI (considered by some investigators as a natural plasmid) |
| pUC8 and pUC9 | 2.8 | Amp$^r$<br><br>lac α-segment | *Eco*RI, *Sma*I, *Xma*I,<br>*Bam*HI, *Sal*I,<br>*Acc*I, *Hind*III |

From Karl Kammermeyer and Virginia L. Clark. 1989. *Genetic Engineering Fundamentals: An Introduction to Principles and Applications.* New York: Marcel Dekker, Inc.

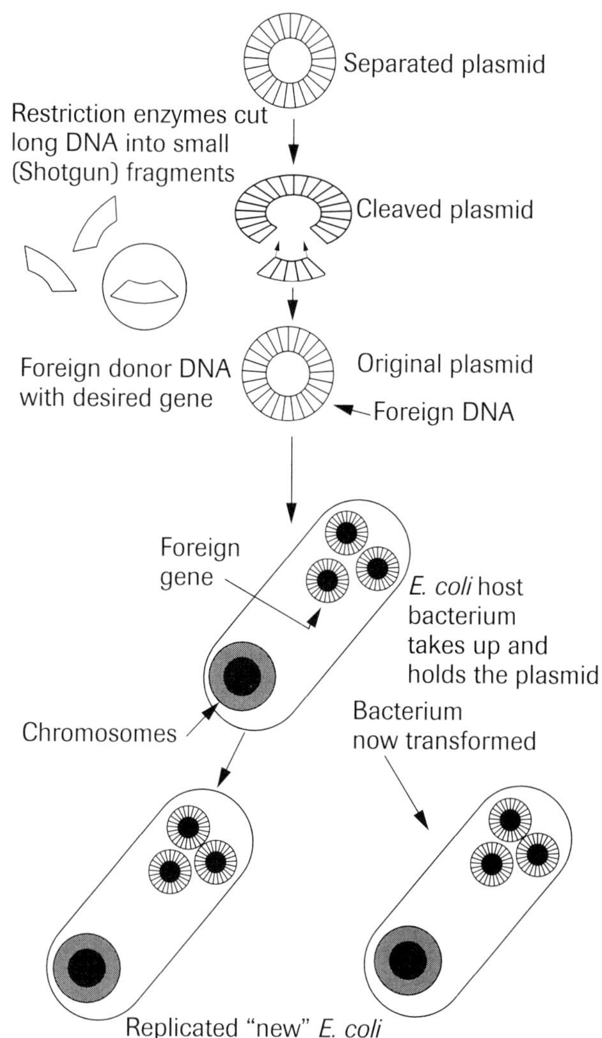


**FIGURE 9.4**  Schematic DNA insertion into plasmid. Plasmid is cleaved and foreign DNA is inserted, recircularized and cloned in host.

From Kammermeyer, Karl and Virginia L. Clark. 1989. *Genetic Engineering Fundamentals: An Introduction to Principles and Applications.* New York: Marcel Dekker, Inc.

 

i. Add 2 volumes of cold 95% EtOH and cool at −80° C 15 minutes to precipitate the plasmid.
j. Centrifuge 10,000 rpm 15 minutes.
k. Wash with 70% EtOH.
l. Centrifuge 10,000 rpm 10 minutes.

### *Pellet*

a. Dry under vacuum in a vacuum oven.
b. Resuspend in 100 µl TE.
c. Hydrate 30 minutes at 60° C.
d. Centrifuge 10,000 rpm 5 minutes.

### *Supernatant*

a. Transfer to a tube and determine DNA concentration.
b. Add 25 µl of sample to 475 µl TE, vortex.

c. Measure absorbance at 260 nm and 280 nm. Corresponds to a 20-fold dilution, and the absorbance can be directly converted into µg ml⁻¹ by multiplying by 1,000.
d. Digest 10 µg plasmid 1 hour with 20 units *Pst*I to free the insert.
e. Load isolated plasmid on an overnight 1% agarose gel; run molecular weight markers of uncut lambda phage, lambda phage cut with *Hind*III and ΦX 174 and with *Hae*III; also run appropriate isolated plant DNA.
f. Characterize DNA quality. Stain and photograph the gel containing isolated plasmid and tomato DNA samples.
g. Calculate the molecular weight insert from your plasmid by plotting it against the molecular weight standards (Figure 9.6).
h. Check the quality of your DNA samples. The DNA should all be in a single band, which is >30 Kb.

## Preparation of ³²P-labeled Probes by the Random Primer

a. Mix 70 mg probe (plasmid-containing plant DNA) with sufficient H₂O to yield 8 µl and heat in a 100° C heat block 10 minutes to determine the DNA.
b. Cool on ice several minutes; add 11 µl of LS and 3 units of 1 µl Klenow (large subunit of DNA polymerase 1). Klenow retains the 5′ → 3′ polymerase activity but lacks 5′ → 3′ exonuclease activity. Klenow can be employed to synthesize continuous strands of uniformly labeled DNA. Requires a synthetic set of primers (6 bp random oligmers).
c. Add 5 µl ³²P dCTP (50µCi) and incubate at 37° C, 1 hour.[9]
d. Terminate the reaction by adding 20 µl stop solution.
e. Pipette the entire mixture onto a G-50 Sephadex column in a 1.5 ml microfuge tube.
f. Add 50 µl 25mM EDTA/1% SDS directly on top.
g. Wait 30 seconds, then add another 50 µl.
h. Centrifuge 1,000 rpm for 20–30 seconds in a microfuge.
i. The blue dextran should be in the tube, the bromphenol blue should still be in the column. If required, add another 50 µl EDTA/SDS.
j. Recentrifuge.
k. Add 100 µl EDTA/SDS to the tube and mix the eluent by tagging the tube.
l. Count 2 µl sample by LS counting.
m. Calculate the spc. act.—it should have between 1 and 10 × 10⁸ cpm/µg (Table 9.5). See Module 10 for the use of liquid scintillation.

## Protocol for DNA Hybridization

a. Remove the weight, towels, and Whatman squares from the Southern blot.

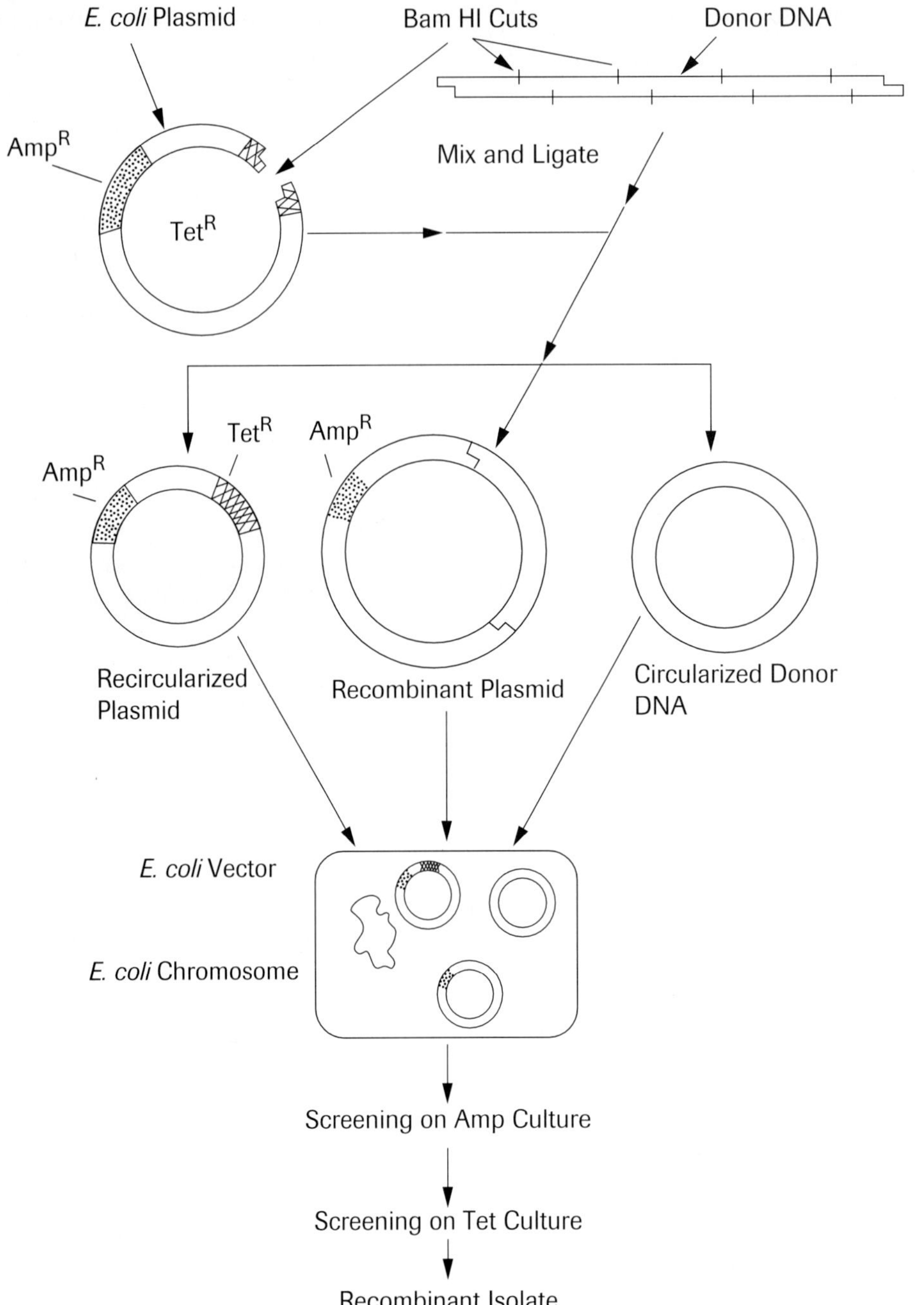


**FIGURE 9.5** Generalized foreign gene insertion. Individual steps to make plasmid receptive to gene insertion, and antibiotic screening for selecting properly transformed cells.

From Kammermeyer, Karl and Virginia L. Clark. 1989. *Genetic Engineering Fundamentals: An Introduction to Principles and Applications.* New York: Marcel Dekker, Inc.

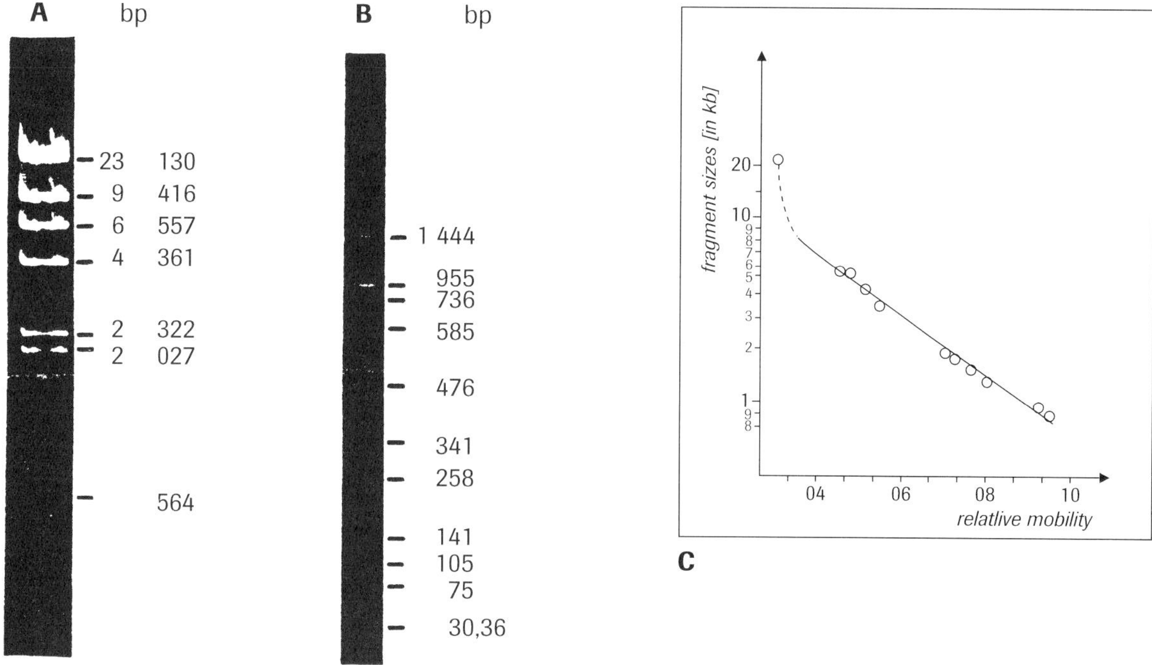


**FIGURE 9.6**   **A)** HMW DNA Ladders (1 µl) were run on a 0.8% agarose gel with TAE buffer at 100V for 1.5 hours. **B)** LMW DNA Ladders (2.5 µl) were run on a 5–15% ISS Mini Gel with TBE buffer at 100V for 3 hours. **C)** Standard curve.

From Integrated Separation Systems. Natick, MA; and Bernatzky, R. and S. D. Tanksley. 1989. *Plant Molecular Biology Reporter.*

**TABLE 9.5**   Specific Activities of Probes Obtained Using Various Amounts of Precursors and Templates[a]

| Labeling Protocol | Amount of DNA Template (µg) | Labeled Precursors[b](µCi) | Label Incorporated (% at plateau)[c] | Specific Activity of Probe (dpm/µg) |
|---|---|---|---|---|
| A | 0.062 | dCTP (50) | 87 | $1.2 \times 10^9$ |
| B | 0.062 | dCTP (100) | 90 | $1.9 \times 10^9$ |
| C | 0.062 | dCTP (150) | 89 | $2.4 \times 10^9$ |
| D | 0.062 | dCTP (200) | 82 | $2.7 \times 10^9$ |
| E | 0.062 | dCTP (100 + dATP (100) | 84 | $3.7 \times 10^9$ |
| F | 0.125 | dCTP (50) | 88 | $0.7 \times 10^9$ |
| G | 0.031 | dCTP (50) | 74 | $1.7 \times 10^9$ |

Etienne, Decant, J. 1988. *Genetic Biochemistry from Genes to Protein.* New York: John Wiley and Sons.

[a] Oligo-labeling reactions were performed as described in using the human gamma globin cDNA insert from plasmid JW 151.

[b] Specific activity of precursor was 3000 Ci/mmol.

[c] Plateau was reached in 3 h or less in protocols A, B, C. E, and F. Plateau was reached in 6–12 h with protocols D and G.

b. Wash the filter in 0.2 M Tris-HCl, pH 7.5/2X SSC, 10 minutes with shaking.

c. Bake the filter at 80° C in a vacuum oven 2 hours. Prehybridization.[2]

d. Calibrate an incubator to 65° C. Warm the hybridization buffer in a 65° C $H_2O$ bath.

e. Put the salmon testes DNA into a $H_2O$ bath 10 minutes.

f. Place on ice.

g. Place 50 ml hybridization buffer in hybridization boxes.

h. Add the filters one by one, completely covering each with buffer before adding the next.

i. Placing the filters DNA-side up is more advantageous later.

j. Cover with a plastic sheet, eliminating air bubbles; cover the box with a lid to prevent evaporation.

k. Place in a 65° C incubator (no shaking for at least 4 hours before adding probe).

## *Hybridization*[6]

a. Denature probe at 90° C or more on the heating block for no less than 10 minutes (Figure 9.7).

b. Obtain the box containing pre-hybridized filters from the incubator.

c. Lift the plastic sheet and filters out of the "incubator," pour the probe into the hybridization buffer, and mix very well. Place filters back into the buffer one by one (DNA-side up), making sure each well is coated with liquid before adding the rest. Replace the plastic sheet and lid and put into a 65° C incubator overnight.[10]

### Molecular Hybridization and Probes

A probe possesses the following characteristics:

1. It is a segment of single-stranded nucleic acid (DNA or RNA).

2. It is complementary to the segment of nucleic acid to be recognized, this recognition being able to take place DNA–DNA or DNA–RNA (or RNA–RNA). Quite clearly in this hybridization reaction, the nucleic acids to be recognized must be in the form of one strand. The probe can cover all or part of the nucleic acid segment to be recognized. It is capable of detecting its complementary copy among thousands of different DNA (or RNA) fragments.

3. The probe should be marked. At present radioactive probes are being used. The positioning of this probe can therefore be easily marked, and consequently the clone to be identified, which will be hybridized with this probe.

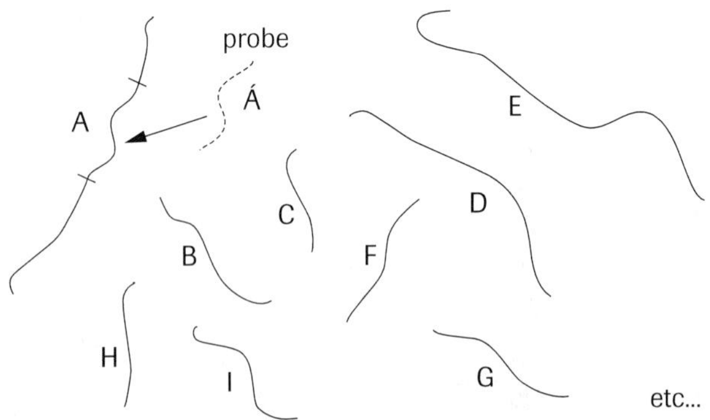


**FIGURE 9.7**   Molecular hybridization: general diagram.
Etienne, Decant, J. 1988. *Genetic Biochemistry from Genes to Protein.* New York: John Wiley and Sons.

## Summary of Procedures for Washing Filters and Autoradiography Applied to RFLPs[6]

a. Prepare 2X SSC, 0.1% SD wash; 20X SSC—100 ml (175.3 g NaCl, 88.2 g; Na$_3$—citrate dihydrate pH 7.0, dH$_2$O to 1 l dH$_2$O to 1 l.

b. Heat wash solutions to 65° C in a microwave. Be certain to stir before the temperature reading, as a temperature gradients will be formed in the solution in the microwave.

c. Remove filters from the hybridization box.

d. Save the probe in a 50 ml Falcon tube. If required for later use; otherwise dispose of it in the high-activity waste bottle.

e. Add 1 l of wash solution per box; do not place anymore than 7 filters per box. Place screens between each filter.

f. Wash on a gyrotory shaker at 65° C, 20–30 minutes.

g. Pour out the solution into the high activity waste bottle.

h. Add 1 l per box of 1X SSC solution. Bring to 60° C before adding filters; shake at 65° C, 20–30 minutes.

i. Discard the solution into the low-activity waste bottle.

j. Wash again as above at 1X SSC for low stringency (heterologous probes) or 0.5X SSC for moderate stringency (homologous probes). Wash to 0.5X SSC, 65° C, anything that is ≥80% homologous will remain hybridized

k. Wrap the filters in plastic wrap and place against film in cassettes at –80° C for overnight exposures. Finally, Figure 9.8 depicts the utilization of the module's procedures as a Southern blot analysis of the PPO gene family.

Table 9.6 summarizes the computer programs[11,12,13] that have been designed for RFLP data analysis.

In addition to the now commonplace RFLPs, the degree of relatedness between individual organisms or inheritance in progeny populations can be assessed with multiple arbitrary amplicon profiling (MAAP)[14] techniques such as random amplified polymorphic DNA (RAPD) analysis (see PCR, Module 8), arbitrarily primed PCR (AP-PCR),[15] and DNA amplification fingerprinting (DAF).[16] These strategies, which were generated independently, employ one or more arbitrary oligonucleotide primers to target specific but unknown sites in the genome. Many of the sites are polymorphic. In this regard, detected amplification fragment length polymorphisms (AFLPs) can be utilized as markers for genetic typing and map applications.[17]

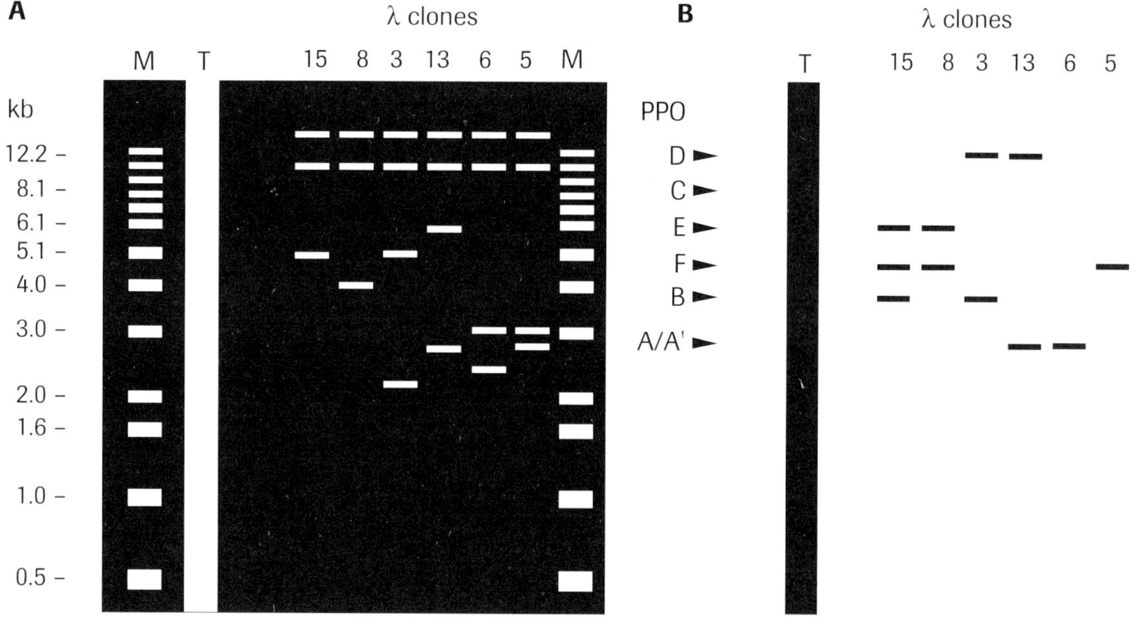


## FIGURE 9.8

Southern blot analysis of PPO gene family. **(A)** Ethidium bromide staining of 12 µg *Hind*III-digested tomato genomic DNA (T) and 1.5 µg *Hind*III-digested λCharon 35 PPO genomic clones 15, 8, 3, 13, 6 and 5 on a 1% agarose gel. Marker lanes (M) are the BRL 1 kb ladder. (Note: in the λCharon35 cloning vector *Hind*III sites are present in the multiple cloning site region immediate adjacent to the *Sau* 3A insertion sites). **(B)** Autoradiogram of the blotted gel probed with the p22 cDNA. The tomato genomic DNA lane was exposed to film for 9 days, the λ clone lanes were exposed for 6 h. From restriction map analysis, the predicted *Hind*III fragment sizes for the PPOs ( in kb) are: A (1.87), B (2.5), D (5.2), E (3.25), and F (2.67). These size bands were generated as expected. λ15 contains PPOs E and F. λ8 contains PPO F only. λ3 contains PPO B and a PPO D fragment 3'-truncated to 5.05 kb. λ13 contains PPOs D and A. λ6 contains PPO A'. λ5 contains a C fragment 5'-truncated. to 2.75 kb. For the PPO C fragment, the 3' *Hind*III site is located in the coding region of PPO C, but the 5' *Hind*III site is located within the vector multiple cloning site region, and thus does not represent the native upstream *Hind*III site. The truncation of λ5's 5' *Hind*III fragment prevents its use for the prediction of the size of the corresponding PPO C fragment in tomato DNA, but the tomato C fragment must be larger than 2.75 kb. Accordingly, the 4.2 kb band is presumed to represent the native *Hind*III fragment for PPO C. PPOs A and A' cannot be differentiated by the 1.87 kb *Hind*III fragment they each generate. (The faint hybridizing fragments associated with PPOs, E and F, seen at 2.41 kb and 3.3 kb, are the downstream *Hind*III fragments of E and F, respectively. The cDNA probe encompassed 180 bp of sequence, 3' of the coding region *Hind*III site, that has homology to these two fragments.)

Newman, Sally M. Nancy T. Eanetta, Haifeng Yu, James P. Prince, M. Carmen de Vicente, Steven D. Tanksley and John C. Steffens. 1993. *Plant Molecular Biology.* 21:1035–51.

**TABLE 9.6**   Computer Procedures Employed to Analyze RFLP Data

| Task | Program | Usefulness |
|---|---|---|
| Calculating genetic distance | Nei's statistic[a]<br><br>$mx$ = total number of fragments in one sample; $my$ = total number of fragments in the other sample; $mxy$ = number of stored fragments<br>$F = (2m_{xy}) / (m_x + m_y)$ | From $F$ values, a phylogenetic tree can be constructed for cultivars |
| Genetic linkage analysis | Mapmaker[b] | Two-point, three-point and multi-point linkage analyses; comparisons between different orders of markers on your map |
| Generate and analyze graphical representations of genotypes | Hyper-gene[c] | Chromosomes can be represented by patterns that indicate the parental origin of each region of the chromosome around RFLP markers and indicate the area in which crossovers have taken place; indicate regions of the genome as targets for selection; select individual plants from a breeding program. |

[a]Nei, M. 1987. *Molecular Evolutionary Genetics,* New York: Columbia University Press.
[b]Lander, E. et al. 1987. *Genomic* 1:174–81.
[c]Young, N. D., A. H. Patteson, and S. D. Tanksley. 1989. *RFLP Training Course Laboratory Manual.* New York: Rockefeller Foundation Program on Rice Biotechnology.

# Review Questions

1. What is restriction fragment length polymorphism and what are its uses?
2. Design an experiment to isolate and purify genomic DNA and assess its quality and purity.
3. What are restriction endonucleases and of what value are they to RFLP analysis? What factors mediate the choice of a restriction endonuclease for RFLP analysis?
4. How can a restriction fragment generated genomic DNA fragment be incorporated into a plasmid? What factors direct the choice of a plasmid?
5. Detail the construction of a Southern Blot. How does it differ from Northern and Western Blots?
6. What function does a Southern Blot serve in RFLP analysis?
7. What is a probe and what are its characteristics? Describe how probes are employed in RFLP analysis.
8. What is the end product of RFLP analysis and how is the analysis quantified?

# REFERENCES


1. Kochert, G. 1992. "RFLP technology." *Advances in Cellular and Molecular Biology of Plants.* Netherlands: Kluwer Academic Publishers.
2. Kochert, G. 1991. "Restriction fragment length polymorphism in plants and its implications." *Subcellular Biochemistry: Plant Genetic Engineering,* vol. 17. New York: Plenum Press. 167–90.
3. Murray, M., and W. F. Thompson. 1980. "Rapid isolation of high molecular weight plant DNA." *Nucl. Acids Res.* 8:4321–25.
4. Rogers, S. O., and A. J. Bendrich. 1980. "Extraction of DNA from plant tissues." *Plant Molecular Biology Manual.* Boston: Kluwer Academic Publishers. 11.
5. Saghai-Maroof, M. A., K. M. Soliman, R. A. Jorgenses, and R. W. Allord. 1984. "Ribosomal DNA spacer-length polymorphisms in barley: Mendelian inheritance, chromosomal location and population dynamics." *Proc. Natl. Acad. Sc. USA* 81:8014–18.

6. Kochert, G., S. Tanksley, and J. P. Prince. 1989. *RFLP Training Course Laboratory Manual.* New York: The Rockefeller Foundation Program on Rice Biotechnology.

7. McCouch, S. R., G. Kochert, Z. H. Yu, Z. Y. Wang, G. S. Khush, W. R. Coffman, and S. D. Tanksley. 1988. "Molecular mapping of rice chromosomes." *Theor. Appl. Genet.* 76:815–29.

8. Kammermeyer, K., and V. L. Clark. 1989. *Genetic Engineering Fundamentals: An Introduction to Principles and Applications.* New York: Marcel Dekker, Inc.

9. Feinberg, A. P., and B. Vogelstein. 1984. "A technique for radiolabeling DNA restriction fragments to a high specific activity." *Anal. Biochem.* 132: 6–13.

10. Miyada, C. G., and R. Bruce Wallace. 1987. *Oligonucleotide Hybridization Techniques: Methods in Enzymology.* New York: Academic Press.

11. Nei, M. 1987. *Molecular Evolutionary Genetics.* New York: Columbia University Press.

12. Lander, E. S., P. Green, J. Abrahamson, A. Barlow, M. J. Daly, S. E. Lincoln, and L. Newburg. 1987. "MAPMAKER: An interactive computer package for constructing primary genetic linkage maps of experimental and natural populations." *Genomics.* 1:174–81.

13. Young, N. D., A. H. Patteson, and S. D. Tanksley. 1989. *RFLP Training Course Laboratory Manual.* New York: Rockefeller Foundation Program on Rice Biotechnology.

14. Caetano-Anolles, G., B. J. Bassam, and P. M. Gresshoff. 1992. "DNA fingerprinting: MAAPing out a RAPD redefinition?" *Biotechnology.* 10:937.

15. Welsh, J., and M. McClelland. 1991. "Fingerprinting genomes using PCR with arbitrary primers." *Nucleic Acids Res.* 18:7213–18.

16. Caetano-Anolles, G., B. J. Bassam, and P. M. Gresshoff. 1991. "DNA amplification fingerprinting: A strategy for genome analysis." *Plant Mol. Biol. Rep.* 9:292–305.

17. Caetano-Anolles, G., B. J. Bassam, and P. M. Gresshoff. 1991. "Amplifying DNA with arbitrary oligonucleotide primers." *PCR Methods and Applications.* 3:85–94.

## Supplementary References

Avers, C. 1986. "Molecular Nature of the Genome." *Molecular Cell Biology.* Reading, MA. 479–527.

Beckmann, J. S., and M. Soller 1983. "Restriction fragment length polymorphisms in genetic improvement: methodologies, mapping and costs." *Theor. Appl. Genet.* 67:35–43.

Beckmann, J. S., and M. Soller. 1986. "Restriction fragment length polymorphisms and genetic improvement of agricultural species." *Euphytica.* 35:111–24.

Beckmann, J. S. 1988. "Oligonucleotide polymorphisms: A new tool for genomic genetics." *Bio. Technology.* 6:1061–64.

Bonnierhale, M. W., R. L. Plaistedl, and S. D. Tanksley. 1988. "RFLP maps based on a common set of clones reveal modes of chromosomal evaluation in potato and tomato." *Genetics.* 120:1095–1103.

Botstein, D., R. L. White, M. Skolnick, and R. W. Davis. 1980. "Construction of a genetic linkage map in man using restriction fragment length polymorphisms." *Am. J. Hum. Genet.* 32:314–31.

Brummer, E. C., J. H. Bouton, and G. Kochert. 1993. "Development of an RFLP map in diploid alfalfa." *Theor. Appl. Genet.* 86:329–32.

Buchko, J., and C. R. Klassen. 1990. "Detection of length heterogeneity in the ribosomal DNA of *Pythium ultimum* by PCR amplification of the intergenic region." *Curr. Genetics.* 18:203–208.

Burger, S. L., and A. R. Kimmel. *Guide to Molecular Cloning Techniques: Methods in Enzymology, 152.* New York: Academic Press.

Cubeta, M. A., E. Echanti, T. Abernathy, and R. Vilgalyu. 1991. "Characterization of anastomosis groups of binucleate *Rhizoctonia* species using restriction analysis of an amplified ribosomal gene." *Phytopathol.* 81:1395–1400.

Eckstein, F., and D. M. J. Lilley. 1987–1992. *Nucleic Acids and Molecular Biology,* vols. 1–6. Berlin: Springer-Verlag.

Etienne, Decant J. 1988. *Genetic Biochemistry from Genes to Protein.* New York: John Wiley and Sons.

Johnson, B. S., and P. H. Thing. 1991. "DNA fingerprinting with [35]S nucleotides." *Nucl. Acids. Res.* 18:7459–60.

Kochert, G, 1992. "RFLP technology." *Advances in Cellular and Molecular Cell Biology.* Reading, MA: Addison Wesley. 479–527.

Kochert, G., T. Halwood, W. D. Brauch, and C. E. Simpson. 1991. "RFLP variability in plant cultivars and wild species." *Theor. Appl. Genet.* 8:566–70.

Krstulovic, A. M. 1987. *CRC Handbook of Chromatography: Nucleic Acids and Related Compounds.* Boca Raton, Florida: CRC Press.

Landry, B. S., and R. W. Michelmore. 1987. "Methods and applications of restriction fragment length polymorphism analysis to plants." *Tailoring Genes for Crop Improvement.* New York: Plenum Press. 25–44.

Maniatis, T. E., T. Fritsch, and J. Sambrook. 1982. *Molecular Cloning: A Laboratory Manual.* New York: Cold Spring Harbor Laboratory Press.

Nevens, P., N. S. Shepherd, and H. Saedler. 1986. "Plant transposable elements." *Advances in Botanical Research.* 12:103–203.

Paterson, A. H., E. S. Landen, J. D. Hewitt, S. Paterson, S. E., Lincoln, and S. D. Tanksley. 1988. "Resolution of quantitative traits into Mendelian factors by using a complete RFLP linkage map." *Nature.* 355:721–26.

Pethe, V., M. Lagu, P. K. Chitnis, V. Gupta, and P. K. Ranjekar. 1989. "Restriction fragment length polymorphisms—A recent approach in plant breeding." *J. Bicohem. Biophys.* 26:285–88.

Rigby, P., M. Dieckmann, C. Rhodes, and P. Berg. 1977. "Labeling deoxyribonucleic acid to high specific activity in vitro by nick-translation with DNA polymerase I." *J. Mol. Biol.* 113:237–51.

Rodriquez, R. L., and R. C. Tait. 1983. "Recombinant DNA Techniques. Restriction endonucleases. Chapter 4. Restriction endonuclease digestion of plasmid and *E. coli* chromosomal DNA. Exercise 6." Menlo Park, CA: Benjamin Cummings. 53–66.

Shure, M., S. Wessler, and M. Federoff. 1983. "Molecular identification of the waxy locus in maize." *Cell.* 35:225–33.

Southern, E. M. 1975. "Detection of specific sequences among DNA fragments separated by gel electrophoresis." *J. Mol. Biol.* 98:530–51.

Weissbock, A., and H. Weissbock. *Plant Molecular Biology Methods in Enzymology.* vol. 118. New York: Academic Press.

White, T. J., N. Arnheim, and H. A. Erlich. 1989. "The polymerase chain reaction." *Trends in Genetics.* 5:185–89.

Zhao, Y., and G. Kochert. 1993. "Phylogenetic distribution and genetic mapping of a (GGC) microsatellite from rice (*Oryza sativa* L.)." *Plant Molecular Biology.* 21:607–14.

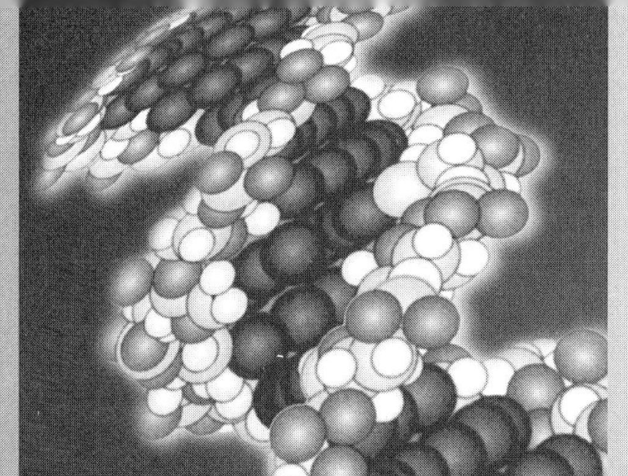



# The Use of Radiolabeled Compounds in Biochemistry: *In Vitro* Translation

## Outline of Module

### THEORY OF RADIOACTIVE DECAY

Overview of decay
  Specific activity
  Some errors in radiotracer assays
Concepts
  Half-life
  The Curie
  Random decay
  Quenching
Quantification of radioactivity
  Gas flow
  Liquid scintillation counting
  Gamma well counting
  Radioautography

### PROTEIN SYNTHESIS

Transcription
  Types of RNA
  Transcription
  Post-transcriptional alteration

## LABORATORY EXERCISES

Growth of *Solanum berthaultii* plants
    *Duration:* Growth to the 1st, 2nd, and 3rd internode stage

Isolation of total RNA and selection of poly (A)-mRNA
    *Duration:* Bulk RNA 1 laboratory period; poly (A)-mRNA 1 laboratory period

*In vitro* translation of leaf-derived poly (A)-mRNA
    *Duration:* One laboratory period

Immunoprecipitation of *in vitro* translation products
    *Duration:* Preparation of *S. aureus* cells in advance of laboratory; immunoprecipitation—one
              laboratory period

SDS-PAGE resolution of *in vitro* translation products
    *Duration:* One laboratory period

Autoradiography and fluorography
    *Duration:* One week exposure in dark

# INTRODUCTION

Many biochemical analyses demand the detection of minute ($10^{-14}$–$10^{-6}$ moles) quantities of material. However, chemical tests are rarely responsive to less than $10^{-7}$ moles. This limitation has been overcome by the development of radiotracer technology through which extraordinarily sensitive detection of many radioisotopically labeled material substances of $10^{-12}$ moles has become routine. In addition, the use of radioactivity has permitted the development of powerful experimental approaches to various types of problems.[1] These approaches employ the double-labeling technique for following two identical substances synthesized at different times: the pulse-chase method,* following the fate of a substance after its synthesis without the interference of material concurrently synthesized; and exchange analysis for measuring participation in reactions.[2]

In the following laboratory exercises, some of these and other techniques will be described, and students will gain an experience on *in vitro* translation of *Solanum berthaultii* poly(A)-mRNA.

## Overview of Decay

*Isotopes* are atoms that contain the same number of protons (have the same atomic number) but different numbers of neutrons (have different atomic weights). Most naturally occurring elements exist as mixtures of isotopes. For example, magnesium exists as $Mg^{24}$, $Mg^{25}$, and $Mg^{26}$, which accounts for about 78.6%, 10.11%, and

---

*Note: The utilization of radionuclides at most institutions is under strict supervision and thus the procurement, use and disposal of $^{35}$S-methionine will require the approval of the institutions' radiation safety officers

11.29%, respectively, of the total magnesium in nature. Because of the mass distribution, the average atomic weight of magnesium is 24.31. Radioactivity is the result of an unstable combination of protons and neutrons in the nucleus, and the attempt to arrive at a more stable combination. This stable combination is frequently attained by the emission of an alpha or beta particle. Such isotopes are called "radioactive isotopes."[1]

## *Specific Activity*

Specific activity refers to the amount of radioactivity per second unit of substance. It is normally expressed in terms of Curies per gram (Ci/g), millicuries per milligram (mCi/mg), millicuries/millimole (mCi/mmole), disintegrations per minute per millimole (DPM/mmole), counts/per minute per micromole (CPM/μmole), or any other convenient way.[3]

## *Some Errors in Radiotracer Assays*

Radiotracer assays must be done with a realization that several inherent experimental errors exist. These are summarized as follows.

### Background Radiation

The detector normally measures a background count because of cosmic radiations, radioactive contaminants in the phototube, and/or contamination from prior spills in the laboratory. The background will vary with the instrument and the laboratory. Therefore, a background count must be made and subtracted from each experimental reading.[3]

### Coincidence Losses

When more than one disintegration occurs at the same time, the instrument records only one count and thus loses the coincident events. This coincident loss is particularly relevant at high counting rates. The error is sometimes included under dead time.

### Dead Time

After every count is recorded, there is a certain time (a fraction of a microsecond to over a hundredth of a microsecond) before the instrument can record the next count. Dead time is considered most important for very active materials. For example, if an instrument with a 100 microsecond dead time recorded 100,000 counts in one minute, then the total dead time was ($100,000 \times 100 \times 10^{-6}$, or 10 sec min$^{-1}$). The total counts that would have been detected using a zero dead-time instrument are

$$100,000 \times \frac{(60)}{60 - 10} = 120,000 \text{ cpm}$$

*Alpha Particles* are essentially energetic helium nuclei (two protons and two neutrons) and are normally emitted by isotopes of the heavier elements (atomic mass greater than 82).

*Beta Particles* are energetic electrons emitted from the nucleus (neutron → electron + proton + V–) of many of the radioisotopes.

*Gamma Rays* are electromagnetic radiation (no mass, no charge) and are almost never emitted alone. These are normally emitted accompanying alpha and beta particles. Most radioisotopes used in biochemical studies are beta and/or gamma emitters.

## Concepts

### *Half-Life*

The activity of each radionuclide is characterized by a constant called the half-life ($t_{1/2}$), which is the time required for half of the original number of atoms to decay. The relationship between $t_{1/2}$ and the disintegration rate is as follows:

$$\lambda = \frac{0.693}{t_{1/2}}$$

or

$$t_{1/2} = \frac{0.693}{\lambda}$$

The average life of a radionuclide is called the mean life ($\overline{T}$) and is essentially the reciprocal of the disintegration constant:

$$T = \frac{1}{\lambda}$$

or

$$\overline{T} = \frac{T}{0.693}$$

***Problem:***   $C^{14}$ has a half-life of 5700 years. Calculate the fraction of the $C^{14}$ atoms that decays a) per year and b) per minute.

***Solution:***   a) Calculate:

$$= \frac{0.693}{5700 \text{ yr}} = \frac{6.93 \times 10^{-1}}{5.7 \times 10^{-3} \text{ yr}^{-1}}$$

Therefore, $1.216 \times 10^{-4}$ atoms per atom decays per year or 1 atom out of $1/1.216 \times 10^{-4}$ atoms decays per year.

$$\frac{1}{1.216 \times 1 \times 10^{-4}} = 0.8225 \times 10^4 = 8.225 \times 10^3$$

Therefore, 1 out of every 8225 radioactive atoms decays per year.

b)   $$\frac{1.216 \times 10^{-4} \text{ yr}^{-1}}{(365)(24)(60)} = 2.31 \times 10^{-10} \text{ min}^{-1}$$

$$\frac{1}{2.31 \times 10^{-10}} = 4.32 \times 10^9$$

Therefore, 1 out of $4.32 \times 10^9$ radioactive atoms decays per minute.

### *The Curie*

The Curie is defined as the quantity of any radioactive substance in which the decay rate is $3.700 \times 10^{10}$ disintegrations per second ($2.22 \times 10^{12}$ DPM). Because the efficiency of most radiation detection devices is less than 100%, a given number of Curies almost always yields a lower than theoretical count rate. Therefore, there is a distinction between DPM and CPM. For example, a sample containing 1 μCi of radioactive material has a decay rate of $2.22 \times 10^{6}$ DPM. If only 30% of the disintegrations are detected, the observed count rate is $6.66 \times 10^{5}$ CPM.[3]

### *Random Decay*

The rate of decay of a radionuclide at any instant is variable. Within a small number of counts, this variation is large; therefore, by taking more counts, the error is reduced. The desired accuracy in any experiment can be obtained within certain limits by calculating the uncertainty due to randomness from the statistical consideration discussed in the biochemical calculation chapter.[1,3]

### *Quenching*

In liquid scintillation counting, the presence of either inert material or color can reduce the recorded scintillation. Normally, this problem can be overcome by adding a small amount of a standard solution of isotope being counted (internal standard).

## Quantification of Radioactivity

### *Gas Flow*

If a gas is contained in a chamber in which there are two charged electrodes, a β particle can pass through this gas and can dislodge an orbital electron from one of the atoms of the gas, which results in the production of an ion pair—the dislodged electron plus the remaining positively charged ion. The β particle has sufficient energy to ionize several atoms successively. If the gas is contained in a chamber in which there are two charged electrodes, the secondary electrons and the positive ions will be attracted to the anode and cathode, respectively, and this can be recorded as a tiny pulse of charge or current. This property of ion chambers is shown in Figure 10.1. At a much higher voltage, the dislodged orbital electrons are accelerated toward the anode at such high velocities that they cause ionization of the gas atoms, producing what is called an avalanche of ions or gas amplification. At even higher voltage, the number of secondary ions collected becomes proportional to the number formed in the original ionization (the proportional region). This is followed by a second region of saturation (the Geiger-Müller region) in which all possible ion pairs are collected; ultimately, the voltage is so high that the gas is ionized by the applied voltage even when no β particles are present.[1,2]

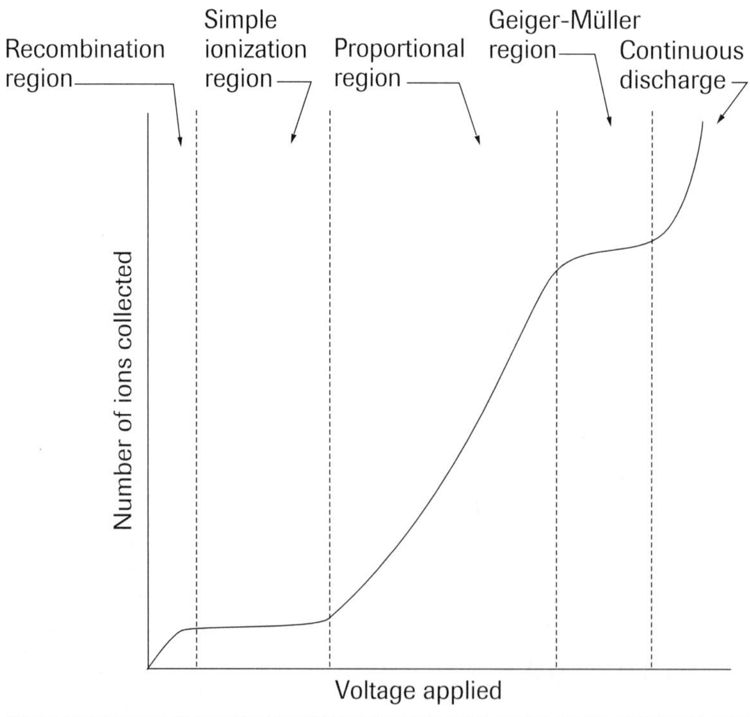


**FIGURE 10.1**   Output (number of ion pairs) of a Geiger-Müller tube as a function of the voltage between anode and cathode.

Adapted from Freifelder, D. 1976. *Physical Biochemistry, Applications to Biochemistry and Molecular Biology.* San Francisco: W. H. Freeman and Company Publisher. 94.

## Liquid Scintillation Counting

In liquid scintillation counting, the sample is either dissolved or suspended in a solvent counting one or more substances that are fluorescent. A scintillation sample vial normally contains the following components:

1. The radioactive sample.
2. A liquid scintillation cocktail normally consisting of the following:
   a. solvent—typically toluene, xylene, pseudocumene, or an alkyl benzene type solvent;
   b. emulsifier—a detergent type molecule (like Triton X-100) that ensures the proper mixing of aqueous samples in organic solvents;
   c. fluor—a fluorescent solute (like PPO). The function of the scintillation cocktail is to convert the energy of the radioactive decay particle into visible light that can be detected by the scintillation counter. The amount of light being emitted from the vial is proportional to the energy of the particle. That is, the higher the energy of a particle, the more solvent molecules it is able to excite, and therefore, more light is generated (Figure 10.2).

Many modern liquid scintillation counters maintain quench curves in memory to correct for quench and convert from cpm to dpm. An important technique in biochemistry/molecular biology is double labeling— e.g., simultaneous administration of a [3H]-labeled compound and a [14C]-labeled compound. Double labeling poses some problems for counting both [3H] and [14C] at the same time because of overlapping β-decay spectra. This results in spillover from one counting channel to another but can be corrected by proper programming of the liquid scintillation counter.

### *Quench*

Quenching within a sample refers to any mechanism that reduces the amount of light being emitted from the vial. Reducing the amount of light reaching the photomultiplier tubes (PMTs) results in a reduction of the pulse height. The three most common mechanisms are chemical, color, and dilution quenching.

### *Chemical Quenching*

Chemical agents (e.g., dissolved oxygen, water, and other solvents) added to the cocktail with the radioactive sample can absorb some of the energy of the β particles without emitting any photons or absorb the photons emitted by the excited solvent molecules without fluorescing.

### *Color Quenching*

Red, green, and yellow colors in the counting vials absorb some of the photons emitted by the secondary fluor in reducing efficiency.[3]

### *Dilution Quenching*

The dilution of the solvent and the fluor by the sample reduces the probability of a scintillation event. This

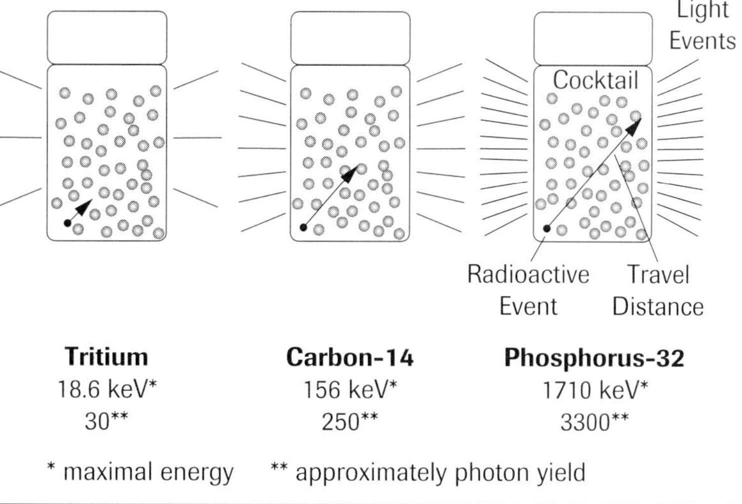


**FIGURE 10.2**  The amount of light being emitted from the vial is proportional to the energy of the particle. The light is emitted from the sample vial in all directions and is "directed" into two photomultiplier tubes which convert the light into a measurable pulse.

Adapted from Burns and Steiner. 1991. *Advanced Technology Guide for LS 6000 Series Scintillation Counters.* Palo Alto, CA: Beckman Instruments, Inc.

can be corrected for in the analysis of data. For more detailed information about quenching, students are advised to read the chapter on quenching by Burns and Steiner in *Advanced Technology Guide*[1] or references 3–5 cited at the end of this module.

### *Gamma-Ray Detection*

Gamma emitters have shown a great value in biochemistry, especially in radioimmunoassay where antibody or other proteins are labeled with the emitter [[126]I]. This radiation can easily be detected and counted by commercially available gamma counters. For more details, students are advised to read certain of the references cited at the end of this module.

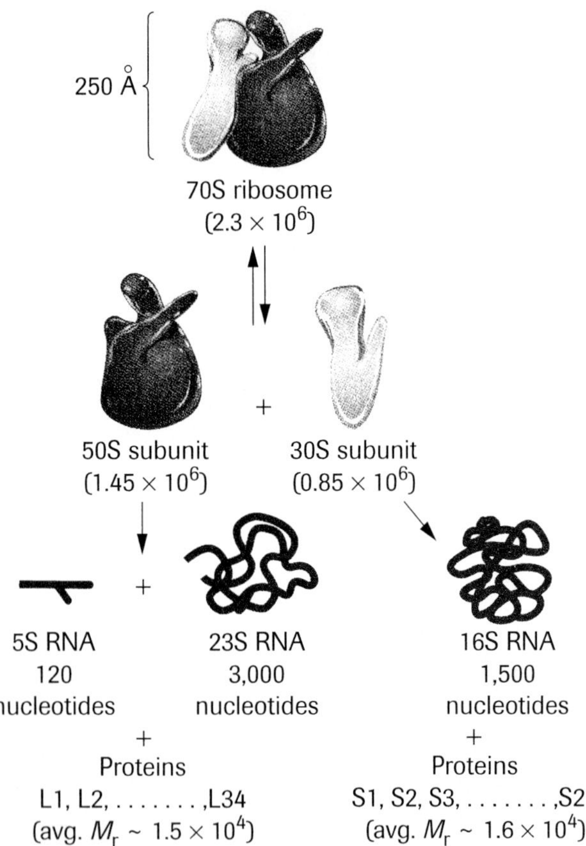



**FIGURE 10.3** Composition of the *E. coli* ribosomes. The 70S ribosome can dissociate into a 50S and a 30S subunit. *In vitro* can be done by lowering the Mg ion concentration. The individual subunits can be dissociated into their constituent RNAs and proteins by exposure to urea denaturant. Molecular weights are given for the subunits and the proteins, and the numbers of nucleotides are given for the RNAs.



### *Radioautography*

The details of this quantitation have already been discussed in the cloning module. Students are advised to read Module 7 in detail.

## Protein Synthesis

Proteins are synthesized possessing amino acid sequences (protein primary structure) via the translation of codons (specific nucleotide triplets) in messenger mRNA by ribosomes, complexes of ribosomonal (r)-RNA and protein (ribonucleoprotein). The (r)-RNA is transiently associated with mRNA, as are initiation, elongation, and termination factors. Translation begins with binding of the ribosome to mRNA.

Whereas bacteria contain 70S ribosomes composed of 50S and 30S subunits (Figure 10.3), eukaryotic ribosomes are larger containing many more proteins than bacterial ribosomes. In addition to free ribosomes, these organelles can exist upon membranes together as the rough endoplasmic reticulum.

The amino acids (Table 10.1) are inserted into the elongating polypeptides by specific amino acyl trans-

**TABLE 10.1** The Genetic Code

| First position (5' end) | | Second position — U | Second position — C | Second position — A | Second position — G | Third position (3' end) |
|---|---|---|---|---|---|---|
| U | UUU | UUU UUC } Phe | UCU UCC UCA UCG } Ser | UAU UAC } Tyr | UGU UGC } Cys | U |
| | | | | | | C |
| | | UUA UUG } Leu | | UAA UAG } STOP | UGA STOP | A |
| | | | | | UGG Trp | G |
| C | | CUU CUC CUA CUG } Leu | CCU CCC CCA CCG } Pro | CAU CAC } His | CGU CGC } Arg | U |
| | | | | CAA CAG } Gin | CGA CGG | C |
| | | | | | | A |
| | | | | | | G |
| A | | AUU AUC AUA } Ile | ACU ACC ACA ACG } Thr | AAU AAC } Asn | AGU AGC } Ser | U |
| | | AUG Met | | AAA AAG } Lys | AGA AGG } Arg | C |
| | | | | | | A |
| | | | | | | G |
| G | | GUU GUC GUA GUG } Val | GCU GCC GCA GCG } Ala | GAU GAC } Asp | GGU GGC GGA GGG } Gly | U |
| | | | | GAA GAG } Glu | | C |
| | | | | | | A |
| | | | | | | G |

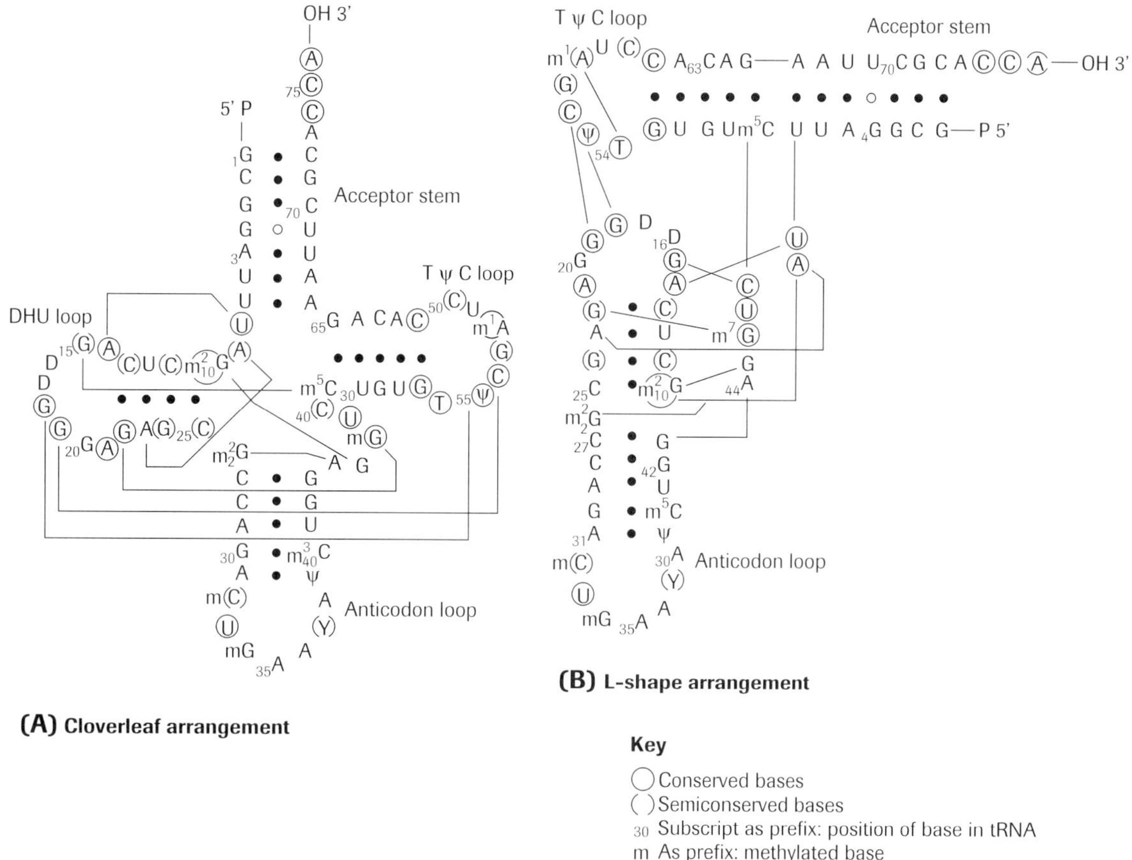



**FIGURE 10.4**   Nucleotide sequence of yeast tRNA[Phe] in the **(A)** cloverleaf and **(B)** L arrangements. The bases that are strongly conserved are circled; less conserved bases are in parentheses. The bases that form tertiary hydrogen bonds are connected by thin lines. Notice that most of the conserved bases are localized in the middle of the L and are involved in forming the tertiary hydrogen bonds. The solid circles indicate hydrogen-bonded base pairs. The open circle is a nonstandard base pair.

From Freifelder, D. 1983. *Molecular Biology.* Portola Valley, CA: Science Books International.

fer t-RNAs (Figure 10.4) formed by amino acyl t-RNA synthetases (Figure 10.5). The tRNAs transfer amino acids to the ribosomes, where the anticodons of tRNA recognize the mRNA codons in non-overlapping arrangement.

Polypeptide elongation is initiated at the amino ($NH_2$) terminal end and occurs through successive additions of new residues toward the carboxyl (COOH) terminus (Figure 10.6). In bacteria, the initiating amino acid, N-formylmethionine, is coded by AUG whose recognition requires a purine-rich (see nucleic acid chemistry) initiating signal on the 5' side of the AUG (Table 10.2).

The genetic code, the aforementioned codons, is "degenerate" in that multiple code words exist for almost all of the amino acids (see Table 10.1). The genetic words appear to be universal in all species except for some differences in mitochondria and a few single-celled organisms.

As for the later stages of protein synthesis (Figure 10.7), polypeptide chain termination occurs via UAA, UAG, and UGA codons. Then polypeptides can undergo enzyme-catalyzed folding into their active three-dimensional forms. Furthermore, some proteins can be modified by post-transitional reactions including attachment of carbohydrates or hybrid moieties or specific side chains.

Finally, many proteins are directed to particular cellular loci following protein synthesis. For example, targeting mechanisms exist involving peptide signal sequences generally found at the amino terminus of newly synthesized proteins. Targeting mechanisms may include a signal recognition particle, certain carbohydrates, or three-dimensional structures of proteins (Figure 10.8).

In-depth discussions of protein synthesis occur within Bosch, Weil and Bogarod, Linderstrom-Lang Conference, Spedding, and Arnstein.[6–10]

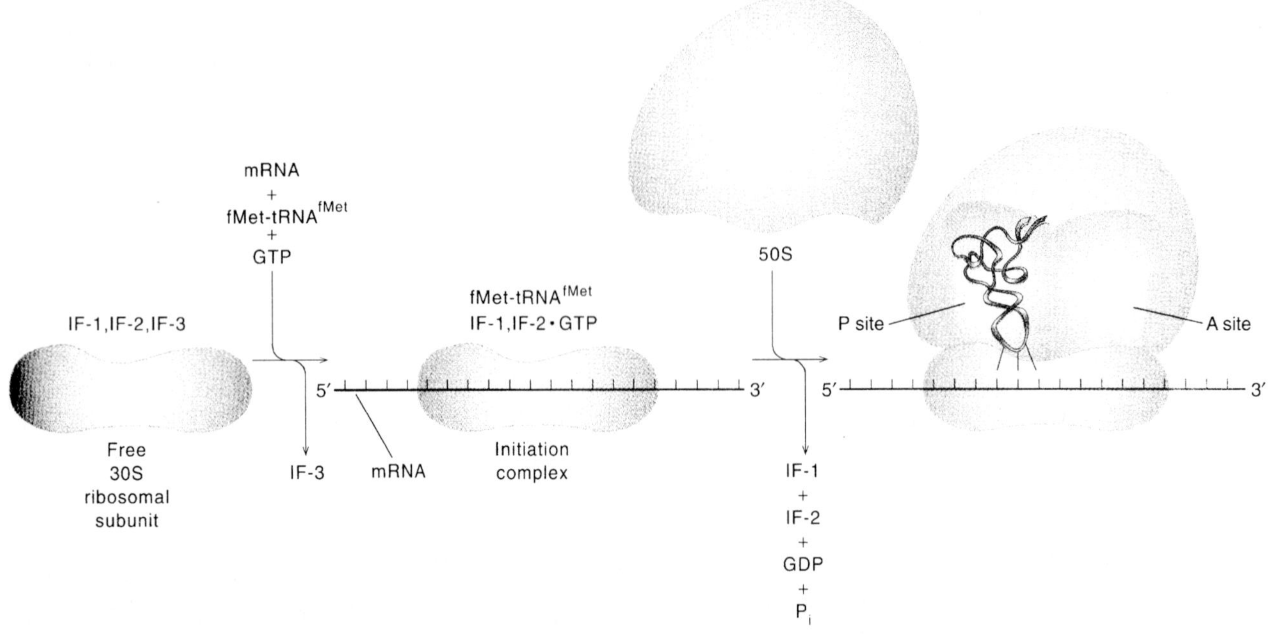



**FIGURE 10.5** Formation of the initiation complex for protein synthesis in prokaryotes. *E. coli* has three initiation factors bound to a pool of 30S ribosomal subunits. One of these factors, IF-3, holds the 30S and 50S subunits apart after termination of a previous round of protein synthesis. The other two factors, IF-1 and IF-2, promote the binding of both fMET-tRNA^fMet and mRNA to the 30S subunit. The binding of mRNA occurs so that its Shine-Dalgarno sequence pairs with 16S rRNA and the initiating AUG sequence with the anticodon of the initiator tRNA. The 30S subunit and its associated factors can bind fMET-tRNA^fMet and mRNA in either order. Once these ligands are bound, IF-3 dissociates from the 30S subunit, permitting the 50S subunit to join the complex. This release the remaining initiation factors and hydrolyzes the GTP, which is bound to IF-2.



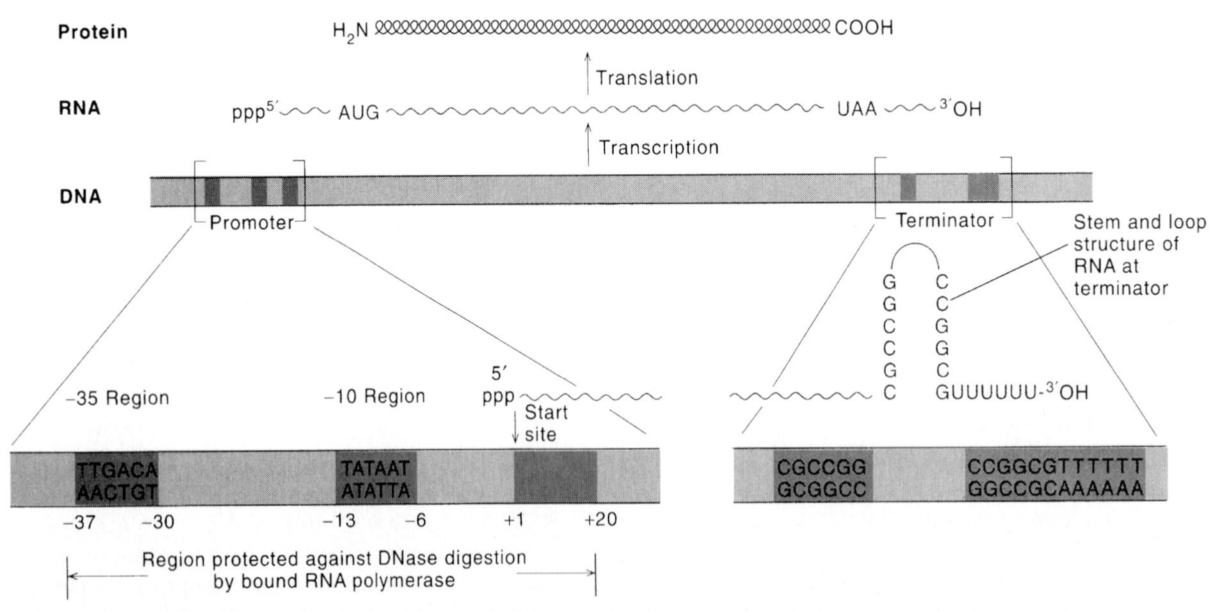



**FIGURE 10.6** Important features of a typical transcription unit. DNA is shown with promoter and terminator regions expanded below. RNA is transcribed starting in the promoter region at +1 and ending after the stem and loop of the terminator. The protein resulting from translation of this RNA is shown above with its N and C terminal indicated.

**TABLE 10.2**  The Sequence of the 3′ End of *E. coli* 16S RNA
and some Shine-Dalgárno Sequences at the 5′ End of Bacterial mRNAs

| | The pyrimidine-rich complement to the Shine-Dalgárno sequence |
|---|---|
| 16S rRNA | 3′ . . . HOAUUCCUCCACUA . . . 5′ |
| *lacZ* mRNA | 5′ . . . ACACAGGAAACAGCUAUG . . . 3′ |
| *trpA* mRNA | 5′ . . . ACGAGGGGAAAUCUGAUG . . . 3′ |
| RNA polymerase β mRNA | 5′ . . . GAGCUGAGGAACCCUAUG . . . 3′ |
| r-protein L10 mRNA | 5′ . . . CCAGGAGCAAAGCUAAUG . . . 3′ |

The purine-rich Shine-Dalgárno sequence      The initiation codon



# LABORATORY EXERCISES

## GROWTH OF POTATO PLANTS

***Supplies and Equipment***
Centrifuge
Liquid scintillation counter
SDS-PAGE electrophoresis cell (see protein
  purification, Module 3)
μl pipettes and tips
Water bath
–20° C freezer
Refrigerator
pH meter
Spectrophotometer
Incubator 65° C

***Chemicals***
Seeds of *Solanum berthaultii* (wild Bolivia
  potato—Hawkes, plant introduction number
  473334)
Department of Plant Breeding and Biometry, 252
  Emerson Hall, Cornell University, Ithaca, NY
  14853
10% v/v chlorox
Sterile water
9 mM gibberellic acid ($GA_3$)
Soil mix Promix BX (Premier Brands)
Fertilizer 20-10-20 NPK
Biweekly application of chelated iron
(Sequesterene 330 Fe)
1000 W metal halide lamps
Growth chamber with timer for controlling
  photoperiod

## Introduction

An overview of the steps comprising *in vitro* translation is illustrated in Figure 10.9. In the following exercises, the student will learn the protocols for the reticulocyte-mediated *in vitro* synthesis of *Solanum berthaultii* PPO. *In vitro* synthesis in mammalian systems will be included in the next edition.

## Procedures

### Growth of Solanum Berthaultii Plants

The procedures for the growth of potato plants are depicted in Figure 10.10. Either the laboratory instructor or the student should grow the plants from seed in advance. These potato plants can be grown from seed within a growth chamber under photoperiodic control.

## ISOLATION OF TOTAL RNA AND POLY (A)-mRNA

***Supplies and Equipment***
Centrifuge
Liquid scintillation counter
SDS-PAGE electrophoresis cell (see protein
  purification, Module 3)
μl pipettes and tips
Water bath

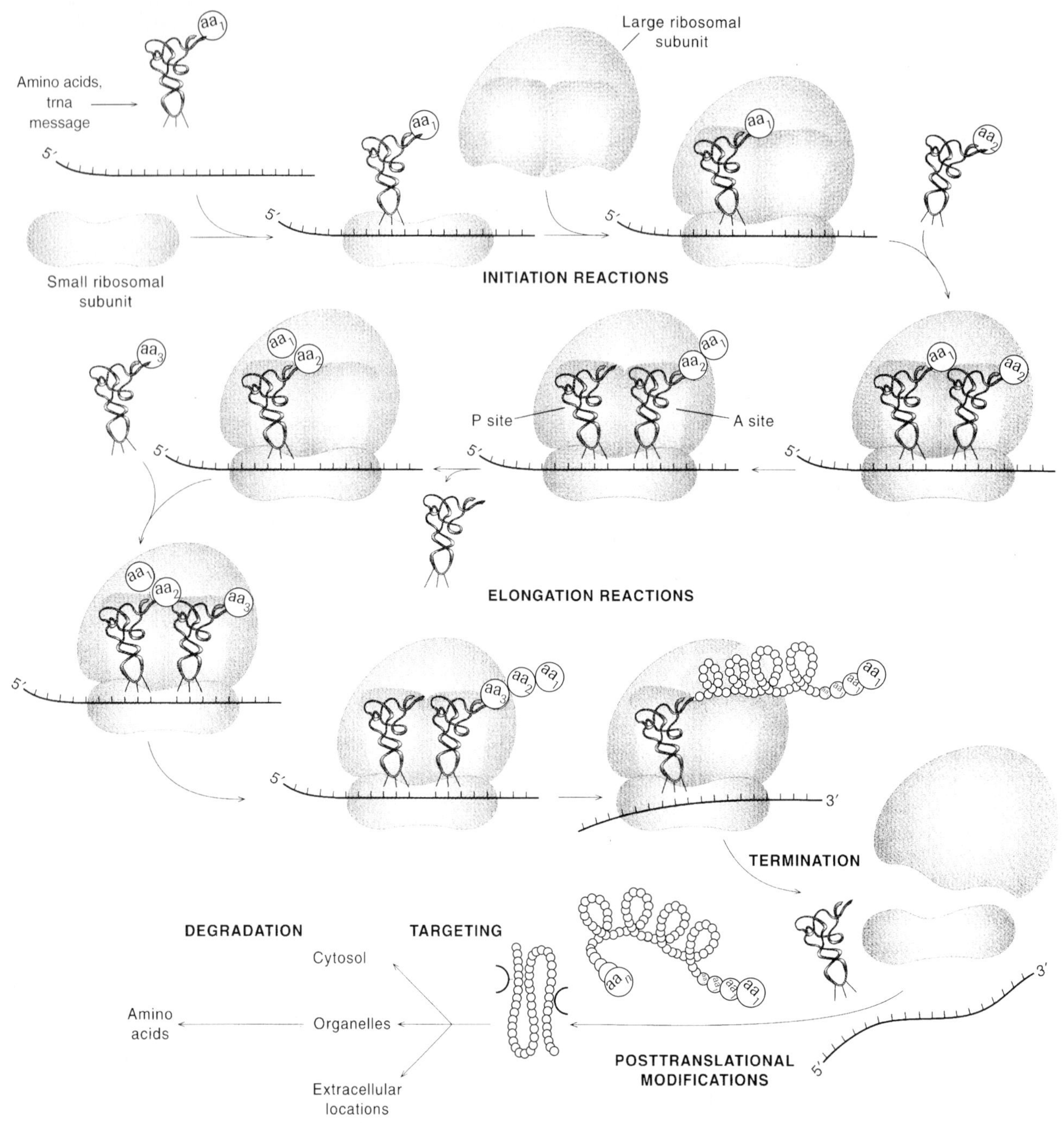


**FIGURE 10.7**   Overview of reactions in protein synthesis. (aa$_1$, aa$_2$ aa$_3$ = amino acids 1, 2, 3). Protein synthesis requires transfer RNAs for each amino acid, ribosomes, messenger RNA, and a number of dissociable protein factors in addition to ATP, GTP, and divalent cations. First the transfer RNAs become charged with amino acids, then the initiation complex is formed. Peptide synthesis does not start until the second aminoacyl tRNA becomes bound to the ribosome. Elongation reactions involve peptide bond formation, dissociation of the discharged tRNA, and translocation. The elongation process is repeated many times until the termination codon is reached, an event that signals the end of polypeptide synthesis. Termination is marked by the dissociation of the messenger RNA from the ribosome and the dissociation of the two ribosomal subunits. The polypeptide chain sometimes folds into its final form without further modifications. Frequently the folded polypeptide chain is modified by removal of part of the polypeptide chain or by addition of various groups to specific amino acid side chains. The completed polypeptide chain will migrate to different locations according to its structure. It may remain in the cytosol or it may be transported into one of the cellular organelles or across the plasma membrane into extracellular space. Proteins have different lifetimes. Eventually a protein will be degraded, usually down to its component amino acids.

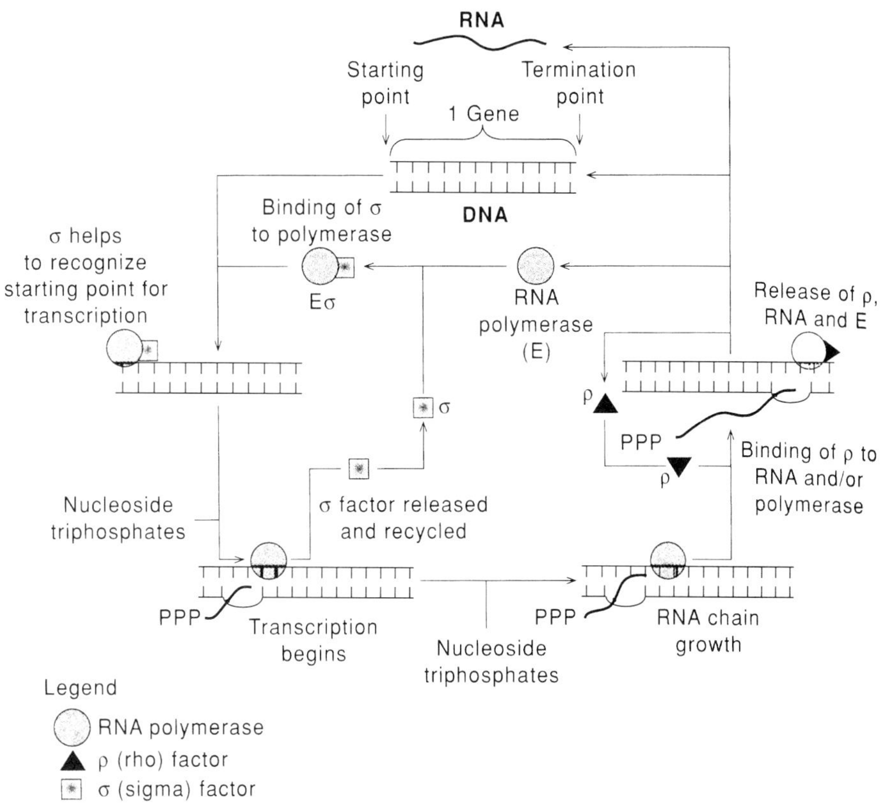



**FIGURE 10.8**   Two successive translocations are required to target proteins such as cytochrome $c_1$ to the intermembrane space. The precursors of cytochrome $c_1$ have two uptake-targeting sequences at its N terminus. The first targets the polypeptide to the matrix. In the matrix the first target sequence is cleaved by a specific protease. The second target sequence is thereby exposed and targets the polypeptide to intermembrane space, where the second target sequence is removed by another protease. The molecule folds and adds heme to become fully functional.

F. U. Hartl et al., in *Cell* 51:1021–1027, 1987. Copyright © 1987. Cell Press, Cambridge, Mass.: and E. C. Hurt and A. P. G. M. van Loon, "How proteins find mitochondria and intramitochondrial compartments." *Trends in Biochemical Sciences* 11:204–207, 1986. Copyright © 1986 Elsevier Trends Journals, Cambridge, England.

---

Plant tissue.
↓
Isolate poly -(A)-mRNA.
↓
Translate poly (A)-mRNA in a rabbit reticulocyte system together with [$^{35}$S]-RNA.
↓
Resolve *in vitro* translation products by SDS-PAGE.
↓
Immunoprecipitate with *S. aureus* cells and antibody.
↓
Autoradiography and fluorography.

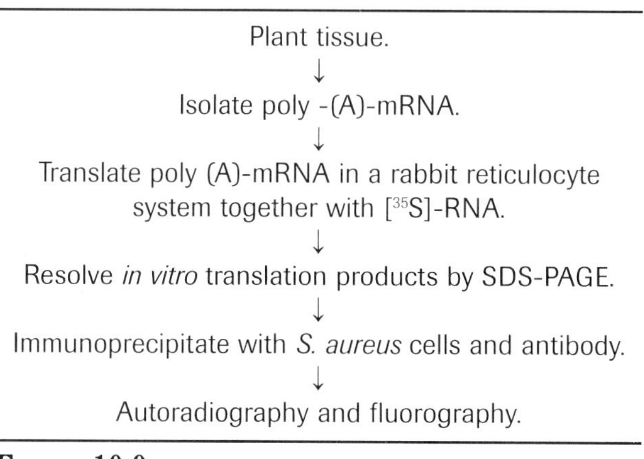

**FIGURE 10.9**   Overview of *in vitro* translation.

---

Treat seeds of *Solanum bethaultii* Hawkes, plant introduction number 473334, with 10% v/v chlorox for 10 seconds.
↓
Wash twice with 200 ml–300 ml deionized sterile water.
↓
Soak in 9 mM GA$_3$ 24 hours prior to sowing.
↓
Sow in Promix BX soil mix.
↓
Fertilize weekly 20–10–20 NPK, plus biweekly application of chelated iron (Sequesterene 330 Fe).
↓
Grow plants at 25° C under 1,000 W metal halide lamps 14–16 hours light 10–8 hours dark photoperiod until the 3rd leaf node stage.

**FIGURE 10.10**   Summary of conditions employed to grow potatoes (to be performed in advance of *in vitro* translation laboratory).

Hunt, M. D., N. T. Eannetta, Y. Haiferg, S. M. Newman, and J. C. Steffens. 1993. "cDNA cloning and expression of potato polyphenol oxidase." *Plant Molecular Biology.* 21:59–68.

–20° C freezer
Refrigerator
pH meter
Spectrophotometer
Incubator 65° C

### Chemicals

Total RNA
Denaturing solution
4M guanidinium thiocyanate
25mM sodium citrate, pH 7.0
0.5% sarcosyl
0.1 M β-mercaptoethanol
Phenol (nucleic acid grade, Bethesda Res. Lab)
   saturated with $H_2O$—maintain at 4° C up to
   1 month
Chloroform
Preparation of denaturing solution
Prepare stock solution of guanidinium thiocyanate
   by dissolving 250 g guanidinium thiocyanate
   (Fluka) in the manufacturer's bottle (without
   weighing) with 293 ml $dH_2O$, 17.6 ml 0.75M
   sodium citrate, pH 7.0, and 26.4 ml 10%
   sarcosyl—65° C; can store at least 3 months
   at room temperature.
Solution D—add 0.36 ml β-mercaptoethanol/50 ml
   stock solution; can store for 1 month at room
   temperature

# Introduction

To obtain *in vitro* poly (A)-mRNA translation products, it is first necessary to isolate bulk RNA and then
select poly (A)-mRNA from it. Whereas Figure 10.11
provides the procedures for isolating bulk RNA[12] from
100 mg of potato leaves excised from the first, second, and third nodes, Figure 10.12 details the protocols for the subsequent selection of poly (A)-mRNA
oligo dT chromatography.[13] The mechanism by which
an oligo dT column separates poly (A)-mRNA takes
advantage of the poly (A) tail of mRNA, which combines with the thymidylic residues of the oligo dT
resin—i.e., A with T (see  DNA chemistry, Chapter 7).

# *IN VITRO* TRANSLATION OF POLY (A)-mRNA IN A RABBIT RETICULOCYTE SYSTEM

### Supplies and Equipment

Centrifuge
Liquid scintillation counter
SDS-page electrophoresis cell (see protein
   purification, Module 3)
μl pipettes and tips
Water bath

–20° C freezer
Refrigerator
pH meter
Spectrophotometer
Incubator 65° C

### Chemicals

Loading buffer—20mM Tris-Cl (pH 7.6, containing
   0.05 M NaCl, 1 mM EDTA, 0.1% SDS)
0.1 M NaOH containing 5mM EDTA
10mM Tris-Cl (pH 7.6), 1mM EDTA, 0.05% SDS
3M sodium acetate, pH 5.2
Ethanol
70% ethanol
Pasteur pipettes
Rabbit Reticulocyte System
[$^{35}$S]-methionine (spc. act. 1,000 Ci/nmol)
Promega reticulocyte system kit
Liquid scintillation counting cocktail

## Derivation of Reticulocyte System

Poly (A)-mRNA can be translated in a Promega,
Biotec (Madison, WI) rabbit reticulocyte system
according to the procedures depicted in Figure 10.13.

Reticulocyte lysate is prepared by inducing reticu-
locytosis in rabbits, obtaining immature red cells,
washing the cells free from serum contaminants, and
lysing the reticulocyte by suspension in $H_2O$ or low
ionic strength buffer. Finally, membrane ghosts are
removed by centrifugation at 10,000 $xg$ for 20 minutes. A review of the use of reticulocyte lysate in
translation of mRNAs can be found in Shafritz.[14] In
addition, Rapoport[15] has discussed the "machinery of
protein synthesis" in the reticulocyte. This machinery
consists of initiation and elongation factors, tRNAs
and mRNAs.

The following three paragraphs are paraphrased
from *Promega Technical Bulletin 127*,[16] Flexi™ Rabbit Reticulocyte Lysate System. The rabbit reticulocyte *in vitro* translation system was originally
described by Pelham and Jackson.[17] The system has
become a routine method for identifying mRNA
species and characterizing their products. In addition,
the system can be employed to post-translationally
alter proteins through proteolysis, phosphorylation,
acetylation, and isoprenylation.

Promega manufacturers the Flexi™ Rabbit Reticulocyte Lysate System as well as a standard rabbit
reticulocyte system. The former offers enhanced flexibility of reaction conditions. Promega adds several
factors to the Flexi system to optimize translation
efficiency. These factors are hemin, phosphocreatine
kinase, and phosphocreatine and calf liver tRNAs.
The latter expand the range of mRNAs that can be
translated. To destroy endogenous RNA, the flexi
lysate is treated with micrococcal nuclease.

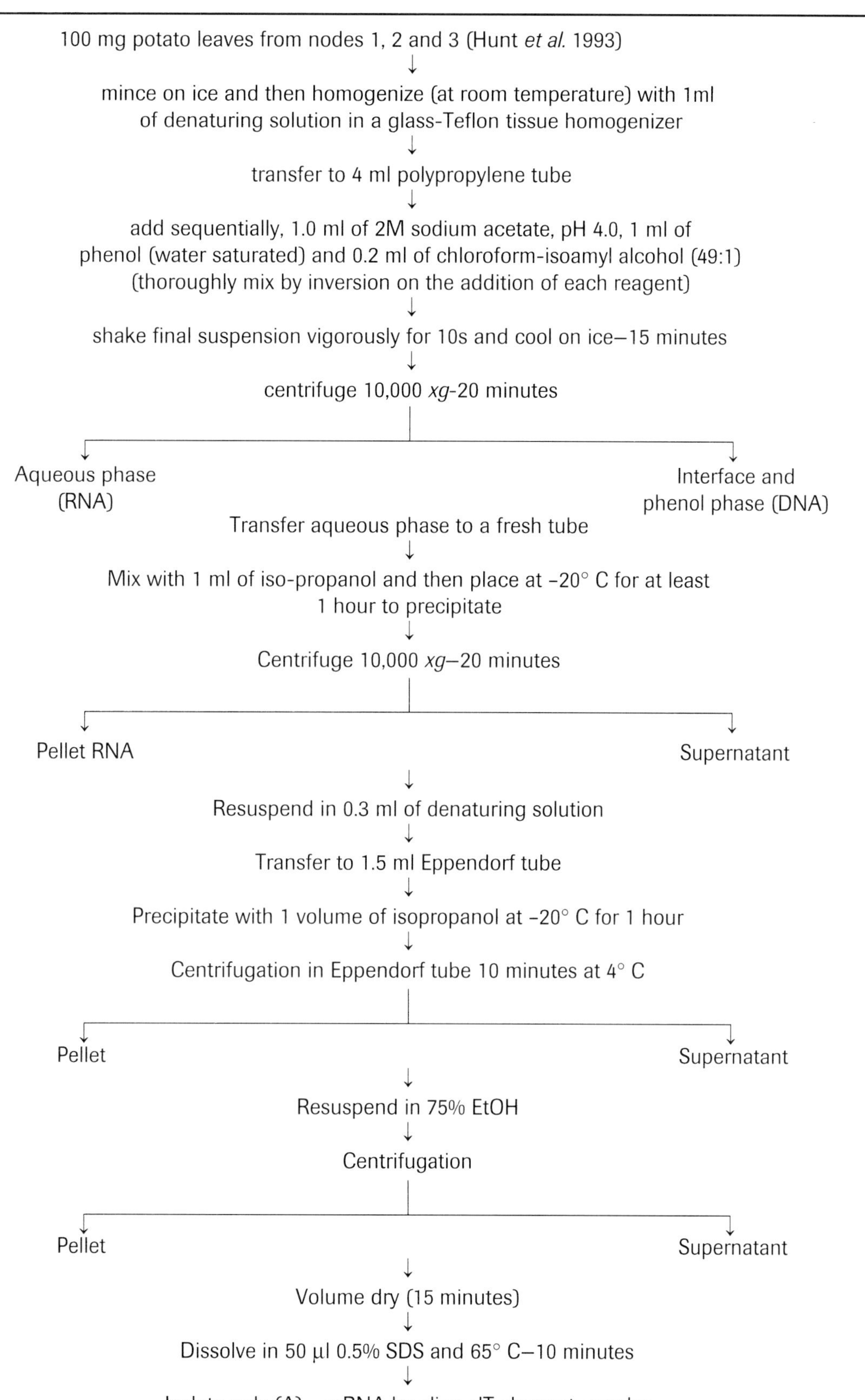


**FIGURE 10.11**   Summary of procedure for isolation of total RNA.

From Chromczynski, P., and N. Sacchi. 1987. "Single-step method for RNA isolation by acid guanidinium thiocyanate-phenol chloroform extraction." *Analytical Biochemistry.* 162:156–59.

Equilibrate the oligo (dT)-cellulose in sterile loading buffer (20mM Tris Cl, pH 7.6, 0.5 M NaCl, 1mM EDTA, 0.1% SDS). The SDS should be added from a 20% stock solution which has been treated at 65° C for 1 hour. As an alternative, the Tris can be replaced with 0.05 M sodium citrate and the loading buffer and SDS treated with diethylpyrocarbonate.

↓

Construct a 1.0 ml column (a Pasteur pipette will suffice); wash the oligo-dt cellulose with 3 column volumes each of:

1. sterile $dH_2O$
2. 0.1 N NaOH and 5 mM EDTA
3. sterile $dH_2O$

Monitor the pH of the column's effluent to ensure that the pH < 8.

↓

Wash the column with 5 volumes of sterile loading buffer.

↓

Dissolve the RNA in sterile $dH_2O$ and heat to 65° C for 5 minutes. Add an equal amount of 2X loading buffer, cool to room temperature, and apply to the column.

↓

Collect the column's effluent, re-heat to 65° C, cool, and reapply to column.

↓

Wash the column with 5–10 column volumes of loading buffer, followed by 4 column volumes of loading buffer containing 0.1 M NaCl.

↓

Elute the poly (A)[+]-mRNA with 2–3 column volumes of sterile 10 mM Tris-Cl pH 7.5 containing 1 mM EDTA and 0.05% SDS.

↓

Monitor each column fraction at 260 nm. (The poly (A)[+]-RNA can be re-selected by a repetition of oligo dT cellulose chromatography.) Add sodium acetate (3M, pH 5.2) to a final concentration of 0.3 M.

↓

Precipitate the RNA with 2.2 volumes of ethanol at –20° C.

↓

Rinse the pellet with 70% ethanol.

↓

Dissolve the pellet in sterile $H_2O$.

↓

Store on ice at 4° C.

**Figure 10.12** Preparation of poly (A)-mRNA by oligo dT-chromatography.

Adapted from Sambrook J., E. F. Fritsch, and T. Maniatis. 1989. *Molecular Cloning: A Laboratory Manual*, 2d ed., Cold Spring Harbor, NY: Cold Spring Harbor Laboratory Press.

Finally, the Flexi™ Rabbit reticulocyte system permits translation reactions to be optimized for a wide range of parameters including $Mg^{++}$ and $K^+$ concentrations and whether to employ DTT. In addition, a control RNA encoding the firefly luciferase gene can be part of the kit.

An example of a standard reaction using [35S]-methionine and the protocol for a translation procedure are presented in Figure 10.13. Students are encouraged to read and thoroughly digest the contents of *Promega Technical Bulletin 127*.[16] In addition to the translation procedure depicted in Figure 10.13, this bulletin contains methods for post-translational analysis, luciferase positive control translation reaction, positive control luciferase assays, and related procedures.

# IMMUNOPRECIPITATION OF *IN VITRO* TRANSLATION PRODUCTS

### *Supplies and Equipment*

Centrifuge
Liquid scintillation counter
SDS-PAGE electrophoresis cell (see protein purification, Module 3)
μl pipettes and tips
Water bath
–20° C freezer
Refrigerator
pH meter
Spectrophotometer
Incubator 65° C

### *Chemicals*

Anti-PPO (rabbit anti-*Solanum* trichome PPO—Kowalski et al., 1992, 1993)
10% *Staphylococci aureus* cells
Protein A Sepharose
Normal rabbit serum
Phosphate buffered saline containing 1% Triton X-100, v/v, and 1% deoxycholate (w/v)

## Introduction

### Importance of Staphylococcal Cells in Immunoprecipitation

The sequential steps for antigen isolation utilizing staphylococcal absorbent involves complexing radio-labeled, solubilized antigen with a molar excess of specific antibody, binding of immune complexes to *Staphylococci*, separation from other cell components by centrifuging and washing, and elution from the absorbent for subsequent analysis.[18]

**Example of Standard Reaction Using
<sup>35</sup>S-Methionine (See Note 3 below for Information Concerning Other
Radiolabeled Amino Acids)**

| | |
|---|---|
| Flexi Rabbit ™Reticulocyte Lysate (see Note 1 below) | 33µl |
| Amino Acid Mixture Minus Methionine, 1mM | 1µl |
| $^{35}$S-methionine (1,200Ci/mmol at 10mCi/ml) (see Notes 2 and 3 below) | 2µl |
| Magnesium Acetate, 25mM | 0–4µl |
| Potassium Chloride, 2.5M | 1.4µl |
| DTT, 100mM | 0–1µl |
| RNasin Ribonuclease Inhibitor (40u/µl) | 1µl |
| RNA substrate (20ng/µl) | 1–12µl |
| Nuclease-free H$_2$O | to a final volume of 50µl |

Notes:

1. The 70% concentration of lysate in the standard reaction is optimal for most applications. If required, though, the lysate can be diluted to 50% without a substantial reduction in translational efficiency. If optimal expression is desired in a 50% reaction, then the reduced endogenous levels of $Mg^{2+}$ and $K^+$ may be adjusted.

2. The addition of RNasin Ribonuclease Inhibitor to the translation reaction is recommended but not required. RNasin Ribonuclease Inhibitor acts to inhibit degradation of sample mRNAs by contaminating RNase.

3. We have found acceptable results using 1µl–4µl of [$^{35}$S]methionine ($^{35}$S-met) (1,200Ci/mmol) at 10mCi/ml. Depending upon the translational efficiency of the experimental RNA and number of methionines present in the protein, the amount of $^{35}$S-met can be adjusted to balance exposure time versus cost of label.

4. It is recommended to use a "translational" grade of label, such as Amersham International [$^{35}$S]methionine (Catalog No. SJ1515 or SJ1015). This grade of $^{35}$S-met does not cause the background labeling of the rabbit reticulocyte lysate 42kDa protein, which can occur using other grades of label. $^{35}$S-labeled amino acids are easily oxidized to translation-inhibiting sulphoxides and should be stored in aliquots with buffer containing DTT or β-mercaptoethanol at –70° C.

5. An unfractionated cytoplasmic RNA preparation is 90%–95% rRNA and, as a result, translates poorly. Usually such preparations yield no better than 20%–30% of the maximum incorporation attainable, and concentrations of 100µg/mg– 200µg/ml (final concentration) of RNA are needed to stimulate translation. In contrast, viral RNAs and poly(A) + mRNAs (including mRNA transcribed *in vitro*) can be used at much lower concentrations (5µg/ml–80µg/ml final concentration).

6. Average preparations of mRNA give a stimulation over background of about 10- to 20-fold. If the translation efficiency of sample RNA is low, the following suggestions may be utilized.

   a. The optimal RNA concentration for translation should be determined prior to performing definitive experiments. In determining the optimal concentration, serially dilute your RNA template *first* and then add the same *volume* of RNA to each reaction to ensure that other variables are kept constant. See also Note 5.

**FIGURE 10.13**   Procedures for the use of a Promega Rabbit Reticulocyte *in vitro* translation kit.

From *Promega Corp. Technical Bulletin 127.* 1992. Madison, WI: Promega.

b. To determine if inhibitors are present in your mRNA preparation, mix your RNA with Luciferase Control RNA and determine if the translation of Luciferase RNA is inhibited relative to a control translation containing only Luciferase RNA. Oxidized thiols, low concentrations of double-stranded RNA, and polysaccharides are typical inhibitors of translation in rabbit riticulocyte lysate.

c. Optimum potassium concentration varies from 80mM–120mM, depending on the mRNA used. Additional potassium can be added if the initial translation results are poor. Similarly, specific mRNAs may also require altered magnesium concentrations. A range between 0.5mM– 2.5mM is generally sufficient for the majority of mRNAs utilized. See Table 10.3 for the concentrations of key components present. For further optimization of salt concentrations, we recommend the use of Promega's Flexi™ Rabbit Reticulocyte Lysate.

**TABLE 10.3**  Final Concentrations of Rabbit Reticulocyte Lysate Components in a 50 µl Translation Reaction

| | |
|---|---|
| Creatin phosphate | 10mM |
| Creatine phosphokinase | 50µg/ml |
| DTT | 2mM |
| Calf liver tRNA | 50µg/ml |
| Potassium acetate | 79mM |
| Magnesium acetate | 0.5mM |
| Hemin | 0.02mM |

d. Avoid adding calcium to the translation reaction. Calcium may reactivate the micrococcal nuclease and result in degradation of the RNA template.

e. Residual ethanol should be removed from mRNA preparations and labeled amino acids before they are added to the translation reaction.

7. Hemin is included in the reticulocyte lysate because it suppresses an inhibitor of the initiation factor elF2a. In the absence of hemin, protein synthesis in reticulocyte lysates ceases after a short period of incubation.

8. The addition of polyamines, such as spermidine (0.1mM–0.5mM) and certain diamines (0.1mM–40mM), have been shown to stimulate translation. The addition of these compounds results in the reduction of the corresponding optimal $Mg^{2+}$ concentration for a given RNA. Therefore, less $Mg^{2+}$ is required for the translation reaction.

9. Except for actual translation incubation, all handling of lysate components should be done at 4° C. Any unused lysate should be refrozen as soon as possible after thawing to minimize loss of translational activity. It is not recommended to freeze/thaw the lysate more than two times.

10. Each batch of lysate contains 100µg/ml–200µg/ml of endogenous protein using BSA as a standard.

11. Use capped vials. This avoids changes in the reaction volume which may affect the concentration of important components.

12. RNase A can be added to the lysate reaction (after completion) to a final concentration of 0.2mg/ml for 5 minutes at 30° C to digest aminoacyl tRNAs, which can sometimes produce background bands

**FIGURE 10.13**  Continued.

Approximately 90% of the *S. aureus* strains possess protein A attached to the cell walls' peptidoglycan; although there appears to be variation in protein density, Cowan I (ATCC 12598, NCTC 8530) is perhaps the best characterized strain possessing a high protein A density. Protein A exhibits a marked affinity for the Fe portion of the immunoglobulin molecule mainly IgG of most mammalian systems. The advantages of utilizing *Staphylococci* for antigen isolation are: 1) specific binding of immune complexes to *Staphylococci* is very rapid (seconds), 2) binding is stable in a variety of solvent systems, 3) non-specific binding to other proteins is minimum, 4) bacteria can be sedimented by low *xg* forces, 5) ability to elute the antigen from *Staphylococci* with several denaturing solvents—e.g., SDS, urea, or lyotropic salts, and 6) versatility and economy.

## Procedures

### Preparation, Harvesting and Maintenance of Staphylococci

The procedures for preparation and harvesting of Staphylococci are presented in Figure 10.14 and 10.15.

To prevent contamination during culturing, streak bacteria from agar slant cultures onto commercially available blood agar plates, inoculate at 37° C, 24 hours.

↓

Employ single, isolated colonies from the streaking to inoculate 30–50 ml (seeding culture); culture at 37° C, 8–12 h.

↓

Pour contents of the seeding cultures into 2 l–4 l Erlenmeyer flasks containing 500 or 1,000 ml sterile broth.

↓

Culture Cowan I (ATCC 12598, NCTC 8330) *Staphylococci aureus* in 4 l–5 l of either CCY broth or Penassay broth (Expect 13 g l$^{-1}$), 37° C 16–17 hours.

↓

Monitor cell density by quantifying absorbance at 525 nm as prolongation of culture in stationary growth phase (720 hours) may result in a significant loss of protein A.

**FIGURE 10.14** Protocol for preparation of *Staphylococci.*

From Kessler, S. W. 1981. "Use of protein a bearing *Stapholococci* for the immune precipitation and isolation of antigens from cells." *Methods Enzymology.* 73:442–59.

---

Harvest cultures in 250 ml polycarbonate centrifuge bottles with firm-fitting screw caps.

↓

Use 7,000 rpm for 5–10 minutes in a 6 × 250 ml Sorvall GSA or Beckman JA-14 rotor or equivalent.

**Pellet**
Refill bottles until all the bacteria have been collected in 6–12 bottles.

**Supernatant**
↓
Autoclave

Resuspend *S. aureus* cells in 40 mM phosphate, 150 mM NaCl, 0.05% NaN3, pH 7.2. Perform resuspension with a wide-base glass pipette to prevent cell fragmentation. Pool suspension into 1 or 2 bottles.

↓

Estimate the yield of *S. aureus* cells by comparing the weight of the bottles containing bacteria against a similar empty bottle. Following resuspension to a 10% (w/v) concentration in PBS-azide, stir the bacteria 90 minutes with 1.5% formaldehyde (kills bacteria).

↓

Centrifuge fixed bacteria and resuspend immediately to the same 10% (w/v) concentration. Swirl bacteria rapidly and continuously for 5 minutes in a large Erlenmeyer flask containing no more than 2 cm liquid depth immersed in an 80° C water bath. Cool contents to room temperature by swirling in a water-ice bath.

↓

Centrifuge bacteria twice more and resuspend in PBS azide to 10% by volume.

---

**FIGURE 10.15** Procedures for harvesting *Staphylococcus aureus.*

From Kessler, S. W. 1981. "Use of a protein bearing *Stapholococci* for the immune precipitation and isolation of antigens from cells." *Methods Enzymology.* 73:442–59.

Staphylococci are stored as frozen aliquots but can be maintained at 4° C for up to 2–3 weeks subsequent to training. Within 24 hours of use, an appropriate volume is centrifuged at 3,000 *xg* 15 minutes and then equilibrated in extraction washing buffer (0.05 M Tris-HCl, pH 7.4, 0.15 M NaCl, 0.005 M EDTA, 0.005M KI, 0.01M appropriate unlabeled amino acid, 0.02% NaN$_3$ containing 0.5% NP-40 detergent). Centrifuged *Staphylococci* are washed again in the same buffer containing 0.05% NP-40 and are then resuspended in this buffer to the original volume.

## Employment of *S. Aureus* Cells in Immunoprecipitation

The procedure employed for the immunoprecipitation of *in vitro* synthesized PPO within translation mixtures are in Figure 10.16.

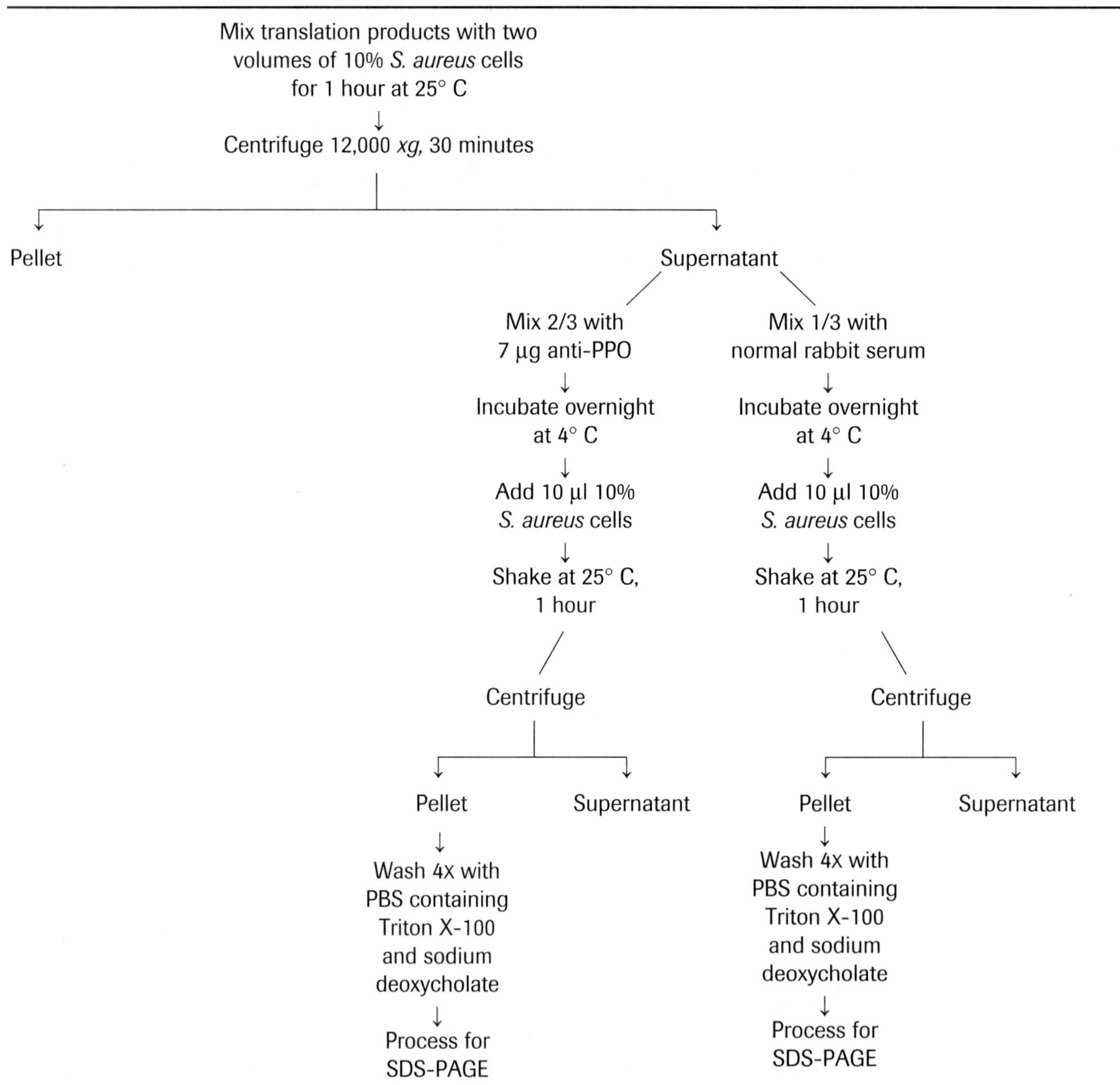


**FIGURE 10.16**   Protocol for immunoprecipitation in *in vitro* translation products.

From Steffens, J. C., S. P. Klowalski, and H. Yu. 1990. "Characterization and cloning of glandular trichome and plastid polypenoloxidase." *Molecular Biology of the Potato*. Barking, Essex, England: C.A.B. International.

# SDS-PAGE RESOLUTION OF *IN VITRO* TRANSLATION PRODUCTS

### *Supplies and Equipment*
Centrifuge
Liquid scintillation counter
SDS-PAGE electrophoresis cell (see protein
purification, Module 3)
μl pipettes and tips
Water bath
–20° C freezer
Refrigerator
pH meter
Spectrophotometer
Incubator 65° C

### *Chemicals*
13% SDS-PAGE gels (see SDS-PAGE Lab)

## Introduction

To resolve the products of *in vitro* translation immunoprecipitates are boiled and processed for SDS-PAGE as in Module 3. Table 10.4 offers the quantities of ingredients to make separating gels of varying percent SDS. A variety of references exist that detail SDS-PAGE procedures and at variants of it—e.g., combined SDS-PAGE and IEF—such as Allen,[19] Dunbar[20] and Anderson.[21]

# AUTORADIOGRAPHY AND FLUOROGRAPHY

### *Supplies and Equipment*
Centrifuge
Liquid scintillation counter

SDS-PAGE electrophoresis cell (see protein
purification, Module 3)
μl pipettes and tips
Water bath
–20° C freezer
Refrigerator
pH meter
Spectrophotometer
Incubator 65° C

### *Chemicals*
X-ray film—Kodak XAR film
Developing solution

## Introduction

Following resolution of the products of *in vitro* translation by SDS-PAGE, the radioactive products can be detected by autoradiography and fluorography (Scintillation autoradiography).

The autoradiography of SDS-PAGE is similar to that for DNA sequencing gels and students are referred to the cloning module for autoradiographic procedures. Finally, Figure 10.17 presents the *in vitro* translation of potato epidermal and leaf mRNA profiles as published by Steffens et al.[22]

## Review Questions

1. What is your understanding of how animals and plants carry out protein synthesis?
2. Design an experiment about the *in vitro* protein synthesis of an enzyme.
3. Differentiate between the isolation of total and poly-(A)-RNAs.
4. What ingredients are components of an *in vitro* protein synthesis system? What does each component do?

**TABLE 10.4**  SDS-Polyacrylamide Separating Gels

| Ingredients | Percent Acrylamide mg Bisacrylamide/ml | | | | | | |
|---|---|---|---|---|---|---|---|
| | 5%:2.6 | 7.5%:1.95 | 10%:1.3 | 12.5%:1.0 | 15.0%:0.86 | 17.5%:0.73 | 20%:0.65 |
| 30% Acrylamide (ml) | 5.0 | 7.5 | 10.0 | 12.5 | 15.0 | 17.5 | 20.0 |
| 2% Bisacrylamide (ml) | 3.9 | 2.9 | 2.0 | 1.5 | 1.3 | 1.1 | 1.0 |
| 1 M Tris–Cl (pH 8.7) (ml) | 11.2 | 11.2 | 11.2 | 11.2 | 11.2 | 11.2 | 11.2 |
| 20% SDS (ml) | 0.15 | 0.15 | 0.15 | 0.15 | 0.15 | 0.15 | 0.15 |
| $H_2O$ (ml) | 9.45 | 8.25 | 6.65 | 4.65 | 2.35 | — | — |
| TEMED (μl) | 25 | 25 | 25 | 25 | 25 | 25 | 25 |
| 10% Ammonium Persulfate (μl) | 100 | 100 | 100 | 100 | 100 | 100 | 100 |

Sambrook J., E. F. Fritsch, and T. Maniatis. 1989. *Molecular Cloning: A Laboratory Manual*, 2d ed., Cold Spring Harbor, New York: Cold Spring Harbor Laboratory Press.

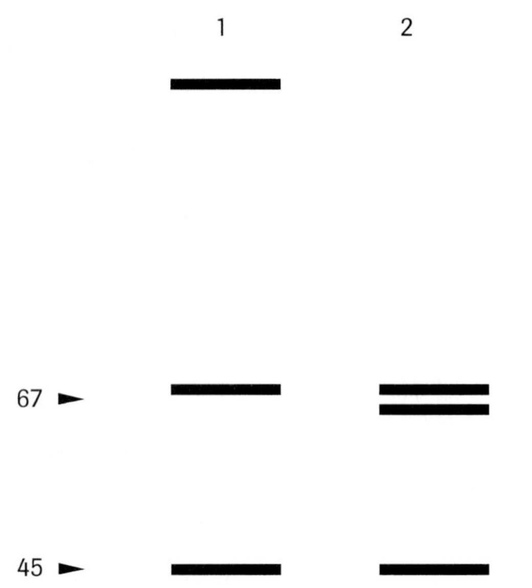



**FIGURE 10.17** *In vitro* translation of epidermal and leaf mRNA. Total RNA was isolated from epidermis by the guanidinium isothiocyanate method (Chromczynski et al., 1987) and mRNA was affinity purified by oligo-dT cellulose chromatography. Polyadenylated RNA was translated in a reticulocyte system in the presence of [35S]-methionine, immunoprecipitated with anti-59 kD trichome PPO and Protein A, and resolved by 10% SDS-PAGE and autoradiography. Lane 1, *in vitro* translation products of epidermal mRNA, and Lane 2, of total leaf mRNA.

From Steffens, J. C., S. P. Klowalski, and H. Yu. 1990. "Characterization and cloning of glandular trichome and plastid polypenoloxidase." *Molecular Biology of the Potato*. Barking, Essex, England: C.A.B. International.

# REFERENCES


1. Burns D. P., and R. Steiner. 1991. *Advanced Technology Guide for LS 6000 Series Scintillation Counters*. Palo Alto, CA: Beckman Instruments, Inc. 3–12.
2. Freifelder, D. 1976. *Physical Biochemistry, Applications to Biochemistry and Molecular Biology*. San Francisco: W. H. Freeman and Company. 88–118.
3. Montgomery, R., and C. A. Swenson. 1969. *Quantitative Problems in the Biochemical Sciences*. San Francisco: W. H. Freeman and Company. 251–62.
4. Segel, H. 1982. *Biochemical Calculations*. New York: John Wiley and Sons. 354–94.
5. Peng, C.-T., and D. L. Horrachs. 1980. *Liquid Scintillation Counting: Recent Applications and Development*. New York: Academic Press.
6. Bosch, L. 1972. *The Mechanism of Protein Synthesis and its Regulation*. Amsterdam: Elsevier.
7. Weil, J. H., and L. Bogorad. 1977. *Nucleic Acids and Protein Synthesis in Plants*. New York: Plenum Press.
8. Linderstrom-Lang Conference. 1983. *Protein Synthesis Translational and Post-Translational Events*. Clifton, NJ: Human Press.
9. Spedding, G. 1990. *Ribosomes and Protein Synthesis: A Practical Approach*. Oxford: Oxford University Press.
10. Arnstein, H. R. V. 1992. *Protein Biosynthesis*. Oxford: IRL Press.
11. Hunt, M. D., N. T. Eannetta, Y. Haiferg, S. M. Newman, and J. C. Steffens. 1993. "cDNA cloning and expression of potato polyphenol oxidase." *Plant Molecular Biology*. 21:59–68.
12. Chromczynski, P., and N. Sacchi. 1987. "Single-step method for RNA isolation by acid guanidinium thiocyanate-phenol chloroform extraction."*Analytical Biochemistry*. 162:156–59.
13. Sambrook J., E. F. Fritsch, and T. Maniatis. 1989. *Molecular Cloning: A Laboratory Manual*, 2d ed., Cold Spring Harbor, NY: Cold Spring Harbor Laboratory Press.
14. Shafritz (see literature in Rapoport, S. M.[15] Promega Technical Bulletin 127).
15. Rapoport, S. M. 1986. *The Reticulocyte*. Boca Raton, FL: CRC Press.
16. *Promega Corp. Technical Bulletins 125 and 127*. 1993 and 1992. Madison, WI: Promega.
17. Pelham, R. B., and R. J. Jackson. 1976. *Eur. J. Biochem*. 67:247. Cited in *Promega Technical Bulletin 127*. Madison, WI: Promega.
18. Kessler, S. W. 1981. "Use of a protein bearing *Stapholococci* for the immune precipitation and isolation of antigens from cells." *Methods Enzymology*. 73:442–59.
19. Allen, R. C. 1984. *Gel Electrophoresis and Isoelectric Focusing of Proteins: Selected Technologies*. New York: de Guyter.
20. Dunbar, B. S. 1987. *Two-Dimensional Electrophresis and Immunological Techniques*. New York: Plenum Press.
21. Anderson, L. 1988. *Two-Dimensional Electrophoresis: Operation of the Iso-Dalt System*. Washington, D.C.: Large Scale Biology Press.
22. Steffens, J. C., S. P. Klowalski, and H. Yu. 1990. "Characterization and cloning of glandular trichome and plastid polypenoloxidase." *Molecular Biology of the Potato*. Barking, Essex, England: C.A.B. International.


## Supplementary References


Beckman Instruments. 1991. *The Counter Solution: Advanced Technology Guide*. Fullerton, CA: Beckman Instruments, Inc.

Bordier, C. 1981. "Phase separations of integral membrane proteins in Triton X-114 solution." *J. Biol. Chem.* 256:1604–07.

Dunn, M. J. 1986. *Gel Electrophoresis of Proteins.* Bristol: John Wright.

Flurkey, W. H., and P. E. Kolattakudy. 1981. "*In vitro* translation of antisense mRNA evidence for a precursor of an extracellular fungal enzyme." *Arch. Biochem. Biophys.* 212:154–61.

Flurkey, W. H. 1985. "*In vitro* biosynthesis of *Vicia faba* polyphenoloxidase." *Plant Physiol.* 79:564–67.

Flurkey, W. H. 1986. "Polyphenoloxidase in higher plants. Immunological detection and analysis of *in vitro* translation products." *Plant Physiol.* 81:614–18.

Flurkey, W. H. 1989. "Polypeptide composition and amino terminal sequence of broad bean polyphenoloxidase." *Plant Physiol.* 91:481–83.

Fox, B. W. 1976. *Techniques of Sample Preparation for Liquid Scintillation Counting Isolectric Focusing.* Amsterdam: Elsevier.

Howard, P. L. 1976. *Basic Liquid Scintillation Counting.* Washington, D.C: American Society of Chemical Pathology.

*International Conference on Liquid Scintillation Science and Technology.* 1976. New York: Academic Press.

Kowalski, S. P., N. T. Eannetta, A. T. Hizel, and J. C. Steffens. 1992. "Purification and characterization of polyphenol oxidase from glandular trichomes of *Solanum berthauttii.*" *Plant Physiol.* 100: 677–84.

Lanker, T., T. C. King, S. W. Arnold, and W. H. Flurkey. 1967. "Active, inactive and *in vitro* synthesized forms of polyphenoloxidase during leaf development." *Physiol. Plantarum:* 69:323–29.

Moore, B. M., and W. H. Flurkey. 1988. "Mini-chromato focusing of plant and fungal polyphenoloxidase." *Analytical Biochem.* 72:504–08.

Murray, M., and W. F. Thompson. 1980. "Rapid isolation of high molecular weight plant DNA." *Nucl. Acids Res.* 8:4321–25.

Rogers, S. O., and A. J. Bendrich. 1980. "Extraction of DNA from plant tissues." *Plant Molecular Biology Manual.* Boston: Kluwer Academic Press.

Ross, H., J. E. Noakes, and J. D. Spalding. 1991. *Liquid Scintillation Counting and Organic Scintillations.* Chelsea, MI: Lewis.

Saghai-Maroof, M. A., K. M. Soliman, R. A. Jorgenses, and R. W. Allord. 1984. "Ribosomal DNA spacer-length polymorphisms in barley: Mendelian inheritance, chromosomal location and population dynamics." *Proc. Natl. Acad. Sci. USA* 81:8014–18.

Yu, H., S. P. Kowalski, and J. C. Steffens. 1992. "Comparison of polyphenol oxidase expression in glandular trichomes of *Solanum* and *Lycopersicon* species." *Plant Physiol.* 100:1885–1900.

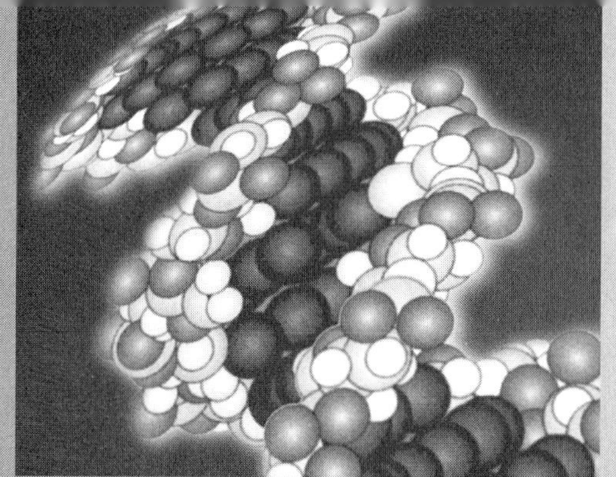



# Bioenergetics: Transformation of Oxidation-Reduction Energy into Bioluminescence

## Outline of Module

LUCIFERASE ASSAY

Stability of firefly luciferase in Tricine buffer and in a commercial enzyme stabilizer
*Duration:* 2.5 hours, one laboratory period (assay at 0, 2, 4, and 6 hours)

Effect of storage at 50° C and –20° C on luciferase activity in the Tricine buffer and AuthentiZyme Stabilizer
*Duration:* 2.5 hours, one laboratory period (assay at 0, 2, 4, and 6 hours)

Effect of temperature in the cold and –20° C after 12, 24, 36, and 48 hours in both luciferase dissolved in AuthentiZyme Enzyme Stabilizer and Tricine, pH 7.8
*Duration:* 2.5 hours, one laboratory period (assay at 12, 24, 36, and 48 hours)

The $FMNH_2$ injection assay for firefly luciferase (substitute for luciferase assay)
*Duration:* 2.5 hours, one laboratory period.

# LABORATORY EXERCISES

## LUCIFERASE ASSAY

### *Supplies and Equipment*
    Luminometer Lumac/3M
    Model 2010A Biocounter for 10 second periods
        or others
    Spectrophotometer
    Fluorescence spectrophotometer

### *Chemicals*
    Sigma Chemical Co.
        Luciferase
        Adenosine triphosphate (ATP)
        Dithiothreitol (DTT)
        Bovine serum albumin (BSA)
        Flavin mononucleotide (FMN)
    Boehringer Manheim Biochemicals
        Luciferin
    Innovative Chemistry (Marshfield, M.A.)
        AuthentiZyme Enzyme Stabilizer

## Introduction

Luminescent organisms include certain bacteria, protozoa, fungi, worms, and crustaceans, and particularly the firefly. In many of these organisms, enzymatic oxidoreductions take place in which the free-energy change is utilized to excite a molecule to a high-energy state. This is followed by a return of the excited molecule to the ground state, a process accompanied by emission of visible light. This phenomenon is called bioluminescence.[1,2]

The molecular components and the mechanism of firefly luminescence have been investigated by D. McElroy and his associates.[3] The two necessary components, a heat-stable heterocyclic phenol, luciferin (Figure 11.1), and a heat labile enzyme luciferase, have been extracted from fireflies and crystallized. Luciferase (mol. wt. 100,000) appears to have no prosthetic group. In the first step, luciferin ($LH_2$) and ATP react to form luciferyl adenylate ($LH_2$–AMP), which remains tightly bound to the catalytic site of luciferase:

$$LH_2 + ATP + E \rightarrow E\text{-}LH_2\text{–}AMP + AMP + PPi$$

When this form of the enzyme is exposed to molecular oxygen, the enzyme-bound luciferyl adenylate is oxidized to yield oxyluciferin (L), which emits light on returning to the ground state:

$$E\text{-}LH_2\text{–}AMP + O_2 \rightarrow L + H_2O + \text{bioluminescence}$$

One quantum of light is emitted for each molecule of luciferin oxidized. A peroxide is presumed to be formed as an intermediate. The color of the emitted light is determined by the enzyme protein, since different species of firefly have the same luciferin but emit light of different colors. This unusual type of energy transformation serves the firefly as a mating signal. The luciferase reaction, followed in a recording spectrophotometer, is often employed as a highly sensitive quantitative assay for ATP.[3–8]

### Effects of Storage on Firefly Luciferase in Tricine Buffer and in a Commercial Enzyme Stabilizer

Many commercial preparations of firefly luciferase contain bovine serum albumin and other additives to stabilize the enzyme.[9,10] The addition of a stabilizing protein can interfere with some uses of firefly luciferase—e.g. as a substrate for the determination of proteolytic activity or derivatization of the luciferase protein. A proprietary preparation called AuthentiZyme Enzyme Stabilizer has been developed by Innovative Chemistry for the stabilization of enzymes in solution. This preparation contains cofactors, buffer, a chelator, metal ions, a carbohydrate, and an antioxidant.[11]

The objectives of this experiment are threefold:

1. Study the effect of storage on firefly luciferase activity in Tricine buffer and AuthentiZyme Enzyme Stabilizer.
2. Determine and compare the activity of firefly luciferase in a Tricine buffer and AuthentiZyme Enzyme Stabilizer.
3. Study the effect of storage at 5° C and –20° C on firefly luciferase in the Tricine buffer and AuthentiZyme Enzyme Stabilizer.

## Procedures

a. Dissolve firefly luciferase prepared in AuthentiZyme Enzyme Stabilizer in 0.05 mol/1 Tricine buffer, pH 7.8, containing 10 mmol/l MgSO4, 1 mmol/l ethylenediamine tetraacetic acid (EDTA), 1 mmol/1 dithiothreitol (DTT), and 100 µg/ml bovine serum albumin.[12] In a separate preparation, dissolve firefly luciferase in 0.05 mol/l Tricine, pH 7.8 only.
b. Divide the preparations (each containing 250 µg of crystalline firefly luciferase) into 16 aliquots of

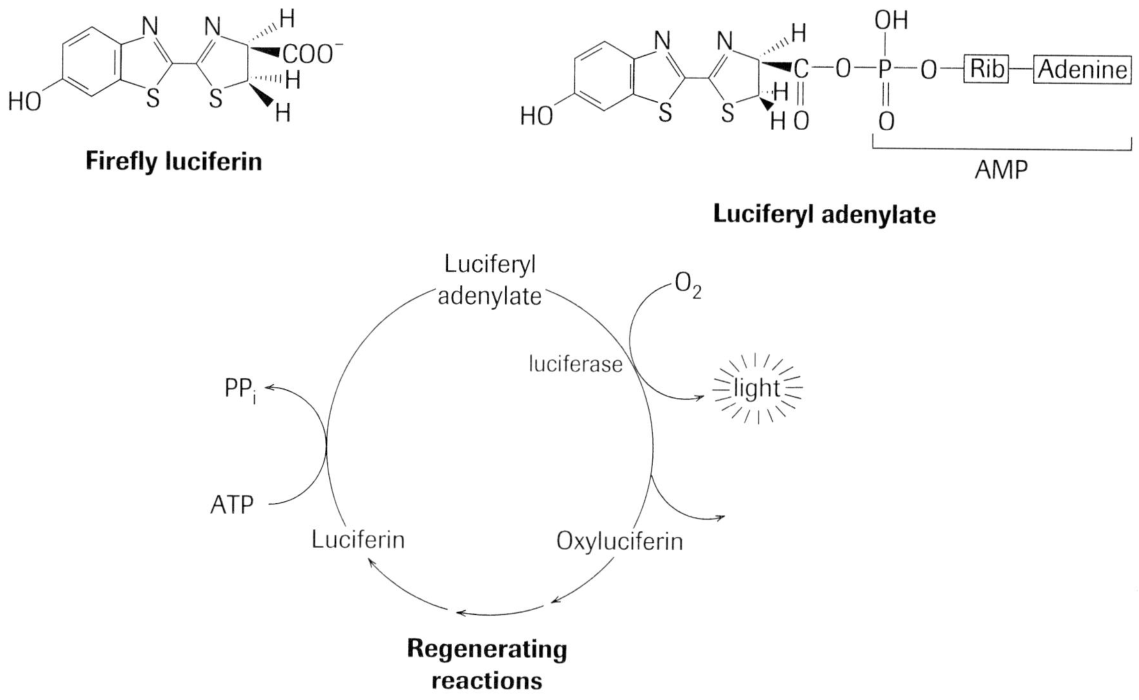

**Firefly luciferin**

**Luciferyl adenylate**

AMP

Luciferyl
adenylate

$O_2$

luciferase

light

$PP_i$

ATP

Luciferin

Oxyluciferin

**Regenerating
reactions**

FIGURE 11.1   Mechanism of firefly luminescence.

each luciferase dissolved in 0.05 M Tricine, pH 7.8, and prepared in AuthentiZyme Stabilizer, and separate them into three groups:

1. Keep four aliquots of each in an ice bath during a 6-hour period. Assay for luciferase activity at time intervals of 0, 2, 4, and 6 hours for each aliquot. Compare the enzyme activity in Tricine buffer versus AuthentiZyme Stabilizer. Plot your data.
2. Keep the second four aliquots of each at 5° C or –20° C over a two-week period. Assay for luciferase activity in both preparations after 12, 24, 36, and 48 hours. Compare the enzyme activity in Tricine buffer versus AutheniZyme Stabilizer stored in a refrigerator and a freezer.
3. Keep the third set of four aliquots of each up to 6 weeks in the freezer so that a solution would have to be thawed only once, and store for up to 6 weeks in sterile 5 ml amber Wheaton tubular serum bottles. The usual temperature for freeze storage is –20° C. At appropriate storage times (fresh, one, two, three, four, and five weeks), assay 6 replicates (50 μl samples for firefly luciferase activity).

## Comparison of Firefly Luciferase Activity in Tricine Buffer and in a Commercial Enzyme Stabilizer

a. Prepare a reaction mixture contained in a total of 200 μl: 25 mmol/l Tricine buffer, pH 7.8, 5 mmol/l MgSO4, 0.5 mmol/l EDTA, .5 mmol/L DTT, 50 μg of Luciferin, 1 ng of ATP, 100 μg of bovine serum albumin, and 10 μl sample of firefly luciferase solution.
b. Measure light production in a Lumac/3M Model 2010 A Biocounter for 10-second periods.[11]

For detailed survey of more than 90 commercially available luminometers and imaging devices, students should refer to Stanley (1992).[13]

c. Determine luciferase activity in AuthentiZyme Enzyme Stabilizer or Tricine pH 7.8 buffer only (see previous page for methods to prepare luciferase in AuthentiZyme Enzyme Stabilizer).
d. Compare luciferase activity under the above conditions and record your results expressed counts per 10 seconds.

Luciferase activity in AuthentiZyme:

Enzyme Stabilizer = Counts per 10 seconds

Luciferase activity in 0.05 M Tricine:

pH 7.8 = Counts per 10 seconds

## Review Questions

1. What is the effect of AuthentiZyme Enzyme Stabilizer on luciferase activity?
2. Discuss the functions of the following:
   a. EDTA
   b. DTT
   c. bovine serum albumin

## Factors Affecting Luciferase Activity

### *The Effect of Storage*
Compare the luciferase activity in AuthentiZyme Enzyme Stabilizer and Tricine, pH 7.8 buffer, at time intervals of 0, 2, 4, and 6 hours. Plot your data expressed in counts per 10 seconds versus hours. From your results, draw your conclusions.

### *The Effect of Temperature*
1. Compare the activity of luciferase stored in Tricine buffer at 5° C and at –20° C after 12, 24, 36, and 48 hours with luciferase dissolved in AuthentiZyme Enzyme Stabilizer and Tricine, pH 7.8.
2. Plot your data expressed as counts per 10 seconds versus time.
3. Compare the luciferase activity for preparations stored at –20° C and at 5° C after 2, 3, 4, 5, and 6 weeks. Plot your data expressed as counts per 10 seconds versus time of storage. What conclusions can be drawn from your graph?

## The FMNH$_2$ Injection Assay for Firefly Luciferase

For departments that are lacking the above equipment facilities, they can follow the FMNH$_2$ injection assay as an alternate.[14]

$$FMNHM_2 + E \rightarrow E.FH_2 \rightarrow O_2 \rightarrow$$
$$E.FHOOH–RCHO–E \rightarrow FHOOH \text{ (complex 1)}$$

$$E \; FMN + H_2O_2$$
$$\downarrow$$

$$E\text{-}FHOH^*$$
$$\downarrow \; hv \; 490 \; nm$$

$$E' \rightarrow E. \; FHOH$$
$$\downarrow$$

$$FMN + H_2O$$

The previous reaction pathway describes FMNH$_2$ injection for luciferase assay. The reaction is initiated by injection of reduced flavin (generally catalytically reduced by H$_2$ over platinized asbestos) into a vial containing luciferase, aldehyde, and O$_2$. The FMNH$_2$ binds first to luciferase enzyme and then reacts with oxygen to form ternary compex 1, which in turn reacts with aldehyde substrate to form an excited flavin species and the carboxylic acid products. The excited flavin species emits a photon and FMN is the product. Luciferase activity is measured as the peak intensity of light emission achieved following the injection of FMN$_2$ and (with saturating substrate concentration) is proportional to the amount of active enzyme present in the reaction.

a. Dissolve firefly luciferase prepared in AuthentiZyme Enzyme Stabilizer as described on page 188. In a separate preparation, dissolve luciferase in 0.05 mol/1 Tricine, pH 7.8.
b. In a separate vial, mix 250 µg luciferase with n-hexanal (20 µl of a 0.01% (v/v) sonicated suspension) in 1.0 ml of air equilibrated buffer solution prepared in step a.
c. Inject 1.0 ml of 0.05 M catalytically reduced flavin mononucleotide (FMNH$_2$) into the vial containing luciferase and n-hexanal prepared in step b.
d. Monitor light intensity as a function of time following injection of FMNH$_2$.
e. Plot your data expressed as light intensity (%) versus time (seconds).
f. Determine luciferase activity by measuring the peak intensity of light emission achieved.

This procedure may substitute for the assay procedure outlined under firefly luciferase assay (page 188). All other experimental procedures, laboratory exercise, stability, effect of storage, etc., should be followed as described in this module.

## Review Questions

1. What is the effect of temperature on luciferase activity?
2. Which luciferase activity is more stable? Why? Explain your results.

Students are encouraged to read References 1–14 with special emphasis to Hall and Leach (1988).[11]

## REFERENCES

1. Chappelle, E. W. 1975. *Laboratory Manual for Luciferase Assay for Adeonsine Triphosphate (ATP)*. Greenbelt, MD: Goddard Space Flight Center.

2. Lehninger, A. L. 1992. *Biochemistry.* New York: Worth Publishers.

3. Deluca, M. A., and W. D. McElroy (eds.). 1986. *Bioluminescence and Chemiluminescence.* New York: Academic Press.

4. DeLuca, M. A. (ed.) 1978. *Bioluminescence and Chemiluminescence.* New York: Academic Press.

5. Leach, F. R. 1981. "ATP determination with firefly luciferase." *J. Appl. Biochem.* 3:473–517.

6. Lurdin, A. 1982. "Analytical applications of bioluminescence: the firefly system." *Clinical and Biochemical Luminescence.* New York: Dekker. 43–74.

7. Slawinski, D., and Slawinski, J. 1985. "Applications of bioluminescence and low level luminescence." *Clin. Biochem. Anal.* 16:533–601.

8. Wulf, K. 1983. "Luminometry." *Methods of Enzymatic Analysis*, 3rd ed. Deerfield Beach, FL: Verlag Chemie. 340–68.

9. Leach, F. R., and J. J. Webster. 1986. "Commercially available firefly luciferase reagents." *Methods in Enzymology*, vol. 133. New York: Academic Press.

10. Webster, J. J., J. C. Chang, J. L. Howard, and F. R. Leach. 1979. "Some characteristics of commercially available firefly luciferase preparations." *J. Appl. Biochem.* 1:471–78.

11. Hall, M. S., and F. R. Leach. 1988. "Stability of firefly luciferase in Tricine buffer and in a commercial enzyme stabilizer." *Journal of Bioluminescence and Chemiluminescence.* 2:41–44.

12. Webster, J. J., and F. R. Leach. 1980. "Optimization of the firefly luciferase assay for ATP." *J. App. Biochem.* 2:469–79.

13. Stanley, P. E. 1992. "A survey of more than 90 commercially available luminometers and imaging devices for low-light measurements of chemiluminescence and bioluminescence, including instruments for manual, automatic and specialized operation, for HPLC, LC, GLC and microtitre plates, part 1: desaysptis." *Journal of Bioluminescence and Chemiluminescence.* 7:77–108.

14. Baldwin, T. O., T. F. Holzman, R. B. Holtzman, and V. A. Riddle. 1986. "Purification of bacterial luciferase by affinity methods." *Methods in Enzymology.* 133:98–102.

## Supplementary References

Ford, S. R., M. S. Hall, and F. R. Leach, 1992. "Comparison of properties of commercially available crystalline native and recombinant firefly luciferase." *Journal of Bioluminescence and Chemiluminescence.* 7:185–93.

Kricka, C. J., P. E. Stanley, G. H. G. Thorpe, and T. B. Whitehead. 1984. *Analytical Applications of Bioluminescence and Chemiluminescence.* New York: Academic Press.

McCapara, F. 1973. "The Chemistry of Bioluminescence." *Endeavour.* 39–145.

Schlomerich, J., R. Andreesen, A. Kapp, M. Ernst, and W. G. Woods. 1987. *Bioluminescence and Chemiluminescence: New Perspectives.* New York: John Wiley and Sons.

Schran, E., and W. Van Witenburg. 1989. Improved ATP methodology for biomass assays. *Journal of Bioluminescence and Chemiluminescence.* 4:390–98.

Simpsons, W. J., and J. R. M. Hammord. 1991. "The effect of detergents on firefly luciferase reactions." *Journal of Bioluminescence and Chemiluminescence.* 6:97–108.

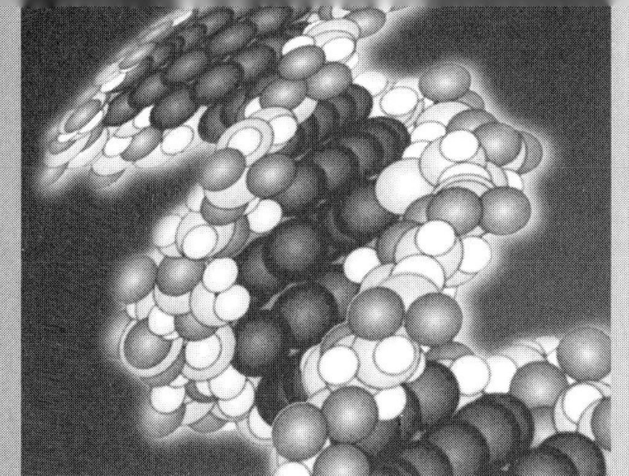



# TLC and GLC of Monosaccharides from Membrane-Bound Glycoproteins

## Outline of Module

Membrane structure/function
    Lipid structure
    Protein structure
        Integral
        Peripheral
    Interactions between lipids and proteins
    Molecular assembly
    Preparation of membrane fractions
        Animal cells
        Plant cells
Solubilization of membrane proteins
    Peripheral proteins
        Mild treatment—low ionic strength solutions
        "Harsh" treatment—guanidinium chloride, urea, dilute acids
    Integral proteins
        Importance of protease inhibitors
        Scheme for interaction of membranes and detergents

## GAS AND THIN-LAYER CHROMATOGRAPHIES OF MONOSACCHARIDES DERIVED FROM DETERGENT-MEMBRANE RELEASED-MEMBRANE GLYCOPROTEINS

Preparation of membrane-bound polyphenol oxidase from *Viva faba* leaflets
   *Duration:*   2 laboratory periods

Detergent-release of membrane-bound polyphenoloxidase from *Vitis vinifera*
   *Duration:*   1 laboratory period

Paper and thin layer chromatographic analyses of monosaccharides
   *Duration:*   1 laboratory period

Gas chromatographic analyses of monosaccharides
   *Duration:*   1 laboratory period

# INTRODUCTION

## Membrane Structure/Function

Lipids comprise approximately 20%–80% of a membrane's mass, and thus are major components of membranes. Proteins may be inserted into the lipid bilayer and/or they may be peripherally associated (Figure 12.1).

Thus, evidence exists for two classes of membrane proteins—i.e., intrinsic (integral) and extrinsic (peripheral). Whereas the former can be released by the action of detergents (Table 12.1), the latter can be solubilized by either low ionic strength solutions or "harsher" treatments such as guanidinium chloride, urea, or dilute acids. The intrinsic proteins are associated with both lipid bilayer "leaflets" and may traverse the membrane more than once. In contrast, the peripheral proteins are associated with only one leaflet—ie., either the inner or outer. In addition to both peripheral and integral membrane proteins, these may be transport proteins that may coincide with either peripheral or membrane proteins and may encompass proteins that may not be anchored within the membrane but nevertheless may interact with it. Examples of transport proteins are pumps, carriers, and channels. An extensive coverage of plant membrane proteins can be found in Leshem et al.[1] The volume edited by Chapman[3] contains papers regarding X-ray diffraction, electron microscopy, Raman spectroscopy, and infrared spectroscopic studies of biomembranes.

For many years, membrane lipids were thought only to form an inert structural framework. However, recent research findings suggest that lipids may mediate cell responses to external stimuli, including hormones, neurotransmitters, and growth factors. The lipids composing membranes in part are the phospholipids and sterols and in certain instances some galactolipids and sulfolipids. The phospholipids can be grouped into the glycerophospholipids and the sphingolipids (Figure 12.2).

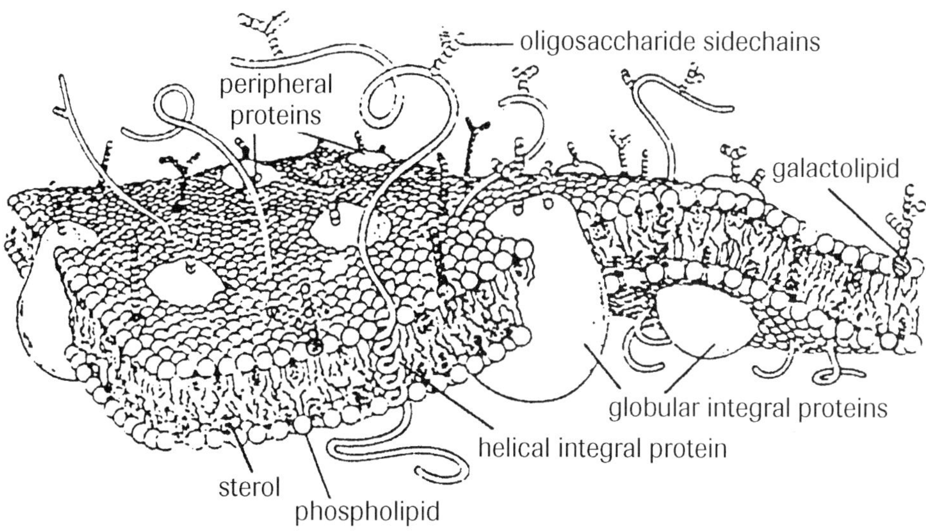


**FIGURE 12.1**  A typical bilayered biological membrane (after Braetscher, 1985). With regard to the membrane proteins this is sometimes referred to as the "apple and pear" model.

Leshem, Y. Y., R. L. Shewfelt, C. M. Willmer, and O. Pantoja. 1991. *Plant Membranes: A Biological Approach to Structure Development and Senescence.* Netherlands: Kluwer Academic Publishers.

**TABLE 12.1**  Chemical and Physical Properties of Commonly Used Detergents

| Detergent[a] | M.P.° C | Mol. wt Monomer | Mol. wt[e] Micelle | CMC[e] % (w/v) | M |
|---|---|---|---|---|---|
| SDS | 206 | 288 | 18000 | 0.23 | $8.0 \times 10^{-3}$ |
| Cholate | 201[b] | 430 | 4300 | 0.60 | $1.4 \times 10^{-2}$ |
| Deoxycholate | 175[b] | 432 | 4200 | 0.21 | $5.0 \times 10^{-3}$ |
| $C_{16}$TAB | 230[c] | 365 | 6200 | 0.04 | $1.0 \times 10^{-3}$ |
| Lyso PC ($C_{16}$) | – | 495 | 9200 | 0.0004 | $7.0 \times 10^{-6}$ |
| CHAPS | 157[c] | 615 | 6150 | 0.49 | $1.4 \times 10^{-3}$ |
| Zwittergent 3–14 | – | 364 | 30000 | 0.011 | $3.0 \times 10^{-4}$ |
| Octyl glucoside | 105[c] | 292 | 8000 | 0.73 | $2.3 \times 10^{-2}$ |
| Digitonin | 235[c] | 1229 | 70000 | – | |
| $C_{12}E_8$ | –[d] | 542 | 65000 | 0.005 | $8.7 \times 10^{-5}$ |
| Lubrol PX | –[d] | 542 | 65000 | 0.005 | $8.7 \times 10^{-5}$ |
| Triton X-100 | –[d] | 650 | 90000 | 0.021 | $3.0 \times 10^{-4}$ |
| Tween 80 | –[d] | 1310 | 76000 | 0.002 | $1.2 \times 10^{-5}$ |

From Jones, O. T., J. P. Earnest, and M. G. McNamee. 1987. "Solubilization and reconstitution of membrane proteins." *Biological Membranes: A Practical Approach.* Oxford: IRL Press.

[a]See REF for structures.
[b]Based on free acid.
[c]Decomposes.
[d]Viscous liquid at room temperature.
[e]Determined at 20–25° C.

The predominant phospholipids in biological membranes are the glycerophospholipids in which phosphate and fatty acids are esterified to glycerol. In contrast, the sphingolipids, e.g. sphingomyelin, possess sphingosine as the alcohol to which the phosphate is esterified. In addition, the second fatty acid chain is linked to an amino group. Table 12.2 presents the various types of phospholipids. Some of the fatty acids that occur in phospholipids are displayed in Table 12.3. For a comprehensive review of the physical principles and models of phospholipid bilayers, readers are referred to the monograph of Cevc and

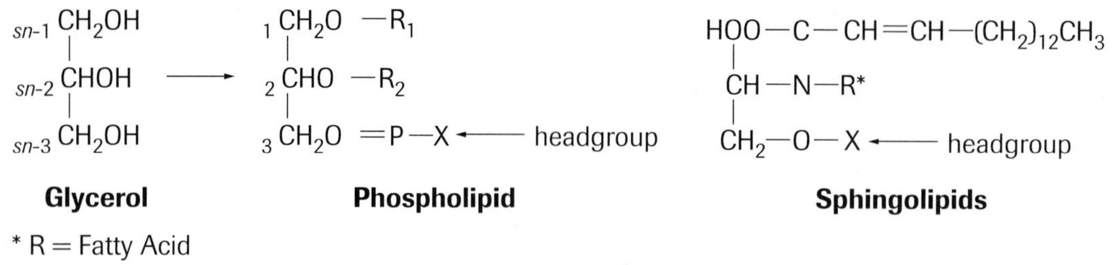

$$sn\text{-}1\ CH_2OH \quad\quad 1\ CH_2O-R_1 \quad\quad\quad HOO-C-CH=CH-(CH_2)_{12}CH_3$$
$$sn\text{-}2\ CHOH \longrightarrow 2\ CHO-R_2 \quad\quad\quad\quad CH-N-R^*$$
$$sn\text{-}3\ CH_2OH \quad\quad 3\ CH_2O=P-X \longleftarrow headgroup \quad\quad CH_2-O-X \longleftarrow headgroup$$

**Glycerol**        **Phospholipid**        **Sphingolipids**

* R = Fatty Acid

**FIGURE 12.2**  Overall glycerol derivation of mem-brane phospholipids.

Leshem, Y. Y., R. L. Shewfelt, C. M. Willmer, and O. Pantoja. 1991. *Plant Membranes: A Biological Approach to Structure Development and Senescence*. Netherlands: Kluwer Academic Publishers.

Marsh.[4] For a more recent review of phospholipids, consult Moore.[5]

As for the amphiphilic sterols, the major sterol found in mammalian membranes is cholesterol, which appears to be absent in plant membranes. Instead, plant membranes may contain stigmasterol, campesterol, and sitosterol. Lanosterol seems to be unique to certain fungi. The structure of stigmasterol is depicted in Figure 12.3.

The polar OH is associated with the polar phospholipid headgroups and appears to be essential for the inclusion of the sterol molecule into the membrane. Both the steroid's skeleton and the aliphatic tail are arrayed parallel to the fatty acyl tailgroups of the membrane phospholipid. The inclusion of sterols into membranes is believed to result in a homeostatic and stabilizing effect.

In addition to lipids, proteins, sterols, and carbohydrates, membranes also contain $H_2O$. Its distribution in membranes is not uniform, and aqueous domains differing in location and physical-chemical properties occur upon the membrane's exterior and at different locales of the membrane's interior.[1]

An exhaustive coverage of membrane chemistry is beyond the scope of this manual, and readers are referred to Leshem et al.[1] for a treatment of plant membrane chemistry. Here will be found treatments of glycolipids and membrane-associated phospholytic and lipolytic enzymes. With regard to animal membranes, refer to Klausner et al.[6] Indeed students may find the 39-volume set of *Current Topics in Membranes*[7] (1979–1991) of value.

## Solubilization of Membrane Proteins

There are two significant steps in the solubilization of membrane proteins: isolation of the membrane fractions enriched in the proteins (see Module 2), and solubilization of the protein with detergents. The scheme for the interaction of detergents with biological membranes is presented in Figure 12.4. With regard to the criteria for solubilization, the most widely employed criterion is the presence of the solubilized protein in a 105,000 $xg$ supernatant.[2] Other means of establishing solubilization include gel filtration, electron microscopy, solution turbidity, and $^{32}P$ nmr. Refer to Jones et al.[2] for a thorough discussion of selection of detergents, factors affecting solubilization, and reconstitution of membrane proteins including detergent removal.

## Purification of Membrane Proteins

As the principles underlying gel filtration, ion exchange, hydrophobic interaction, and affinity chromatographies have already been reviewed, students will here become acquainted with reverse phase chromatography, isoelectric focusing, and chromatofocusing.

Isoelectric focusing is an electrophoretic technique that utilizes a pH gradient rather than a constant pH. The pH gradient is formed between two electrodes and is maintained via ampholytes.[8] During isoelectric focusing, proteins migrate to their isoelectric points (pI), where the protein lacks an overall net charge. Isoelectric focusing is the electrophoretic technique possessing the highest resolution. A thorough discussion of both the theory and application of the technique can be found in Pharmacia's fine chemicals booklet regarding the principles and methods of isoelectric focusing (Pharmacia, Uppsala, Sweden).

Reverse-phase HPLC is an HPLC technique that utilizes a nonpolar stationary phase and a polar phase with an organic solvent gradient. It appears that RP-HPLC of polypeptides involves an absorption-desorption process. There are certain limitations to RP-HPLC. These include variability in protein recovery, loss of biological activity in protein recovery, and loss of biological activity. A thorough discussion of this technique can be found in Henschev.[10]

**TABLE 12.2**  Commonly Occurring Phospholipids

| Headgroup | Systematic Name | Name |
|---|---|---|
| 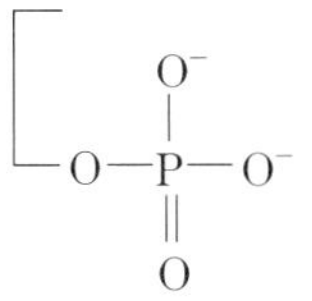 (structure) | 1,2-Diacyl-sn-glycerol-3-phosphoric acid | sn-3-Phosphatidic acid |
| 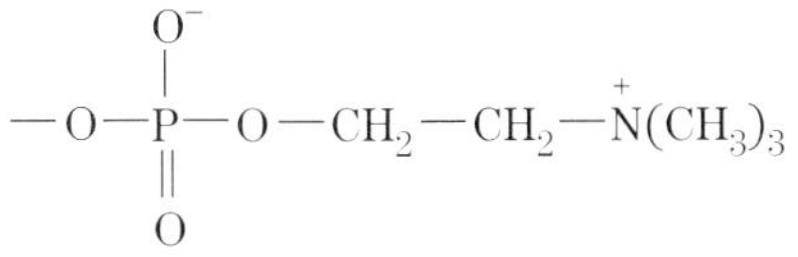 (structure) | 1,2-Diacyl-sn-glycerol-3-phosphocholine | sn-3-Phosphatidycholine |
| 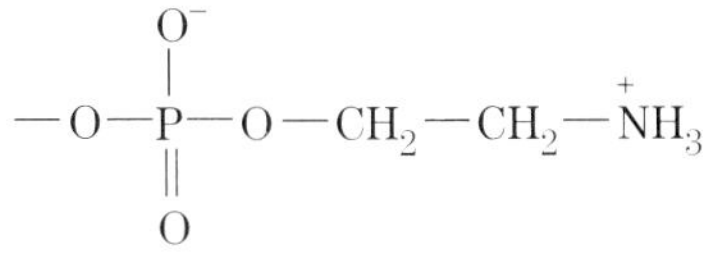 (structure) | 1,2-Diacyl-sn-glycerol-3-phosphoethanolamine | sn-3-Phosphatidylethanolamine |
| 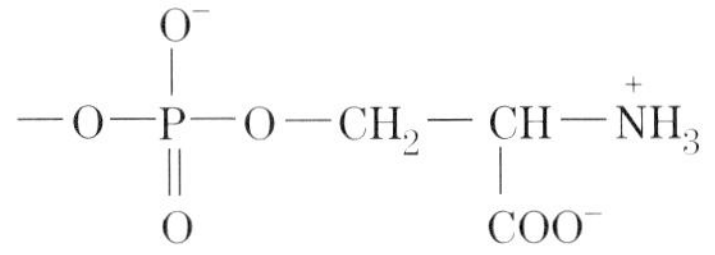 (structure) | 1,2-Diacyl-sn-glycerol-3-phosphoserine | sn-3-Phosphatidylserine |
| 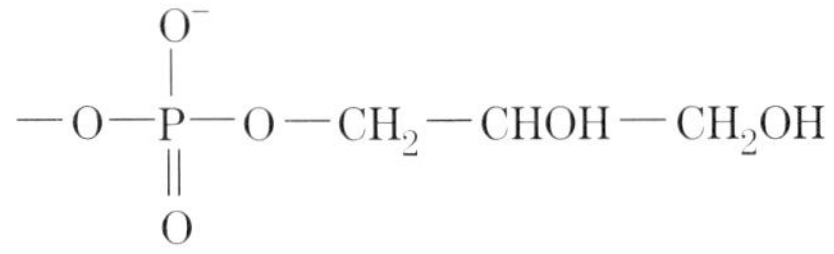 (structure) | 1,2-Diacyl-sn-glycerol-3-phospho-1-sn-glycerol | sn-3-Phosphatidyglycerol |
| 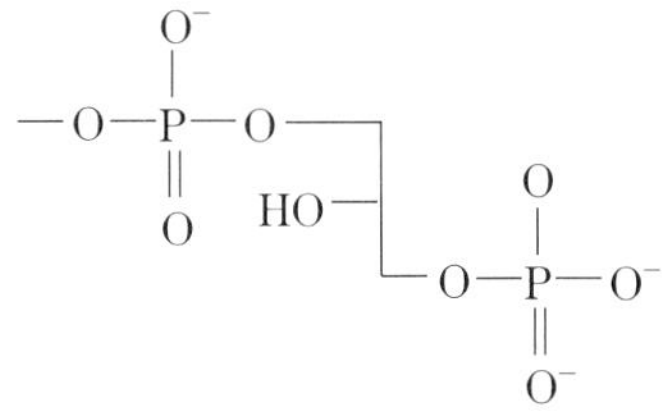 (structure) | 1,2-Diacyl-sn-glycerol-3-phospho-1′ sn-glycero-3′-phosphoric acid | sn-3-phosphatidyglycerol phosphate |
| 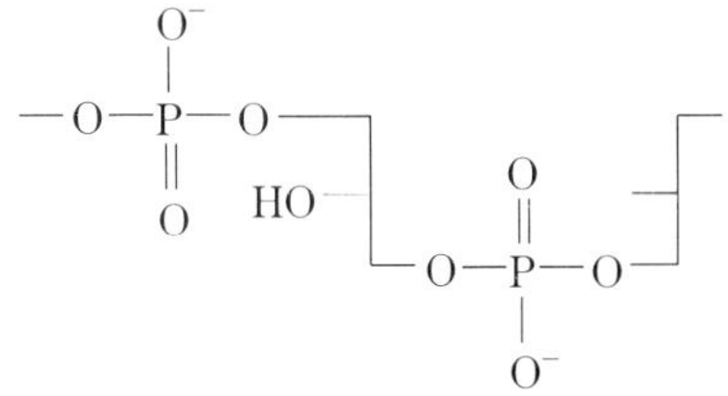 (structure) | 1,2-Diacyl-sn-glycerol-3-phospho-1′-sn-glycero-3′-phospho-1″,2″-diacyl-sn-glycerol | sn-3,3′-Diphosphatidyglycerol |
| (structure) $A_t = OH,\ A_2 = OH$ | 1,2-Diacyl-sn-glycerol-3-phospho-1′-myo-inositol | sn-3′-phosphosphatidylinositol |
| $A_t = PO_4^{2-},\ A_2 = OH$ | 1,2-Diacyl-sn-glycerol-3-phospho-1′-myo-inositol-4′-phosphate | sn-3-phosphatidylinositol phosphate |
| $A_t = PO_4^{2-},\ A_2 = PO_4^{2-}$ | 1,2-Diacyl-sn-glycerol-3-phospho-1′ myo-inositol-4′,5′-diphosphate | sn-3-phosphatidylinositol diphosphate |

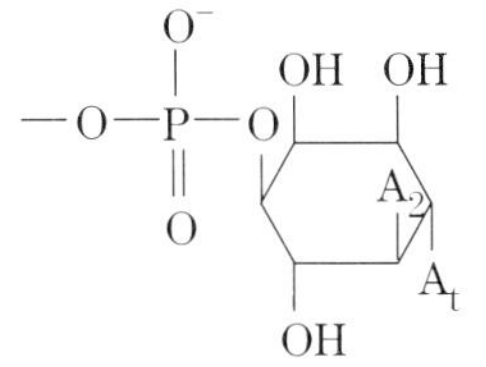

**TABLE 12.3**  Types of the Major Long-Chained Fatty Acids Occurring in Phospholipids

| Biological Name | *Abbreviation | Bond Positions |
|---|---|---|
| Palmitic acid | 16:0 | |
| Palmitoleic acid | 16:1 (*n*-7) | $\Delta^9$ |
| | 16:1 (*n*-10) | $\Delta^6$ |
| Stearic acid | 18:0 | |
| Oleic acid | 18:1 (*n*-9) | $\Delta^9$ |
| Vaccenic acid | 18:1 (*n*-7) | $\Delta^{11}$ |
| Petroselenic acid | 18:1 (*n*-12) | $\Delta^6$ |
| Elaidic acid | *t*-18:1 (*n*-9) | *t*-$\Delta^9$ |
| Linoleic acid | 18:2 (*n*-6) | $\Delta^{9,12}$ |
| Linoelaidic acid | | |
| α-Linolenic acid | 18:3 (*n*-3) | $\Delta^{9,12,15}$ |
| γ-Linolenic acid | 18:3 (*n*-6) | $\Delta^{6,9,12}$ |
| Arachidic acid | 20:0 | |
| Gadoleic acid | 20:1 (*n*-11) | $\Delta^9$ |
| Gondoic acid | 20:1 (*n*-9) | $\Delta^{11}$ |
| Dihomo-γ-linolenic acid | 20:3 (*n*-6) | $\Delta^{8,11,14}$ |
| Mead acid | 20:3 (*n*-9) | $\Delta^{5,8,11}$ |
| Arachidonic acid | 20:4 (*n*-6) | $\Delta^{5,8,11,14}$ |
| Timnodonic acid | 20:5 (*n*-3) | $\Delta^{5,8,11,14,17}$ |
| Behenic acid | 22:0 | |
| Centoleic acid | 22:1 (*n*-11) | $\Delta^{11}$ |
| Erucic acid | 22:1 (*n*-9) | $\Delta^{13}$ |
| Adrenic acid | 22:4 (*n*-6) | $\Delta^{7,10,13,16}$ |
| Docosapentaenoic acid | 22:5 (*n*-6) | $\Delta^{4,7,10,13,16}$ |
| Clupanodonic acid | 22.5 (*n*-3) | $\Delta^{7,10,13,16,19}$ |
| Cervonic acid | 22:6 (*n*-3) | $\Delta^{4,7,10,13,16,19}$ |
| Lignoceric acid | 24:0 | |
| Nervonic acid | 24:1 (*n*-9) | $\Delta^{15}$ |

*Unless indicated with a *t* (-*trans*), all bonds have a *cis* formation.

From Cevc, G., and D. Marsh. 1987. *Phospholipid Bilayers Physical Principles and Models.* New York: John Wiley and Sons. Reprinted by permission.

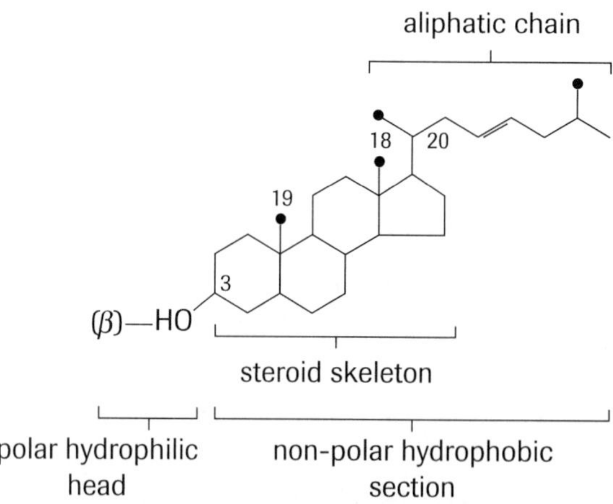


**FIGURE 12.3**  The structure of stigmasterol—a typical plant sterol. The mammalian cholesterol differs from stigmasterol only by the lack of an unsaturated double bond (–C=C–) in the aliphatic sidechain. As here depicted the plane of the –OH group faces upwards, at right angles to the plane of the steroid skeleton, this configuration being termed β-OH, and being the normal biological form. If the –OH group faces downwards the resulting configuration termed α–OH is biologically inactive. ⏽ indicates methyl groups facing the same way as the –OH.

Leshem, Y. Y., R. L. Shewfelt, C. M. Willmer, and O. Pantoja. 1991. *Plant Membranes: A Biological Approach to Structure Development and Senescence.* Netherlands: Kluwer Academic Publishers.

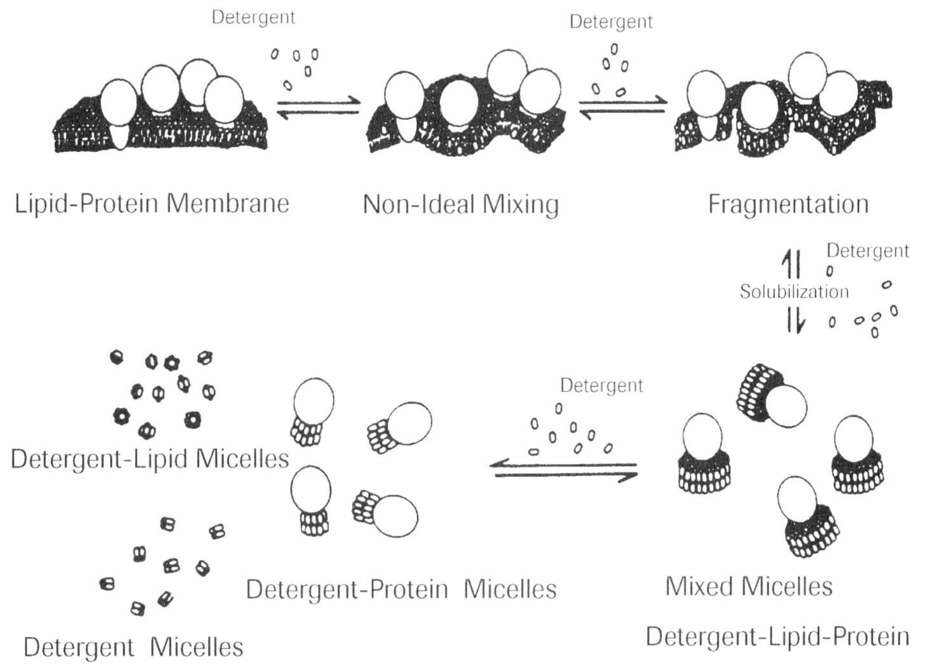


**FIGURE 12.4**   Solubilization and reconstitution of membrane proteins.

From Jones, O. T., J. P. Earnest, and M. G. McNamee. 1987. "Solubilization and reconstitution of membrane proteins." *Biological Membranes: A Practical Approach.* Oxford: IRL Press.

# LABORATORY EXERCISE

## GAS AND THIN LAYER CHROMATOGRAPHIES OF MONOSACCHARIDES DERIVED FROM MEMBRANE GLYCOPROTEINS

### Supplies and Equipment

Amicon ultrafiltration device
Assay tubes
Balances (top loader and analytical)
Bronson sonifier or equivalent
Cheesecloth
Cuvettes, quartz and plastic
Fraction collector
Freezer, –20° C
Gradient maker
Hair dryer
Hydrolysis ampules
Ice bucket
Microcaps
Pasteur pipettes
Pipettes
Preparative centrifuge—i.e., Beckman, International, or Sorvall
Spectrophotometer
Test tubes
Thin layer chromatography chamber
Trip balance
Ultracentrifuge and centrifuge tubes for 105,000 $xg$,—i.e., Beckman or International
Vacuum grease
Waterbath
Whatman 3MM chromatography paper
Ym 10 Amicon ultrafilters

### Chemicals

100mM acetate buffer, pH 5.0
Arabinose 10 µg/100 µl
100mM catechol in 100mM acetate buffer, pH 5.0
DEAE-cellulose
Ethyl acetate:pyridine:H$_2$O (8:2:1, v/v)

Fucose 10 µg/100 µl
Galactose 10 µg/100 µl
Glucose 10 µg/100 µl
Hydroxylapatite bio-gel HTP
Mannose 10 µg/100 µl
0.35 M Potassium phosphate, pH 7.7
Sephadex G-100
Silver reagent
Sodium chloride
100 mM sodium phosphate buffer containing
   10mM sodium ascorbate
1.5% TX-114 in 100 mM, pH 7.3, sodium phosphate
50 mM Tris-HCl, pH 7.9
2N Trifluoroacetic acid
Xylose 10 µg/100 µl
Monosaccharide standards 101–148 µg ml⁻¹ Merck
   Darmstadt
Dichloromethane (pro analysis quality)
Purm grade n-hexane
Thin layer plates—silica gel Kieselguhr or silica
   gel G1
Trifluoroacetic anhydride alumina (1:1 w/w)
   (TFAA), 99% pure (Pierce, Rockford, IL)
50 mm × 0.34 mm
SE-54 column G. C. Labar H. & I (Jaeggi,
   Trogen, Switzerland) for Perkin Elmer Gas
   Chromatograph

### Facilities
Coldroom
Greenhouse
Pots
Soil

### Organisms
Grape berries (*Vitis vinifera* L. cv. Monastrell)
Broadbean (*Vicia faba*)

## Introduction

This module is concerned with the release of membrane-bound polyphenol oxidase and subsequent separation and characterization of the carbohydrate moieties of this glycoprotein. Prior to performing gas chromatography, the student should read an introduction to the subject (see References).

## Procedures

Two procedures are provided for obtaining membrane-bound polyphenol oxidase (PPO). Figure 12.5 shows the flowchart for the isolation of membrane-bound PPO from the chloroplasts of *Vicia faba*[12] leaflets through sonication. Figure 12.6 presents a flowchart for the temperature-induced phase separation of membrane-bound PPO from grape berries.[11] Either procedure may be employed. The carbohydrates released from the purified PPO upon acid hydrolysis[13] may be characterized by TLC and/or GLC.[14]

Perform gas chromatography according to the following conditions (Figures 12.7 and 12.8):

a. stationary phase 50 m × 0.34 mm SE-54 column
b. helium carrier gas flow rate: 0.79 ml min⁻¹
c. scavenger gas flow rate: 24 ml min⁻¹
d. temperature program 70° C–130° C at 1 minute
e. injection temperature 200° C
f. detection temperature 250° C
g. manifold temperature 200° C

### Data Analysis

Compare weight of peaks (cut-out paper containing peak for each monosaccharide) to those for standards or alternatively use a disc integrator if your gas chromatograph is so equipped. Figure 12.9 presents a chromatograph of nine monosaccharides.

## Review Questions

1. What are the differences between integral and peripheral membrane proteins?
2. What are the various classes of detergents and what factors mediate the choice of detergent for release of membrane glycoproteins?
3. State the similarities and dissimilarities between thin layer and gas chromatographies of monosaccharides.
4. How are phosopholipids, sterols, proteins, and $H_2O$ integrated into the structure of biological membranes?

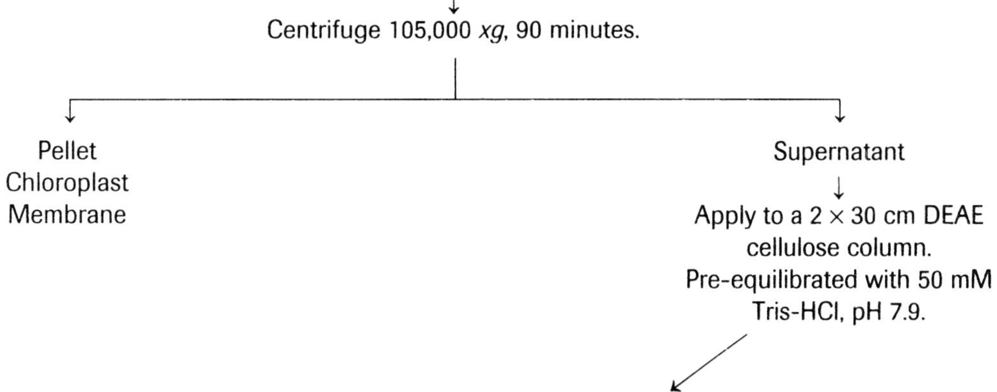


**FIGURE 12.5**  Protocol for the preparation of membrane-bound polyphenol oxidase from *Vicia faba* leaflets—sonication release.

Hutcheson S. W., and B. B. Buchanan. 1966. "Polyphenol oxidation by *Vicia faba* chloroplast membranes studies on the latent membrane-bound polyphenol oxidase and on the mechanism of photochemical polyphenol oxidation." *Plant Physiol.* 66:1150-1154

Grape berries (*Vitis vinifera* L. cv. Monastrell); store at –20° C until use.
↓
Defrost 25 g in 12.5 ml, 100 mM pH 7.3 sodium phosphate buffer containing 10 mM sodium ascorbate; homogenize in a blender 15 seconds.
↓
Filter through 8 layers of cheesecloth and centrifuge 4,000 *xg*, 15 minutes.
↓
Extract pellet with 20 ml 1.5% TX-114 (w/v) in 100 mM, pH 7.3 sodium phosphate, 30 minutes, 4° C.
↓
Centrifuge 60,000 *xg*, 15 minutes.

Pellet                                                                Supernatant

Add TX-114 detergent, 4° C yielding a final detergent % of 4, 15 minutes.
↓
Warm 35° C, 10 minutes.
↓
Centrifuge 5,000 *xg*, 10 minutes.

Pellet                                                                Supernatant
                                                                      (PPO)

**FIGURE 12.6**   Summary of procedure for detergent-release of membrane-bound polyphenol oxidase—temperature induced.

Sanchez-Feuer, A., R. Bru, and F. Garcia-Carmona. 1987. "Novel procedure for extraction of a latent grape polyphenoloxidase using temperature-induced phase separation in Triton X-114." *Plant Physiol.* 91:1481–87.

---

Lyophilyze purified polyphenol oxidase.
↓
Mix 100 mg with 2 ml trifluoroacetic acid; autoclave 1 hour in sealed hydrolysis tubes.
↓
Filter through a single layer of Whatman No. 1 filter paper; wash with 5 ml $H_2O$.
↓
Evaporate filtrates to dryness.
↓
Reconstitute in 200 µl $H_2O$.
↓
Spot onto Whatman 3MM chromatography paper together with 10 µg each arabinose, fucose, galactose, glucose, xylose, and mannose.
Develop in ethyl acetate:pyridine:$H_2O$ (8:2:1 v/v/v),
or
spot onto mixed silica Gel G/alumina G layer or silica gel/Kieselguhr (6%–100%).
Develop in ethyl acetate/dimethylformamide/$H_2O$ (30:60:2, v/v/v).
↓
Following air-drying of papers or plates, develop in acetone-saturated $AgNO_3$ (100 ml:6 drops), followed by NaOH:EtOH (0.5 g in 0.5 ml $H_2O$ plus 9.5 ml EtOH).

**FIGURE 12.7**   Procedure for paper and thin-layer chromatographic analyses of monossacharides.

Albersheim. P., D. J. Nevins, P. D. English, and A. A. Karr. 1967. "A method for the analysis of sugars in plant cell wall polysaccharrides by gas liquid chromatography." *Carbohydr. Res.* 5:340–45.

Transfer 500 μl aliquots (or the appropriate aliquots) of standards and reconstituted
polyphenol oxidase to 1 ml glass vials (Reacti vials, Pierce, or alternative).
↓
Freeze dry.
↓
Add 100 μl TFAA and 100 μl dichloromethane under a stream of nitrogen.
↓
Close vials with PTFE-lined screw caps or alternative, and heat at 130° C, 2 hours.
↓
Cool and evaporate under dry nitrogen with gentle heating from an IR lamp to ~10 μl.
↓
Add 100 μl of n-hexane containing 1% TFAA.
↓
Evaporate to 10 μl.
↓
Add 100 μl n-hexane-TFAA mixture.

Inject into a gas chromatograph equipped with an electron capture device—e.g., a Perkin Elmer
(modify below to be consistent with other companies—check with company regarding use of your chromatograph).

**FIGURE 12.8**   Procedure for gas chromatography of sugars.

Eklund, G., B. Josefsson, and C. Roos. 1977. "Gas liquid chromatography of monosaccharides at the picogram level using glass capillary column, trifluoroacetyl derivitization and electron-capture detection." *J. Chromatogr.* 142:575–85.

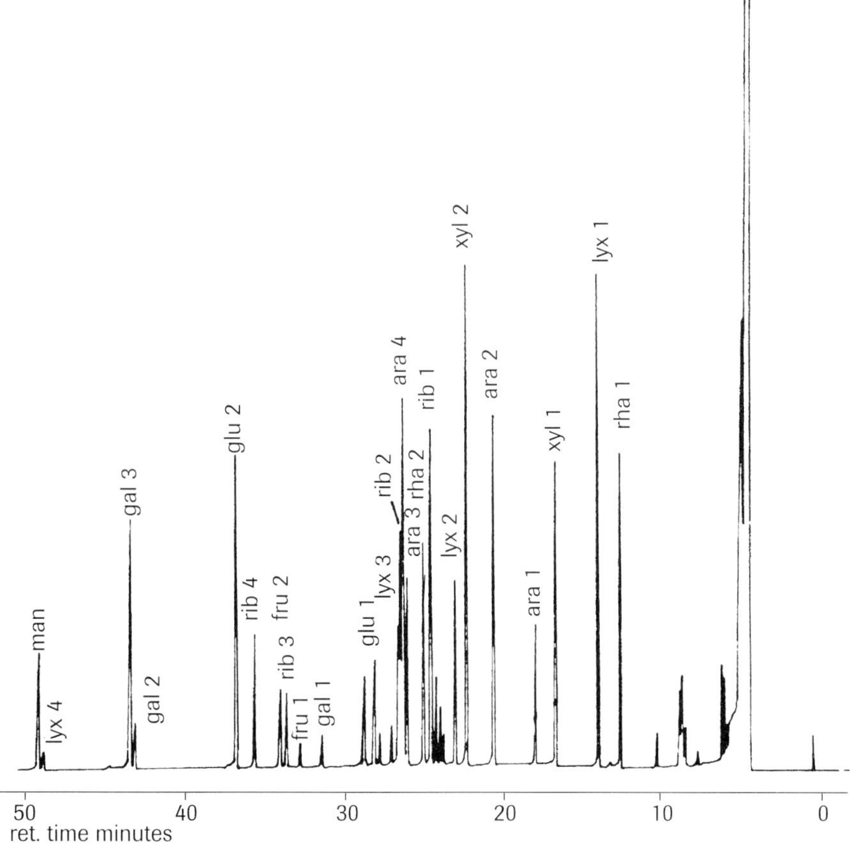


**FIGURE 12.9**   Chromatogram of nine mono-saccharides as their TFA derivatives: approximately 1 ng of each on the column. Carrier gas flow-rate, 0.7 ml min; scavenger gas flow-rate, 24 ml min; temperature programme, 74 (4 min) to 130 at 1 min; stationary phase, SF-54; column, 50 m 0.34 mm I.D.: 1 μl injected, spitless.

Eklund, G., B. Josefsson, and C. Roos. 1977. "Gas liquid chromatography of monosaccharides at the picogram level using glass capillary column, trifluoroacetyl derivitization and electron-capture detection." *J. Chromatogr.* 142:575–85.

# REFERENCES


1. Leshem, Y. Y., R. L. Shewfelt, C. M. Willmer, and O. Pantoja. 1991. *Plant Membranes: A Biological Approach to Structure Development and Senescence.* Netherlands: Kluwer Academic Publishers.
2. Jones, O. T., J. P. Earnest, and M. G. McNamee. 1987. "Solubilization and reconstitution of membrane proteins." *Biological Membranes: A Practical Approach.* Oxford: IRL Press.
3. Chapman, D. 1984. *Biomembrane Structure and Function: Topics in Molecular and Structural Biology.* Deerfield Beach, FL: Verlag Chemie.
4. Cevc, G., and D. Marsh. 1987. *Phospholipid Bilayers Physical Principles and Models.* New York: John Wiley and Sons.
5. Moore, T. S. 1990. "Phospholipids." *Methods in Plant Biochemistry.* 47–70.
6. Klausner, R. O., C. Kempf, and J. Van Rensworde. 1987. "Membrane structure and function." *Current Topics in Membranes and Transport.* 29: 1–297.
7. Hui, S.-W. 1987. "Ultrastructural studies of the molecular assembly in biomembranes: Diversity and similarity." *Current Topics in Membranes and Transport.* 30:29–65.
8. Righetti, P. G. 1983. *Isoelectric Focusing Theory, Methodology and Applications.* New York: Elsevier. 386.
9. Findlay, J. B. C., and W. H. Evans. 1987. *Biological Membranes: A Practical Approach.* Oxford: IRL Press.
10. Henschev, A. 1985. *High Performance Liquid Chromatography in Biochemistry.* New York: VCH Publishers.
11. Sanchez-Feuer, A., B. Roque, J. Cabaner, and F. Garcia-Carmona. 1988. "Characterization of catecholase and cresolase activities of monastrell grape polyphenol oxidase." *Phytochem.* 27:319–21.
12. Hutcheson S. W., and B. B. Buchanan. 1966. "Polyphenol oxidation by *Vicia faba* chloroplast membranes studies on the latent membrane-bound polyphenol oxidase and on the mechanism of photochemical polyphenol oxidation." *Plant Physoil.* 66:1150–1154
13. Albersheim. P., D. J. Nevins, P. D. English, and A. A. Karr. 1967. "A method for the analysis of sugars in plant cell wall polysaccharides by gas liquid chromatography." *Carbohydr. Res.* 5: 340–45.
14. Eklund, G., B. Josefsson, and C. Roos. 1977. "Gas liquid chromatography of monosaccharides at the picogram level using glass capillary column, trifluoroacetyl derivitization and electron-capture detection." *J. Chromatogr.* 142:575–85.


## Supplementary References


Aloia, R. C., C. C. Curtain, and L. M. Gordon. 1988–1990. *Advances in Membrane Fluidity*, 4 vol. New York: John Wiley and Sons.

Bandurski, R. S. 1986. *Gas Chromatography/Mass Spectrometry.* New York: Springer-Verlag. 304.

Biermann, C. J., and G. D. McGinnis. 1989. *Analysis of Carbohydrates by GLC and MS.* Boca Raton, FL: CRC Press.

Bordier, C. 1981. "Phase separations of integral membrane proteins in Triton X-114 solution." *J. Biol. Chem.* 256:1604–07.

Churms, S. S. 1992. *Carbohydrates.* Heftman, E. J. (ed.) of Chromatography, Vol. 51B. Chromatography 5th Edition. Fundamentals and applications of chromatography and related differential migration methods, Part B applications 32:630.

Clement, R. 1990. *Gas Chromatography, Biochemical, Biomedical and Clinical Applications.* New York: John Wiley and Sons. 393.

Eiceman, G. A., R. E. Clement, and H. H. Hill. 1992. *Gas Chromatography: Analytical Chemistry.* 64:170–80.

Elson, E., W. Frazier, and L. Glaser. 1983–1987. *Cell Membranes: Methods and Reviews.* New York: Plenum Press.

Evans, W .H. 1987. "Organelles and membranes of animal cells." *Biological Membranes. A Practical Approach.* Oxford: IRL Press.

Harmon, R. E. 1978. *Cell Surface Carbohydrate Chemistry.* New York: Academic Press.

Higgins, J. A. 1987. "Separation and analysis of membrane components." *Biological Membranes: A Practical Approach.* Oxford: IRL Press. 103–37.

Jennings, W. 1987. *Analytical Gas Chromatography.* Orlando: Academic Press. 259.

Jones, O. T., J. P. Earnest, and M. G. McNamee. 1987. Solubilization and Reconstitution of Membrane Proteins. *Biological Membranes: A Practical Approach.* Oxford: IRL Press.

Martonosi, A. 1985. *The Enzymes of Biological Membranes*, 4 vols. New York: Plenum Press.

Morré, D. J., A. O. Brightman, and A. S. Sandelius. 1987. "Membrane fractions from plant cells." *Biological Membranes: A Practical Approach.* Oxford: IRL Press.

Pfuffenberger, C. D. "Analysis of sugars, sugar alcohols and related compounds." *Glass Capillary Chromatography in Clinical Medicine and Pharmacology.* New York: Marcel Dekker, Inc.

Schomburg, G. 1990. *Gas chromatography: A Practical Course.* New York: Weinheim.

Sherma, J., and B. Fried. 1991. *Handbook of Thin-Layer Chromatography.* New York: Marcel Dekker, Inc.

Vaughn, K., A. R. Lax, and S. O. Duke. 1988. "Polyphenol oxidase. The chloroplast oxidase with no established function." *Physiol. Plantarum.* 72:659–65.

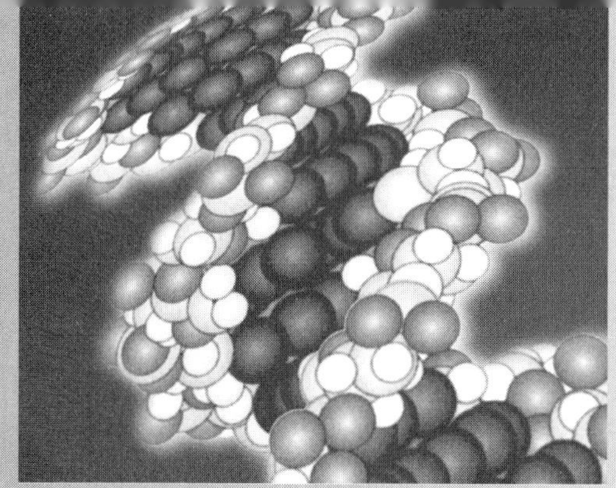

---

# Appendixes

---

## Outline of Appendixes

# Appendix 1
# Metric Terms and Abbreviations

kg = kilogram
g = gram
mg = milligram
µg = microgram
ng = nanogram
pg = picogram
wt = weight
fwt = formula weight
mol = mole
µmol = micromole
nmol = nanomole
ppm = parts per million
ppb = parts per billion
M = molar
mM = milimolar
µM = micromolar

nM = nanomolar
N = normal
l = liter
ml = milliliter
cc = cubic centimeter
µl = microliter
nl = nanoliter
vol = volume
m = meter
cm = centimeter
mm = millimeter
µm = micrometer
nm = nanometer
°C = degrees centrigrade
K = degrees Kelvin

# APPENDIX 2

# DENSITY, % SOLUTION, AND MOLARITY CONVERSION TABLES FOR SUCROSE AND $CsCl_2$ GRADIENT CENTRIFUGATIONS

Density, Refractive Index, and Concentration Data—Cesium Chloride at 25° C. Molecular Weight = 168.37

| Density (g/cm³)[*] | Refractive Index, $\eta_D$ | % by Weight | mg/ml of Solution[**] | Molarity | Density (g/cm³)[*] | Refractive Index, $\eta_D$ | % by Weight | mg/ml of Solution[**] | Molarity |
|---|---|---|---|---|---|---|---|---|---|
| 1.0047 | 1.3333 | 1 | 10.0 | 0.056 | 1.336 | 1.3657 | 34 | 454.2 | 2.698 |
| 1.0125 | 1.3340 | 2 | 20.2 | 0.119 | 1.3496 | 1.3670 | 35 | 472.4 | 2.806 |
| 1.0204 | 1.3348 | 3 | 30.6 | 0.182 | 1.363 | 1.3683 | 36 | 490.7 | 2.914 |
| 1.0284 | 1.3356 | 4 | 41.4 | 0.244 | 1.377 | 1.3696 | 37 | 509.5 | 3.026 |
| 1.0365 | 1.3364 | 5 | 51.8 | 0.308 | 1.391 | 1.3709 | 38 | 528.6 | 3.140 |
| 1.0447 | 1.3372 | 6 | 62.8 | 0.373 | 1.406 | 1.3722 | 39 | 548.3 | 3.257 |
| 1.0531 | 1.3380 | 7 | 73.7 | 0.438 | 1.4196 | 1.3735 | 40 | 567.8 | 3.372 |
| 1.0615 | 1.3388 | 8 | 84.9 | 0.504 | 1.435 | 1.3750 | 41 | 588.4 | 3.495 |
| 1.0700 | 1.3397 | 9 | 96.3 | 0.572 | 1.450 | 1.3674 | 42 | 609.0 | 3.617 |
| 1.0788 | 1.3405 | 10 | 107.9 | 0.641 | 1.465 | 1.3778 | 43 | 630.0 | 3.742 |
| 1.0877 | 1.3414 | 11 | 119.6 | 0.710 | 1.481 | 1.3792 | 44 | 651.6 | 3.870 |
| 1.0967 | 1.3423 | 12 | 131.6 | 0.782 | 1.4969 | 1.3807 | 45 | 673.6 | 4.001 |
| 1.1059 | 1.3432 | 13 | 143.8 | 0.854 | 1.513 | 1.3822 | 46 | 696.0 | 4.134 |
| 1.1151 | 1.3441 | 14 | 156.1 | 0.927 | 1.529 | 1.3837 | 47 | 718.6 | 4.268 |
| 1.1245 | 1.3450 | 15 | 168.7 | 1.002 | 1.546 | 1.3852 | 48 | 742.1 | 4.408 |
| 1.1340 | 1.3459 | 16 | 181.4 | 1.077 | 1.564 | 1.3868 | 49 | 766.4 | 4.552 |
| 1.1437 | 1.3468 | 17 | 194.4 | 1.155 | 1.5825 | 1.3885 | 50 | 791.3 | 4.700 |
| 1.1536 | 1.3478 | 18 | 207.6 | 1.233 | 1.601 | 1.3903 | 51 | 816.5 | 4.849 |
| 1.1637 | 1.3488 | 19 | 221.1 | 1.313 | 1.619 | 1.3920 | 52 | 841.9 | 5.000 |
| 1.1739 | 1.3498 | 20 | 234.8 | 1.395 | 1.638 | 1.3937 | 53 | 868.1 | 5.156 |
| 1.1843 | 1.3508 | 21 | 248.7 | 1.477 | 1.658 | 1.3955 | 54 | 895.3 | 5.317 |
| 1.1948 | 1.3518 | 22 | 262.9 | 1.561 | 1.6778 | 1.3973 | 55 | 922.8 | 5.481 |
| 1.2055 | 1.3529 | 23 | 277.3 | 1.647 | 1.699 | 1.3992 | 56 | 951.4 | 5.651 |
| 1.2164 | 1.3539 | 24 | 291.9 | 1.734 | 1.720 | 1.4012 | 57 | 980.4 | 5.823 |
| 1.2275 | 1.3550 | 25 | 306.9 | 1.823 | 1.741 | 1.4032 | 58 | 1009.8 | 5.998 |
| 1.2387 | 1.3561 | 26 | 322.1 | 1.913 | 1.763 | 1.4052 | 59 | 1040.2 | 6.178 |
| 1.2502 | 1.3572 | 27 | 337.6 | 2.005 | 1.7846 | 1.4072 | 60 | 1070.8 | 6.360 |
| 1.2619 | 1.3584 | 28 | 353.3 | 2.098 | 1.808 | 1.4093 | 61 | 1102.9 | 6.550 |
| 1.2738 | 1.3596 | 29 | 369.4 | 2.194 | 1.831 | 1.4115 | 62 | 1135.8 | 6.746 |
| 1.2858 | 1.3607 | 30 | 285.7 | 2.291 | 1.856 | 1.4137 | 63 | 1167.3 | 6.945 |
| 1.298 | 1.3619 | 31 | 402.4 | 2.390 | 1.880 | 1.4160 | 64 | 1203.2 | 7.146 |
| 1.311 | 1.3631 | 32 | 419.5 | 2.492 | 1.9052 | 1.4183 | 65 | 1238.4 | 7.355 |
| 1.324 | 1.3664 | 33 | 436.9 | 2.595 | | | | | |

Adopted from "Density, Refractive Index and Concentration Data–Cesium Chloride at 20° C." *Techniques of Preparative, Zonal and Continuous Flow Ultracentrifugation.* 5th ed. Palo Alto, CA: Beckman Instruments, Inc. *Computed from the relationship $\rho^{25} = 10.2402\ \eta_D^{25} - 12.6483$ for densities between 1.00 and 1.38, and $\rho^{25} = 10.8601\ \eta_D^{25} - 13.4974$ for densities above 1.37 (Bruner and Vinograd, 1965).

** Divide by 10.0 to obtain % w/v.

Density data are from International Critical Tables.

Density Conversion for Cesium and Rubidium Salts at 20° C

| % w/w | CsCl | CsBr | CsI | $Cs_2SO_4$ | $CsNO_3$ | RbCl | RbBr | RbI | $Rb_2SO_4$ | $RbNO_3$ |
|---|---|---|---|---|---|---|---|---|---|---|
| 1 | 1.00593 | 1.00612 | 1.00608 | 1.0061 | 1.00566 | 1.00561 | 1.00593 | 1.00591 | 1.0066 | 1.0053 |
| 2 | 1.01374 | 1.01412 | 1.01402 | 1.0144 | 1.01319 | 1.01307 | 1.01372 | 1.01370 | 1.0150 | 1.0125 |
| 4 | 1.02969 | 1.03048 | 1.03029 | 1.0316 | 1.02859 | 1.02825 | 1.02965 | 1.02963 | 1.0322 | 1.0272 |
| 6 | 1.04609 | 1.04734 | 1.04707 | 1.0494 | 1.04443 | 1.04379 | 1.04604 | 1.04604 | 1.0499 | 1.0422 |
| 8 | 1.06297 | 1.06472 | 1.06438 | 1.0676 | 1.06072 | 1.05917 | 1.06291 | 1.06296 | 1.0680 | 1.0575 |
| 10 | 1.08036 | 1.08265 | 1.08225 | 1.0870 | 1.07745 | 1.07604 | 1.08028 | 1.08041 | 1.0864 | 1.0731 |
| 12 | 1.09828 | 1.10116 | 1.10071 | 1.1071 | 1.09463 | 1.09281 | 1.09817 | 1.09842 | 1.1052 | 1.0892 |
| 14 | 1.11676 | 1.12029 | 1.11979 | 1.1275 | 1.11227 | 1.11004 | 1.11661 | 1.11701 | 1.1246 | 1.1057 |
| 16 | 1.13582 | 1.14007 | 1.13953 | 1.1484 |  | 1.12775 | 1.13563 | 1.13621 | 1.1446 | 1.1227 |
| 18 | 1.15549 | 1.16053 | 1.15996 | 1.1696 |  | 1.14596 | 1.15526 | 1.15605 | 1.1652 | 1.1401 |
| 20 | 1.17580 | 1.18107 | 1.18112 | 1.1913 |  | 1.16469 | 1.17554 | 1.17657 | 1.1864 | 1.1580 |
| 22 | 1.19679 | 1.20362 | 1.20305 | 1.2137 |  | 1.18396 | 1.19650 | 1.19781 | 1.2083 | 1.1763 |
| 24 | 1.21849 | 1.22634 | 1.22580 | 1.2375 |  | 1.20379 | 1.21817 | 1.21980 | 1.2309 | 1.1952 |
| 26 | 1.24093 | 1.24990 | 1.24942 | 1.2643 |  | 1.22421 | 1.24059 | 1.24257 | 1.2542 | 1.2146 |
| 28 | 1.26414 | 1.27435 | 1.27395 |  |  | 1.24524 | 1.26380 | 1.26616 | 1.2782 | 1.2346 |
| 30 | 1.28817 | 1.29973 | 1.29944 |  |  | 1.26691 | 1.28784 | 1.29061 | 1.3028 | 1.2552 |
| 35 | 1.35218 | 1.36764 | 1.36776 |  |  | 1.32407 | 1.35191 | 1.35598 | 1.3281 | 1.2764 |
| 40 | 1.42245 | 1.44275 | 1.44354 |  |  | 1.38599 | 1.42233 | 1.42806 |  |  |
| 45 | 1.4993 | 1.52626 | 1.52803 |  |  | 1.45330 | 1.50010 | 1.50792 |  |  |
| 50 | 1.58575 | 1.61970 | 1.62278 |  |  | 1.52675 | 1.58639 | 1.59691 |  |  |
| 55 | 1.68137 | 1.72492 |  |  |  |  | 1.68254 | 1.69667 |  |  |
| 60 | 1.78859 |  |  |  |  |  |  | 1.80924 |  |  |
| 65 | 1.90966 |  |  |  |  |  |  | 1.93722 |  |  |

Adapted from "Density, Refractive Index and Concentration Data–Cesium Chloride at 20° C." *Techniques of Preparative, Zonal and Continuous Flow Ultracentrifugation.* 5th ed. Palo Alto, CA: Beckman Instruments, Inc.

# Appendix 3

## References for Preparation of Cells and Tissues for Cytochemical-Histochemical Localizations of Proteins, Carbohydrates, Lipids, and Nucleic Acids

There are many monographs concerned with the preparation of cells and tissues for the light microscopic localization of carbohydrates, lipids, nucleic acids, and proteins. The list presented here is not comprehensive but is representative.


Baker, R. J. 1989. *Autoradiography: A Comprehensive Overview.* Oxford: Oxford Univ. Press.

Gahan, P. B. 1984. *Plant Histochemistry and Cytochemistry: An Introduction.* London: Academic Press.

Galigher, A. E., and E. N. Kosloff. 1964. *Essentials of Practical Microtechnique.* Philadelphia: Lea and Febiger.

Hayat, M. A. 1993. *Stains and Cytochemical Methods.* New York: Plenum Press.

Horubin, R. W. 1988. *Understanding Histochemistry: Selection, Evaluation and Design of Biological Stains.* Chichester: Hopwood.

Jensen, W. 1952. *Botanical Histochemistry.* San Francisco: Freeman.

Kiernan, J. A. 1990. *Histological and Histochemical Methods: Theory and Practice.* Oxford: Pergamon Press.

Lillie, R. D., and H. M. Fuller. 1976. *Histopathology Technique and Practical Histochemistry.* New York: McGraw-Hill.

Quello, A. C. 1983. *Immunohistochemistry.* Chichester: John Wiley and Sons.

Pearse, A. G. E. 1964. *Histochemistry Theoretical and Applied.* Boston: Little Brown.

Sannes, P. L. 1988. *The Histochemical and Cytochemical Localization of Proteases.* Stuttgard: Deerfield BEA.

Sheehan, D. C. 1980. *Histotechnology—Theory and Practice.* St. Louis: Mosby.

Slayter, E. M. 1992. *Light and Electron Microscopy.* Cambridge: Cambridge Univ. Press.

Smith, R. F. 1990. *Microscopy and Photomicrography: A Working Manual.* Boca Raton, FL: CRC Press.

Sumner, B. E. H. 1988. *Basic Histochemistry.* Chichester: John Wiley and Sons.

# APPENDIX 4

# REFERENCES FOR PREPARATION OF CELLS AND TISSUES FOR ELECTRON MICROSCOPY


Representative monographs—early literature prior to 1985 not presented.

Bozzola, J. J., and L. D. Russell. 1992. *Electron Microscopy Techniques for Biologists.* Boston: Jones and Bartlett Publishers.

De Bruijin, W. C., C. W. J. Sorber, E. S. Gelsema, A. L. D. Beckers, and J. F. Jongkind. 1993. "Energy filtering transmission electron microscopy of biological specimens." *Scanning Microscopy.* 7:693–709.

Dykstra, M. J. 1993. *A Manual of Applied Techniques of Biological Electron Microscopy.* New York: Plenum Press.

Egerton, R. F. 1986. *Electron Energy Loss Spectroscopy in the Electron Microscope.* New York: Plenum Press.

*Electron Microscopy Reviews.* 1988–1992. Oxford: Pergamon Press.

Griffin, R. L. 1990. *Using the Transmission Electron Microscope in the Biological Sciences.* New York: Ellis Horwood.

Hader, D. P. 1992. *Image Analysis in Biology.* Boca Raton, FL: CRC Press.

Hall, J. L. 1978. *Electron Microscopy and Cytochemistry of Plant Cells.* Amsterdam: Elsevier.

Hall, J. L., and C. Haves. 1991. *Electron Microscopy of Plant Cells.* New York: Academic Press.

Harris, J. R. 1981–1987. *Electron Microscopy of Proteins.* New York: Academic Press.

Harris, R. 1991. *Electron Microscopy in Biology: A Practical Approach.* Oxford: IRL Press.

Hawkes, P., and V. Valdie. 1990. *Biological and Electron Microscopy.* New York: Academic Press.

Hayat, M. A. 1986. *Basic Techniques for Transmission Electron Microscopy.* New York: Academic Press.

Hayat, M. A. 1990. *Negative Staining.* New York: McGraw-Hill.

Hodges, G. M., and R. C. Hallanes. 1979. *Biomedical Applications of Scanning Electron Microscopy.* New York: Academic Press.

Morgan, J. A. 1985. *X-ray Microanalysis in Electron Microscopy for Biologists.* Oxford: Oxford University Press.

Newbury, D. E. 1986. *Advanced Scanning Electron Microscopy and X-ray Microanalysis.* New York: Plenum Press.

Plattner, H. 1989. *Electron Microscopy of Subcellular Dynamics.* Boca Raton, FL: CRC Press.

Polack, J. M., and S. Van Noorden. 1984. *An Introduction: Current Techniques and Problems.* Oxford: Oxford Univ. Press.

Reed, S. J. B. 1993. *Electron Microprobe Analysis.* Cambridge: Cambridge Univ. Press.

Robinson, D. G. 1987. *Methods of Preparation for Electron Microscopy.* Berlin: Springer-Verlag.

*Scanning Electron Microscopy, X-ray Microanalysis, and Analytical Electron Microscopy: A Laboratory Workbook.* New York: Plenum Press.

Scott, V. D., and G. Love. 1985. *Electronprobe Microanalysis.* Chichester: E. Horwood.

Sigee, D. C. 1993. *X-ray Microanalysis in Biology: Experimental Techniques and Applications.* Cambridge: Cambridge Univ. Press.

# APPENDIX 5
# BIOCHEMISTRY JOURNALS

## Biochemistry Journals and Monographs

1. *Analytical Biochemistry*
2. *Archives of Biochemistry and Biophysics*
3. *Biochemical Archives*
4. *Biochemical Journal*
5. *Biochemistry*
6. *Biochemistry and Molecular Biology International*
7. *Biochemical Biophysical Research Communications*
8. *Biochemie et Biologic Cellulaire*
9. *Current Topics in Plant Biochemistry and Physiology*
10. *Comparative Biochemistry and Physiology*
11. *Critical Reviews in Biochemistry and Molecular Biology*
12. *Enzyme*
13. *European Journal of Biochemistry*
14. *International Journal of Biochemistry*
15. *Journal of Biochemical and Biophysical Methods*
16. *Journal of Biological Chemistry*
17. *Journal of Cellular Biochemistry*
18. *Macromolecules*
19. *Nucleic Acids Reagent*
20. *Nucleosides and Nucleotides*
21. *Phytochemical Analysis*
22. *Phytochemistry*
23. *Plant Physiology and Biochemistry*
24. *Preparative Biochemistry*
25. *The FASEB Journal—Recent Advances in Biochemistry*

## Molecular Biology and Genetics Journals

1. *Biochemical Genetics*
2. *Biotechnology and Bioengineering*
3. *Cellular and Molecular Biology Research*
4. *Current Genetic*
5. *Cytogenetics and Cell Genetics*
6. *DNA and Cell Biology*
7. *Eukaryotic Gene Expression*
8. *Gene*
9. *Genetic Analysis*
10. *Genetics*
11. *Genetical Research*
12. *Geneticha*
13. *Genome*
14. *Genomics*
15. *Heredity*
16. *Human Genetics*
17. *Human Molecular Genetics*
18. *Journal of Genetics*
19. *Molecular Biology of the Cell*
20. *Molecular Biology and Evolution*
21. *Molecular and General Genetics*
22. *Molecular Biology Reports*
23. *Plant Molecular Biology Reporter*
24. *Progress in Biophysics and Molecular Biology*
25. *Somatic Cell and Molecular Genetics*
26. *Theoretical and Applied Genetics*

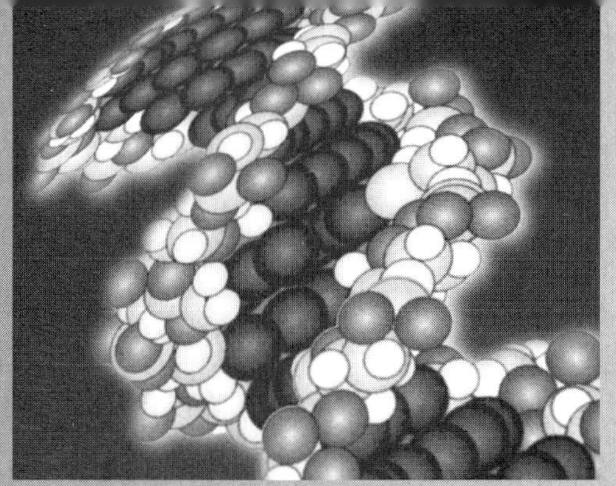

# Index